International Scientific Conference Interstroymeh (ISM 2020)

Held online due to COVID-19

IOP Conference Series: Materials Science and Engineering
Volume 1103

Samara, Russia
17 - 18 December 2020

ISBN: 978-1-7138-3397-0
ISSN: 1757-8981

Printed from e-media with permission by:

Curran Associates, Inc.
57 Morehouse Lane
Red Hook, NY 12571

Some format issues inherent in the e-media version may also appear in this print version.

Institute of Physics
Dirac House, Temple Back
Bristol BS1 6BE UK

Phone: 44 1 17 929 7481
Fax: 44 1 17 920 0979

techtracking@iop.org

Additional copies of this publication are available from:

Curran Associates, Inc.
57 Morehouse Lane
Red Hook, NY 12571 USA
Phone: 845-758-0400
Fax: 845-758-2633
Email: curran@proceedings.com
Web: www.proceedings.com

TABLE OF CONTENTS

IOP Conf. Series: Materials Science and Engineering 1103 (2021) 011001 doi:10.1088/1757-899X/1103/1/011001

International Scientific Conference Interstroymeh – 2020 was successfully held at Samara State Technical University (Samara, Russia) on December 17-18, 2020. ISM 2020 was dedicated to the memory of Doctor of Technical Sciences, Professor Stanislav Yakovlevich Galitskov.

In order to avoid the risk of crowd gathering, the Organizing Committee decided to hold ISM 2020 online, via Zoom platform. This gave scientists and practitioners from all over the country the opportunity to present their research despite the travel restrictions. Professors M.V. Nenashev (SamSTU), E.M. Kudryavtsev (Moscow State University), V.A. Zorin (MADI), B.G. Kim (VISU), V.A. Zhulai (VSTU) made plenary reports at the opening of the conference. There were five sections, where the speakers discussed various issues of construction, mechanical engineering, road building, materials science. Two days of the conference became a time of great insights and cooperation for all the participants.

After careful reviewing, almost 40 papers were selected to be published in the ISM 2020 proceedings. They cover a range of topics such as automation in construction, production and maintenance of technological machines, materials manufacturing, road construction, industrial design. We express our gratitude to all the participants, members of the Organizing Committee, and editors for their valuable contribution in preparing the ISM 2020 proceedings.

Appendix 1. Organizing committee of Interstroymeh – 2020

Chairman

- Bykov Dmitry Evgenievich, Rector of SamSTU, Ph.D., Professor, Samara.

Cochairmen:

- Nenashev Maxim Vladimirovich, First Vice-Rector for Scientific Work of SamSTU, Ph.D., Professor, Samara.
- Kudryavtsev E.M. - Head of the Department "Construction and Lifting Vehicles" NIU MGSU, Dr. Techn. Sciences, prof., Moscow;
- Zorin V.A. - Head. Department "Production and Repair of Cars and Road Machines" Moscow State Automobile and Road Technical University, Dr. Techn. Sciences, prof., Moscow;
- Galitskov K.S. - Head. Department "Mechanization, Automation and Energy Supply of Construction," SamSTU, Cand. tech. of Sciences, Associate Professor, Samara.

Members:

- Sakhapov R.L. - Head. Department "Road Construction Machines" KGASU, Dr. Techn. sciences, prof., Kazan.
- Sinelshchikov A.V. - Head. Department "Lifting and Transport Machines, Production Logistics and Machine Mechanics" AGTU, Cand. tech. sciences, Astrakhan;
- Sharapov R.R. - Head of the Department of Construction Mechanization, Moscow State University, Dr. Techn. Sciences, prof., Moscow;

- Vavilov A.V. - Head. Department "Construction and Road Machines," BNTU, Dr. Techn. Sciences, prof., Minsk;
- Dotsenko A.I. - Head. Department "Technology, Mechanization and Operation of Roads," MGAKKiS, Dr. Techn. Sciences, prof., Moscow;
- Efremov I.M. - Head. Department "Lifting and Transport, Construction, Road Machinery and Equipment," BrSU, Cand. tech. sciences, doc., Bratsk;
- Kim B.G. - Head. Department of "Construction Production" VlSU, Dr. Techn. sciences, prof., Vladimir;
- hulai V.A. - Head. Department "Construction Engineering and Engineering Mechanics," VGASU, Dr. Technician. Sciences, prof., Voronezh;
- Panichkin A.V. - Head. Department "Lifting and Transportation, Construction and Road Vehicles," GUNPK, Cand. tech. sciences, Oryol;
- Starkov S.V. - Head. Department "Transport Machines and Transport and Technological Processes," FEFU, Cand. tech. Sciences, Doctor, Vladivostok;
- Baranchik V.P. - Head. Department "Construction and Road Machines," IzhGTU, Dr. Techn. sciences, prof., Izhevsk;
- Nizhny Novgorod A.I. - Head. Department "Construction and Road Machines and Hydraulic Systems," IrGTU, Dr. Techn. Sciences, prof., Irkutsk;
- Vakhidov U. S. - Head. Department "Construction and Road Machines," NSTU, Dr. Techn. Sciences, prof., Nizhny Novgorod;
- Salmin V.V. - Head. Department "Transport Machines," PSU, Dr. Techn. Sciences, prof., Penza;
- Svitala F. - Professor, Lublin Catholic University, Lublin, Poland;
- Pugin K.G. - Head. Department "Cars and Technological Machines," PNIPU, Cand. tech. sciences, doc., Perm;
- Yevtyukov S.A. - Head of the Department "Ground Transport and Technological Machines," SPbGASU, Dr. Techn. Sciences, prof., St. Petersburg;
- Martyuchenko I.G. - Head. Department "Construction and Road Machines and Equipment," SSTU, Cand. tech. sciences, prof., Saratov;
- Galdin N.S. - Head of the Department "Lifting and Transport, Traction Machines and Hydraulic Drive," SibADI, Dr. Techn. sciences, prof., Omsk;
- Kondratyev A.V. - Head of Department "Construction, Road Machinery and Equipment," TVGTU, Dr. Techn. sciences, prof., Tver;
- Antsev V.Yu. - Head. Department "Lifting and Transportation Machines and Equipment," TulSU, Dr. Techn. sciences, prof., Tula;
- Shavyraa C.D. - Head of the Department "Transport and Technological Means," TuvSU, Cand. tech. sciences, Kyzyl;

- Isakov V.S. - Head. Department "Lifting and Transport, Construction and Road Machines," YURGPU (NPI), Dr. Techn. sciences, prof., Novocherkassk;
- Bondar V.N. - Head. Department "Wheeled, Tracked Cars and Cars," SUSU, Cand. tech. Sciences, prof., Chelyabinsk;
- Prusov A.Yu. - Head. Department "Construction and Road Machines," YaSTU, Cand. tech. sciences, doc., Yaroslavl.

Secretary

- Nazarov M.A., Ph.D., associate professor of the Department of MAES SamSTU.

Appendix 2. Photos from the conference

Opening of the conference

Galitskov K.S. - Head. Department "Mechanization, Automation and Energy Supply of Construction," SamSTU, Cand. tech. of Sciences, Associate Professor, Samara

Kudryavtsev E.M. - Head of the Department "Construction and Lifting Vehicles" NIU MGSU, Dr. Techn. Sciences, prof., Moscow

Kim B.G. - Head. Department of "Construction Production" VlSU, Dr. Techn. sciences, prof., Vladimir

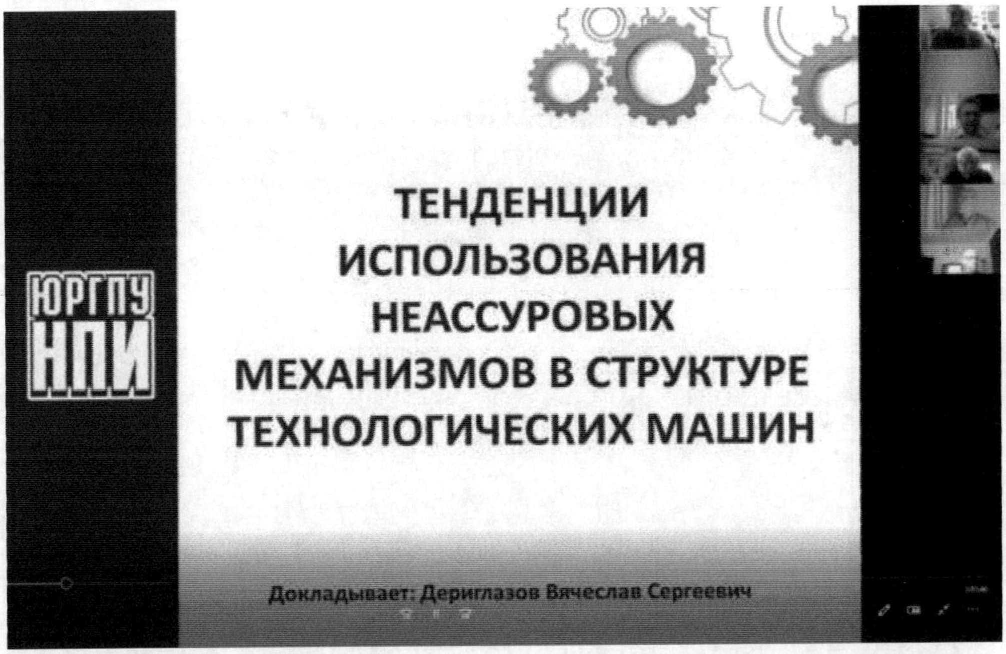

Report made by V.S. Deriglazov

IOP Conf. Series: Materials Science and Engineering 1103 (2021) 011001 doi:10.1088/1757-899X/1103/1/011001

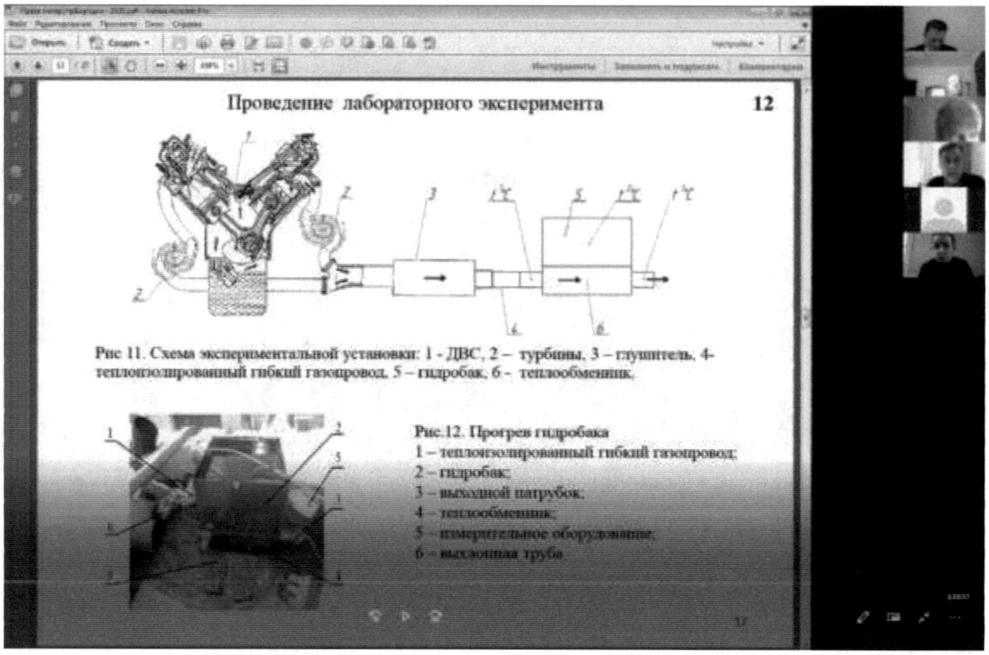

Report made by D.M. Borodin

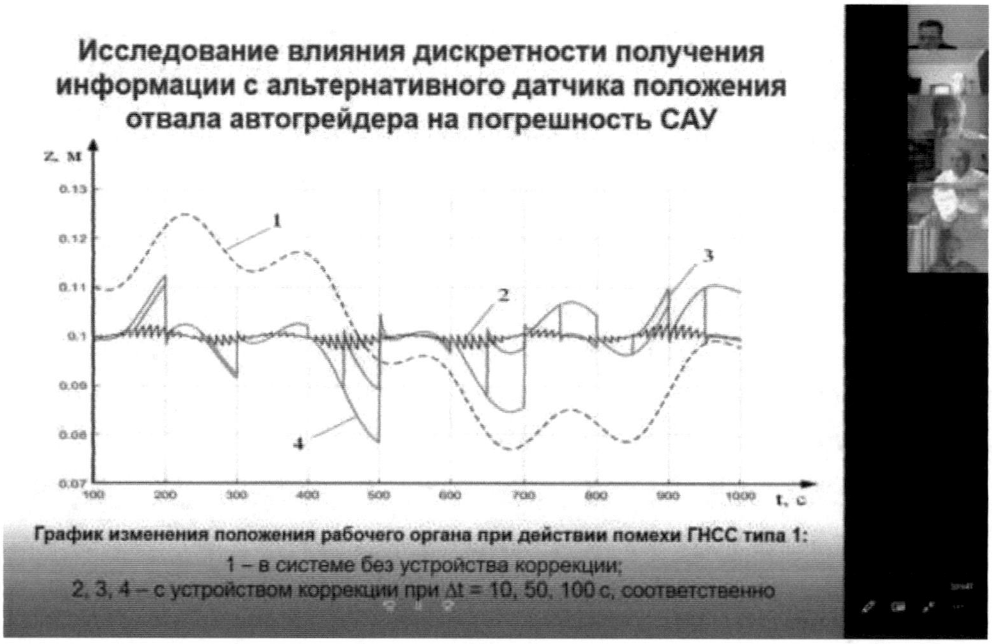

Report made by A.S. Lukyanov

Appendix 3. The link to the conference program: http://sib-publish.ru/pages/en/tech18/Conf.programm_2020_eng.pdf

IOP Conf. Series: Materials Science and Engineering 1103 (2021) 011002 doi:10.1088/1757-899X/1103/1/011002

Peer review declaration

All papers published in this volume of *IOP Conference Series: Materials Science and Engineering* have been peer reviewed through processes administered by the Editors. Reviews were conducted by expert referees to the professional and scientific standards expected of a proceedings journal published by IOP Publishing.

- **Type of peer review:** Single-blind peer review (by two committee members – session head and volume editor).
- **Conference submission management system:** submissions were received and handled via e-mail (caest@ortum-publish.ru)
- **Number of submissions received:** 72
- **Number of submissions sent for review:** 57
- **Number of submissions accepted:** 39
- **Acceptance Rate (Number of Submissions Accepted / Number of Submissions Received X 100):** 54%
- **Average number of reviews per paper:** 2
- **Total number of reviewers involved:** 8
- **Any additional info on review process:** the review pipeline was organized as follows:

1. secretary of the committee handled the correspondence via email and checked submissions formally;
2. all qualified submissions were sent to one of the editors;
3. editors reviewed papers by themselves and assigned an extra referee to each paper;
4. the language editor made style and spell checks of each accepted paper.

- **Contact person for queries: Konstantin Stanislavovich Galitskov – Head of the Department "Mechanization, Automation and Energy Supply of Construction," SamSTU, Cand. of tech. sciences, Associate Professor, Samara. E-mail: ksgal@yandex.ru**

IOP Conf. Series: Materials Science and Engineering 1103 (2021) 012001 doi:10.1088/1757-899X/1103/1/012001

Controlling motion of metal bascule structures by fluid power system (exemplified by lifting of bascule bridge span)

A V Ashcheulov

Peter the Great St. Petersburg Polytechnic University, 29, Polytekhnicheskaya str., St. Petersburg, 195251, Russia

E-mail: atsheulov_av@spbstu.ru

Abstract. This article discusses the dynamics of power processes upon motion of bascule metal structures peculiar for many lifting heavy-duty mechanisms. Due to the forces of inertia, during the acceleration of the mechanism with a load, fluctuations in power parameters occur: efforts; stress; pressures that can lead to the destruction of the metal structures of machines. Due to high alternating load oscillations, the working capacity of machinery decreases about 3 times, herewith, this problem is traced for machinery in various industries. Using the motion of large-tonnage span of bascule bridge as the example, at the design stage it is proposed to apply simulation modeling in order to determine optimum law of drive control, in particular: the law of variation of flow rate in fluid power systems. The article presents the results of simulation of hydraulic lifting mechanisms of a single-wing bascule bridge. Two basic algorithms of proportional control are considered: by the form of the function in the form of a trapezoid and by the form of the cosine function. The results indicate the possibility of significantly reducing the amplitude of fluctuations in load parameters during acceleration of mechanisms and, thereby, to increase the service life of heavy machines. To achieve such results, it is recommended to change the existing algorithms for automatic control of the speed of the mechanisms of heavy machines. Despite the fact that this issue has been studied quite well in cybernetics, it still remains relevant in applied technical specialties, in the design of modern heavy machines. In project organizations, along with constructive design methods, digital, computational design methods should be actively developed.

1. Introduction

Numerous hoisting and road building machines are equipped with mechanisms with bascule metal structures and the fluid power system [1]. Observation of operation of these mechanisms has demonstrated that motion at a constant speed results in increased dynamics upon acceleration and deceleration. In the spaces of hydraulic cylinders, the pressure drops reach high amplitudes. The coefficient of dynamicity K_{dyn} [2]:

$$K_{dyn} = \frac{\Delta p_{\max}}{\Delta p_{stat}},$$

where $\Delta p_{\max}$ is the maximum pressure drop in hydraulic cylinders upon acceleration, Δp_{stat} is the pressure drop in hydraulic cylinders in the static state. According to the author's results it can reach 1.4 and higher. Figure 1 illustrates a program window of the metering system (monitoring system) of bascule bridge. The window shows plots of pressure variations in spaces of hydraulic cylinders and

IOP Conf. Series: Materials Science and Engineering 1103 (2021) 012001 doi:10.1088/1757-899X/1103/1/012001

plot of the rotation angle of the lifting mechanism of the bascule bridge. The presented transition processes are the responses to consecutive activation (deactivation at the stroke end) of two pumps in the hydraulic system: boosting and main pumps. This is the relay stepwise volumetric (machine) control, when it is possible to vary only the steps and time of activation of the second step [3]. The first pressure surge in the discharge line is caused by activation of the boosting pump, the second surge occurs after activation of the main pump. In this case, the oscillation amplitude of the second pressure surge is higher than the first one, since there appear resonant effects of oscillation superposition. However, opposite situations are possible, this depends on the time of activation of the boosting and main pumps.

These oscillations in the range of operation pressure are not dangerous for the fluid power system [4]. However, the pressures in spaces of hydraulic cylinders are equivalent to forces, thus, the pressure oscillations evidence alternating loads in mount fittings of hydraulic cylinders and metal structures, which in the case of high oscillation amplitude can influence significantly their strength [5]. The following facts were detected in practice at some bridges: in the gaps between the rotation axle and the center of counterweights under the layers of old paint, the cracks in metal structures were detected. One significant accident in the 1980-s should be mentioned, when a metal structure at one bridge was completely destroyed and all large-tonnage counterweight felt into a navigable waterway [6].

In order to prevent such accidents, the guide for bridge design includes specified coefficients of load dynamicity, which should not exceed 1.2 [7]. It is possible to achieve the specified coefficient of dynamicity with relay stepwise control by further decrease in motion speed (decrease in flow rates of boosting and main pumps of the hydraulic system). However, this is a dead end, since it would increase the time of bridge lifting, which is unallowable. For bridges of the main waterway, the predetermined lifting time is 2 min, and for the reserved waterway, it is 4 min. The time is also specified for hoisting mechanism of hydraulic facilities (gate drives). The same problem exists for other machines, where increase in performances is highly important.

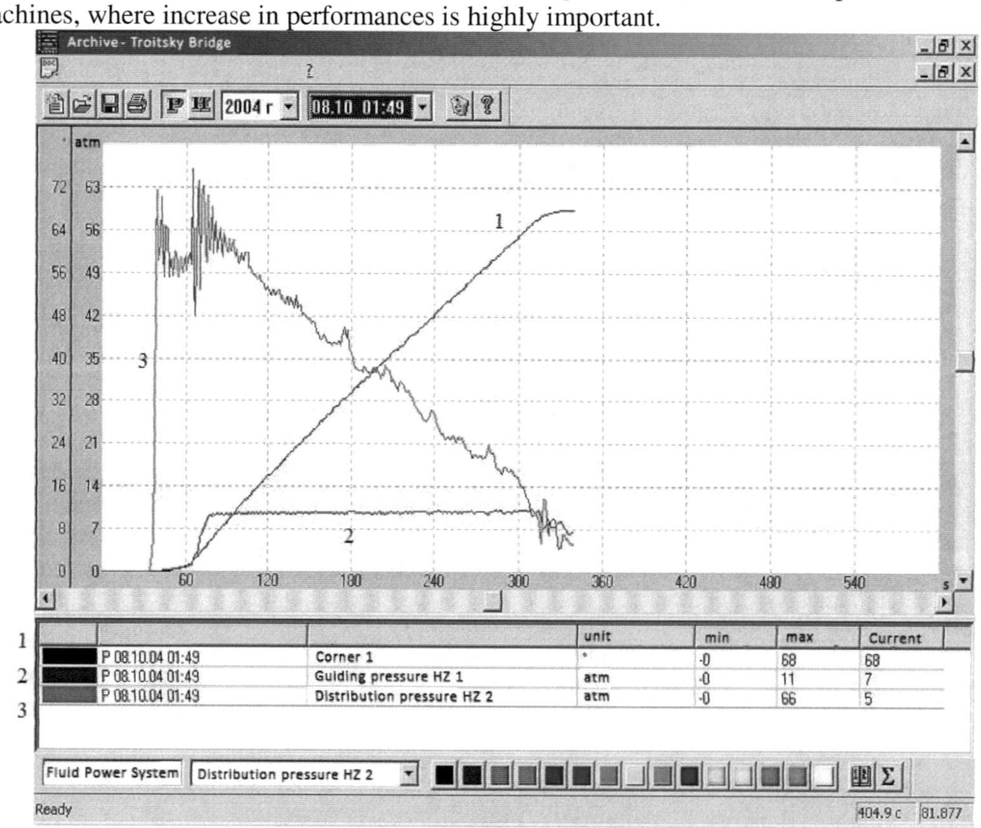

Figure 1. Major specifications of fluid power of Troicky Bridge monitoring system.

2. Materials and methods

A distinctive feature of the research is the use of the author's methodology, which provides for the use of experimental data not of local bench tests of individual machine units, but of data from diagnostic systems (monitoring systems) of full-fledged machines that are directly in operation. We are talking about methods of full-scale tests, but not experimental machines, but industrial, operating models of machines. At the same time, the researcher does not need to write test methods, it is not required to install little reliable non-standard sensors, but it is necessary to study the design of monitoring systems and adapt digital information from the so-called on-board controllers, incl. and "black boxes", which are currently equipped with most critical machines. It should be noted that, first of all, information from monitoring systems is intended to ensure the safe operation of machines. But it also contains a large amount of information that can and should be used by design organizations. First of all, through this information, a feedback is established between the stage of machine operation and the stage of its design. In particular, this information reveals deficiencies in the operation of machines. The accumulation of this information allows you to form design standards and make forecasts for the durability of machine components. Using this information allows us to develop adequate digital models of complex machines.

The data of the monitoring systems installed on movable bridges allowed the author to identify and analyze the problem of the dynamics of the acceleration of the lifting mechanism considered in the article. The same data were used in simulation mathematical modeling for comparison with the calculated dependencies, for correcting the digital model, as well as for clarifying the values of "inaccurate" initial data.

Simulation mathematical modeling was carried out by traditional methods of composing and solving mixed systems of differential equations.

Solution to such sets of differential equations could be performed in general form by numerical integration using any certified application software. The author used the Model Vision proprietary software of StPPU.

3. Concept of motion

Solution to the mentioned problem should be searched at the machinery design stage. Proportional control of the pump flow rate of the fluid power system should be applied. However, it is insufficient to select an adjustable pump and a regulator type for the design of the hydraulic system, it is required to determine a mathematical equation, according to which the pump regulator should operate, it is referred to as the control law by experts in the field of automatic control. The control law is determined by solution to the optimization problem of motion control, it is obtained by mathematical simulation modelling of machinery, and this is beyond the frames of designing. Therefore, numerous Russian designing bureaus do not provide simulation modelling of their machinery due to insufficient competence in this field. It should be mentioned that the time for designing activities is restricted and sometimes additional time for complicated verifications is not stipulated. On the contrary, leading foreign companies, using CAD tools, carry out simulation modelling of hoisting mechanisms at the design stage, sometimes gaining competitive advantages in the form of smooth motion of their mechanisms without hydraulic shocks.

Simulation modelling begins with development of the mathematical model of the machine and stipulates for determination of a set of differential and algebraic equations of parameter interrelation. For instance, for a bascule bridge, it is possible to write the differential equation of motion of cylinder rod in the form of Lagrange equation of the second order. Herewith, the weight of the metal structure is reduced to the weight of movable cylinder rod, and to use equations of mechanism kinematics and equations of hydraulics, accounting for elastic properties of working fluid and the body of the hydraulic cylinder. In addition, it would be reasonable to consider for the volume of the total hydraulic system, that is, the equations of major pipelines. The equations of controlling hydraulic devices should also be considered, if they participate in control of the motion. The order of mathematical model depends on the complexity of the considered system. A general procedure of development of such models is described in the Doctoral thesis by the author. At the second step, the solution to the set of equations is programmed. At the third step, the adequacy of predicted model is verified. If the adequacy is not high, then it should be provided by adjustment of initial data, in particular, by varying inaccurate parameters of the mathematical model. The procedure of adequacy achievement for the predicted model is described in other works by the author. Only a high degree of adequacy of the predicted model allows one to consider the results of simulation modelling as reliable.

$$\begin{cases} m_{red}\cdot\ddot{x}_{hc} = P_{hc}^{A}\cdot S_{hc}^{A} - \left(F_{\mathrm{fr}}^{hc} + \dfrac{M_{\mathrm{fr}}^{\mathrm{bs}}}{h}\right)\cdot signV - \dfrac{M_{L}(\psi)}{h} - \dfrac{M_{W}(\psi,t)\cdot n}{h} \\ \dot{x}_{hc} = \dfrac{Q_{hc}^{A}}{S_{hc}^{A}} = V \\ x_{hc} = L_{hc} - x \\ S_{hc}^{A} = S_{hc}^{W} = \dfrac{\pi}{4}\cdot\left(D_{piston}^{2} - d_{rod}^{2}\right) \\ D_{piston}^{conv} = \sqrt{\dfrac{4\cdot\sum S_{hc}^{A}}{\pi}} \\ \dot{P}_{hc}^{A} = \dfrac{Q_{hc}^{A} - S_{hc}^{A}\cdot\dot{x}_{hc} - Q_{hc}^{y}}{K_{cont}^{A}} \\ K_{cont}^{A} = \dfrac{\Delta V_{A} + S_{hc}^{A}\cdot x_{hc}}{E_{red,hc}} \\ E_{red,hc} = \dfrac{E_{fluid}}{1 - \dfrac{D_{piston}}{\delta_{hc}}\cdot\dfrac{E_{fluid}}{E_{hc}}} \\ x = \sqrt{r^{2} + c^{2} + 2\cdot r\cdot c\cdot\cos(\gamma-\psi)} \\ h = \dfrac{r\cdot c\cdot\sin(\gamma-\psi)}{x} \end{cases}$$

where m_{red} is the weight of moving parts including the metal structure reduced to the rod of the conventional hydraulic cylinder; x_{hc} is the rod stroke; P_{hc}^{A} is the pressure in the work space; S_{hc}^{A} is the surface area of the working space of the conventional hydraulic cylinder; L_{hc} is the length of the hydraulic cylinder with the extended rod; x is the current length of the hydraulic cylinder; Q_{hc}^{A} is the flow rate of working fluid supplied to the working space; V is the rod linear speed; D_{piston} is the piston diameter; d_{rod} is the rod diameter; F_{fr}^{hc} is the dry friction force in the hydraulic cylinder; $M_{\mathrm{fr}}^{\mathrm{bs}}$ is the moment of friction of mechanical units of the bascule span; $M_{L}(\psi)$ is the moment of forces from positional loads (weight, icing, etc.); $M_{W}(\psi,t)$ is the moment of forces of wind loads; n is the coefficient of the wind direction; ψ is the rotation angle of the metal structure; h is the arm of force action on the rod with respect to the rotation axis; δ_{hc} is the wall thickness of the hydraulic cylinder;

ΔV_A is the dead capacity in the hydraulic cylinder; E_{hc} is the elasticity module of the hydraulic cylinder wall, E_{fluid} is the module of fluid bulk elasticity; r, c, γ are the kinematic dimensions of the mechanism.

4. Results

At the first stage of predictions by the simulation model, the pressure was determined in the spaces of hydraulic cylinders according to the existing algorithm (relay stepwise volumetric control), however, at the bridge lifting time equaling to 2 min. Such lifting time can be achieved by increase in the flow rates of boosting and main pumps. The predicted results are illustrated in Fig. 2. The results of analysis are described below.

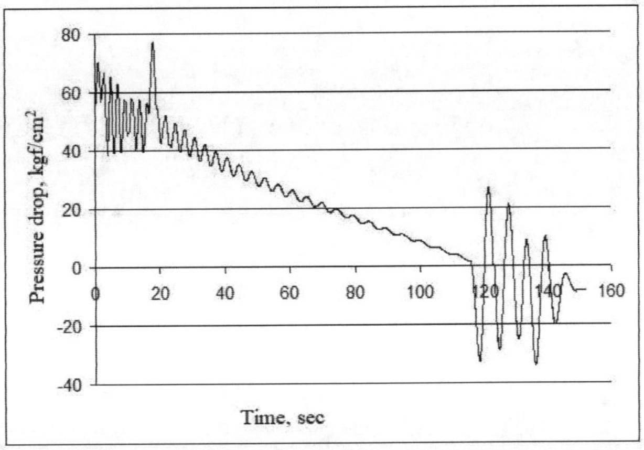

Figure 2. Predicted pressure drop in spaces of hydraulic cylinders of Trinity Bridge as a function of time.

At the second stage of predictions, the simulation model was used for determination of pressure in the spaces of hydraulic cylinders according to the trapezoid law and according to the cosine law of pump flow rates for two lifting times: 4 min and 2 min:

$$\begin{cases} Q = K_Q \cdot t \\ Q = \dfrac{Q_0}{2}\left(1 - \cos\left(\pi \cdot \dfrac{t}{t_{acc}}\right)\right) \end{cases} \qquad at\ 0 \le t \le t_{acc},$$

where K_Q is the coefficient of proportionality for the trapezoid law of the flow rate; t_{acc} is the time of acceleration.

The respective predicted dependences are illustrated in Figures 3 and 4.

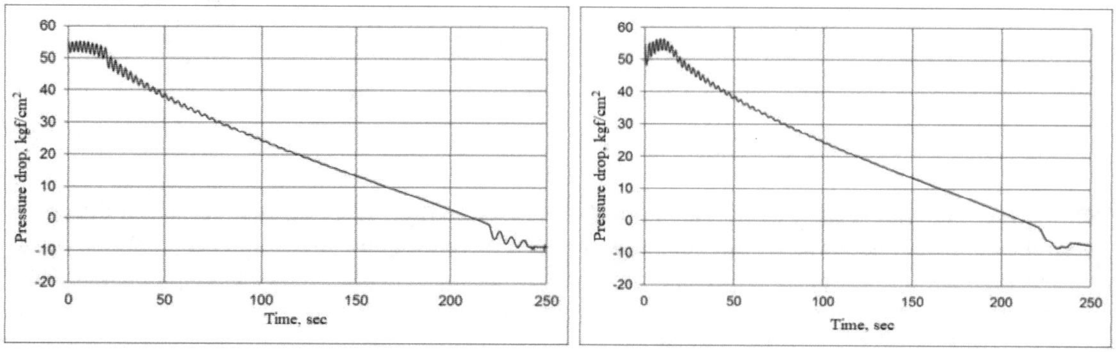

a) at $Q = K_Q \cdot t$; b) at $Q = \frac{Q_0}{2}\left(1 - \cos\left(\pi \cdot \frac{t}{t_p}\right)\right)$

Figure 3. Predicted pressure drop in spaces of hydraulic cylinders of Trinity Bridge as a function of time at 4 min lifting time.

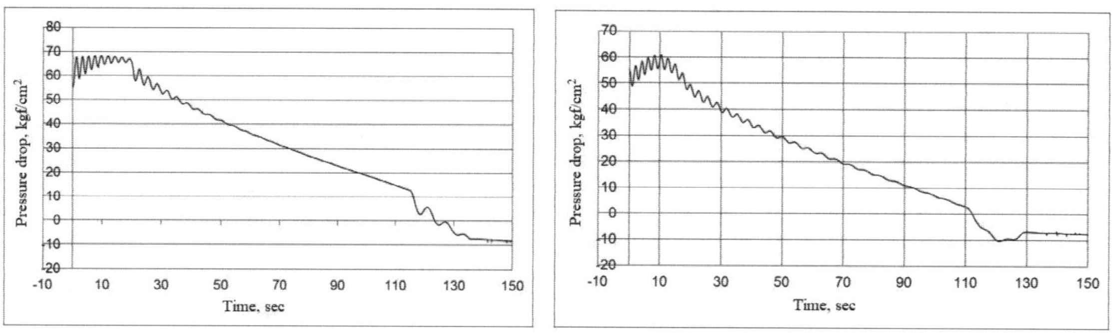

a) b)

Figure 4. Predicted pressure drop in spaces of hydraulic cylinders of Trinity Bridge as a function of time at 2 min lifting time: a) at $Q = K_Q \cdot t$; b) at $Q = \frac{Q_0}{2}\left(1 - \cos\left(\pi \cdot \frac{t}{t_p}\right)\right)$.

5. Discussion

Comparative analysis of Figures 1 and 2 demonstrated that with two-fold decrease in the total lifting time, the coefficient of dynamicity K_{dyn} increased by 20% from 1.33 to 1.7. The oscillation amplitudes of pressure drop reached the presets of safety valves of the hydraulic system (80 kg/cm^2). Oscillations of pressure (forces) were so high that they had no time to decay during overall lifting time.

Analysis of predictions in Figures 3 and 4 demonstrated that in the case of lifting time of 4 min, the trapezoid law of flow rate variations resulted in the coefficient of dynamicity equaling to 1.08, and the cosine law resulted in the coefficient of dynamicity equaling to 1.10. In the case of the lifting time of 2 min, the trapezoid law of flow rate variations resulted in the coefficient of dynamicity equaling 1.16, and the cosine law resulted in the coefficient of dynamicity equaling to 1.19.

6. Conclusion

Conventional relay control of motion speed of the heavy metal structure of the bascule type results in increased dynamics of the process, since acceleration coincides with overriding of maximum loads. The coefficients of dynamicity significantly exceed norms, which increases risks of destruction of metal structures. In order to provide an uninterrupted operation, the machinery works at lower rates, which decreases machinery capacities (according to statistics, the rates of different machines are reduced three times!). A similar situation can be observed not only upon the reciprocal, but also upon rotational motion of mechanisms [8].

It is required to design drives, the fluid power system in particular with proportional control and to carry out additional dynamical predictions at the design stage: simulation modelling of mechanism motion aiming at determination of optimum laws of speed control. At present, these predictions are supported by verified software and prediction procedures [9]. In addition, it is possible to involve specialized researching companies in simulation modeling.

An optimization criterion should be presented by the minimum coefficient of dynamicity of load on the mechanism upon the specified time of mechanism motion.

For the fluid power system the flow rates can be varied either according to trapezoidal law or according to cosine law. Both laws are implemented by modern pump regulators and/or hydraulic valves with proportional control [10].

References

[1] Alexanderov M P 2000 *Hoisting machines*, textbook for universities (Moscow: MGTU named after Bauman, Vysshaia shkola) 552 p

[2] 2013 GOST 32579.1-2013 *Hoisting cranes. Principles for the formation of design loads and load combinations. Part 1. General provisions*

[3] Bashta T M 1972 *Hydraulic drive and hydropneumatic automation* (Moscow: Mashinostroenie) p 320

[4] Galdin N S 2009 *Hydraulic machines, volumetric hydraulic drive:* a tutorial (Omsk: SibADI) 272 p

[5] Sokolov S A 2011 *Structural mechanics and metal structures of machines:* a textbook (St Petersburg.: Politekhnika) 450 p

[6] 55th anniversary of the Alexander Nevsky Bridge - SPb GBU Mostotrest. Retrieved from: https://mostotrest-spb.ru/jubilees/most-aleksandra-nevskogo-91.

[7] 1991 *Movable bridge design*: Guide book (Lengiprotransmost; Transport, Moscow)

[8] Ashcheulov A V, Shestopalov A A, Khoroshanskii A E and Kochetkov A V 2015 Operation of drill rigs with top drive systems *Chemical and Oil-and-Gas Engineering* **2** 14-17

[9] Ashcheulov A V, Filipovskii V M and Meshkovskii E O 2016 Optimization of Regulator Parameters of the Automatic Regulation System of the Rotation Frequency of the Valve Engine of the Top Drive System of Bore Installation *Int. J. of Applied Engineering Research* **11(22)** 11055-11059

[10] Sveshnikov V K (ed) 1999 *Hydraulic equipment of mobile machinery*: Guidebook (Parker Hannifin Corporation)

Organic-inorganic composition for stone masonry walls of ancient tower and necropolis buildings in the North Caucasus

D K-S Bataev[1,2], S-A J Murtazaev1[1,2,4], Kh N Mazhiev[2,4], M M Maloroev[1], G K-S Bataev[1], R U Goitemirov[1,3], M A Gaziev[1,4] and P D Bataeva[1,4]

[1] Department of Materials Science, Kh. Ibragimov Complex Institute of the Russian Academy of Sciences, 21a Staropromyslovoe hw., Grozny, 364051, Russian Federation

[2] Academy of Sciences of the Chechen Republic, 13 M. Esambaev, Grozny, 364024, Russian Federation

[3] Faculty of Technology and Economics Specialty, Chechen State Pedagogy University, 62 Isaeva, Grozny, 364068, Russian Federation

[4] Institute of Construction, Architecture and Design, Grozny State Oil Technical University named after Academican M.D. Millionshikov, 100 Isaeva, Grozny, 364051, Russian Federation

E-mail: kniiran@mail.ru

Abstract. Search works were performed based on archival and bibliographic research, which allowed us to get an idea of the monuments of archeology, history, and culture. For laboratory physical and physicomechanical studies, samples collection of the masonry wall solution was taken. Based on the conducted laboratory studies, a conclusion was made about the necessary grade strength of the masonry mortar of 25-75 kg/cm2 with the following physical and mechanical properties of the solution are as follows: average compressive strength-8.3 MPa; water absorption-11.7%; porosity-18.3%; volume weight-1.56 g / cm^3.stress-related properties: average compression strength – 8.3 MPa; water absorption – 11.7%; porosity – 18.3%; volume weight – 1.56 g/cm^3. The testing of new experimental compositions was performed in places accessible for observation and documented. Local inert, reinforcing, and binding materials were used to select the composition of the stone masonry solution for the walls of ancient tower and necropolis buildings. The research was carried out using existing methods, adjusted to take into account the peculiarities of the historical and cultural monuments under consideration. Based on the study results were developed organic-inorganic compositions of solutions of stone masonry walls of ancient tower and necropolis buildings in the North Caucasus.

1. Introduction

Research works (RW) are an integral part of the investment project for the repair and restoration of archaeological, historical, and cultural monuments. Research works begin at the preliminary studies stage and continue throughout the entire investment process, including its production stages.

Depending on the significance of the monument, the goals and scope of the proposed work, the availability of documentation on previously performed research, as well as the level of their professional performance, research work can be carried out in different volumes and with different degrees of elaboration of individual issues: either in the form of comprehensive research work, or in the form of separate studies that are not allocated to an independent stage, the need for which arises in the process of construction, repair, and restoration. In this work was performed a full-scale proportioning of mortar for stone masonry walls of ancient tower and necropolis buildings in the North Caucasus.

This research work includes the following preliminary studies:
- archival and bibliographic researches;
- development and selection of binders, fillers;
- physical and physicomechanical destructive and non-destructive laboratory studies;
- the complex of search and accompanying works.

2. Problem statement

Research work on the study and proportioning of mortar for stone masonry walls of ancient tower and necropolis structures in the North Caucasus was carried out by experienced specialists of CRDC named after Kh.I. Ibragimov RAS and the Construction Testing Laboratory LLC "Complex" (STL LLC "Complex") in the field of chemical technology, physicochemical, and physicomechanical research methods.

Before the start of complex research and experimental works was appointed a scientific supervisor of these works, who prepared a research program and selected responsible performers for the main sections of complex research and experimental works.

The scientific supervisor had experience in scientific and restoration work on this type of monuments, had the necessary professional qualities, and was responsible for the scientific completeness (validity) and quality of research and experimental work.

Archival and bibliographic research [1-5] allowed us to get an idea of the time of creation and the history of the existence of the considered monuments of archeology, history, and culture.

The composition of archival and bibliographic research included:
- photographs, copies, illustrations in printed publications, etc.;
- testimony, scientific publications, annalistic and other archival materials about monuments;
- selection of analogs;
- analytical processing of the collected material.

Archival and bibliographic studies [6-10] have shown that towers were built in the North Caucasus everywhere on the lime-stone mortar, and the top was covered with a thick layer of yellow or yellow-white plaster, the inside of the masonry joints were covered with mortar. This was a characteristic feature of all architectural monuments: military and residential towers, crypts, and shrines. According to legend were added milk or whey and chicken eggs to the limestone mortar.

Nondestructive (optical-analytical) and laboratory destructive and non-destructive (physical, physicomechanical) research types were used in the course of research work.

3. Materials and methods

For laboratory physical and physicomechanical studies, samples collection of the masonry wall solution was taken (Figure 2). Samples of the masonry wall solution were presented in the original.

Based on the conducted laboratory examination, a conclusion was made about the necessary grade strength of the masonry mortar of 25-75 kg/cm^2. According to the results of laboratory studies, it is clear that the masonry mortar is made on a lime binder in the ratio binder: filler (B/F)-1:2.5÷3.0.

Masonry mortar of white color of a dense structure of uniform mixing with separate coarse particles on the ground lime. The masonry mortar surface of the seam is bumpy with separate large shells, the lateral area is dirty. Lime contains natural crystalline gypsum in an amount of 5.0-10.0%.

To prepare the masonry mortar was used as a filler in the form of a transparent fine of quartz sand of a rounded shape with a size of 0.1-0.2 mm, angular dark gray (black), and light gray particles of metamorphic rock of limy shales with a size from 0.3 to 5.0 mm. There are a lot of opus signinum particles in the size of 0,1-0,2 mm and crushed bricks in the size of 0,5-1,0 mm.

The physical and mechanical properties of the solution are as follows: average compressive strength-8.3 MPa; water absorption-11.7%; porosity-18.3%; volume weight-1.56 g / cm^3. The physical and mechanical properties of the masonry mortar are satisfactory, the solution strength is high and corresponds to the M-75 brand.

The testing of new experimental compositions was performed in places accessible for observation and documented. Recipes and methods for performing experimental work are described below.

The research was carried out using existing methods, adjusted to take into account the peculiarities of the historical and cultural monuments under consideration. Based on the study results were developed organic-inorganic compositions of solutions of stone masonry walls of ancient tower and necropolis buildings of the Republic of Ingushetia.

Local inert, reinforcing, and binding materials were used to select the composition of the stone masonry solution for the walls of ancient tower and necropolis buildings (Figure 1).

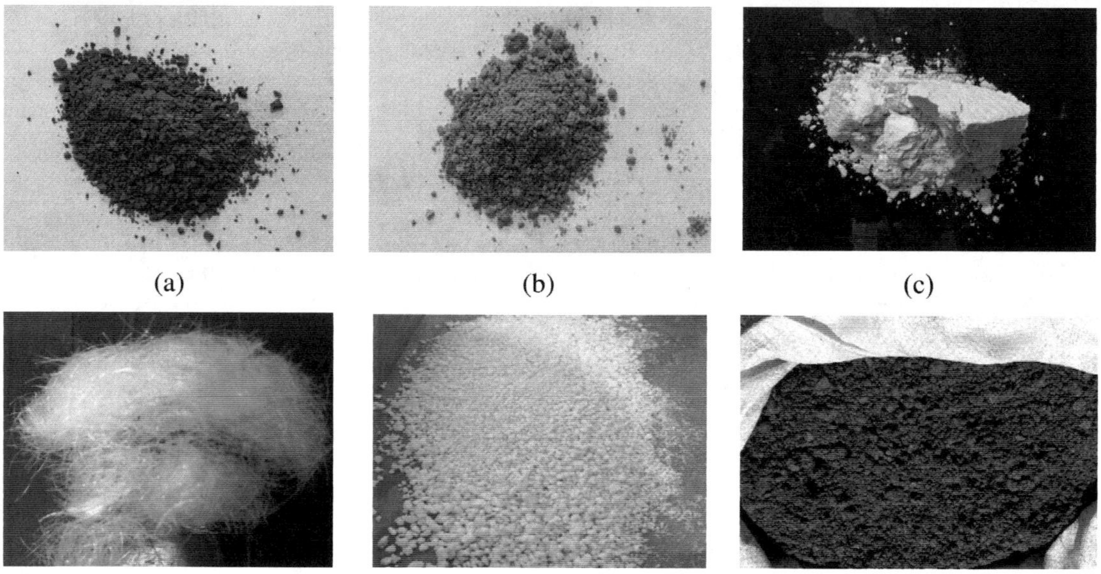

Figure 1. (a) - - slate powder; (b) - dolomitic sand; (C) - hydrated lime; (d) - fibre optic; (e) - technical casein; (f) - opus signinum fraction of 0-5 mm.

The proportioning of mortar of stone masonry walls of ancient tower and necropolis buildings was selected based on the following normative documents:

- AUSS 28013-98 construction solutions. General specifications;
- AUSS 5802-86 construction solutions. Test method;
- AUSS R557338-2016 construction solutions for stone masonry. Methods for determining the ultimate compressive and bending strength;
- AUSS 25818-91 flue ash of thermal power plants for concrete;
- AUSS 3056-90 casein glue in powder. Technical conditions;
- PSA 2007.3 Recommendations for conducting restoration work on monumental painting-cultural properties;
- BG 82-101-98 Preparation and application of construction solutions.

Figure 2. Samples of solutions selected from the masonry joints of the walls of ancient tower and necropolis buildings of the Chechen Republic and the Republic of Ingushetia.

4. Results and discussion

Experimental compositions for research (Table 1, Table 2) have been developed by studying the data of literary sources concerning the experience of construction, restoration, repair, and restoration of an ancient tower and necropolis structures, testimonies of mountain residents, as well as speculative analysis.

Table 1. Experimental compositions of mortar for stone masonry walls of an ancient tower and necropolis buildings on 1,0 m^3 (set No.1)

Composition No.	Lime paste, m^3	Slate powder, m^3	Dolomitic sand, m^3	Water, l	Casein glue, kg	Ash, kg	Egg mass, pcs	Serum, kg
1	0.43	0.91	-	182	-	-	-	-
2	0.38	0.994	-	197	-	-	-	-
3	0.33	-	1.06	156	-	-	-	-
4	0.33	1.06	-	133	-	-	-	-
5	0.29	0.995	-	140	70.0	79.4	-	-
6	0.31	1.00	-	150	-	85.0	-	-
7	0.30	0.88	-	55	58.0	62.0	-	-
8	0.30	0.87	-	95	-	62.0	90	
9	0.17	0.95	-	-	-	117	156	115
10	0.28	1.03	-	-	-	117	-	136
11	-	1.0014	-	-	-	120	330	-
12	0.17	1.0037	-	-	-	125	-	144
13	-	1.03	-	-	-	117	-	226
14	-	1.03	-	-	-	-	203	

At the initial stage were determined the humidity and dry bulk density of inert and matrix materials.

Lime paste:

-water-lime ratio-1/1;

- volume weight (p) - p = 1370 kg / m^3.

Shale sand:

- humidity (W) – W=3.3%;

-bulk density (rd. d) - rd.d=1414 kg / m^3.

Dolomitic sand:

- humidity (W) – W=3.6%;

- bulk density (rd. d) - rd. d=1496 kg / m^3.

Wood ash:

- bulk density (rd. d) - rd. d=500 kg / m^3.

Egg mass:

- density ($r_{e.m}$) – $r_{e.m}$=1050 kg / m^3.

Milk whey:

- density ($r_{m.w}$) - $r_{m.w}$=1024 kg/m^3.

Technical casein:

-physical properties were not determined because the preparation of the test mix requires casein glue ("Ordinary", "Extra").

In laboratory conditions based on experimental compositions of a solution for stone masonry walls of ancient tower and necropolis buildings in standard forms were made:

- cubes 70, 7x70,7x70, 7 3 PCs. for each test to determine the compressive strength (cube strength) (Figure 3, a);

- 4x4x16 beams of 5 pcs. for each test to determine the prism strength and bending tensile strength (Figure 3, b);

- paired and glued on mortar compositions samples of quarry stone and concrete to determine the adhesion of the solution in rubble and stone masonry (Figure 3, c).

(a) (b) (c)

Figure 3. Experimental samples: (a) - cubes 70,7x70,7x70,7; (b) - beams 4x4x16; (c) - paired and glued on mortar compositions of concrete and quarry stone.

Table 2. Experimental compositions of mortar for stone masonry walls of an ancient tower and necropolis buildings on a single beam 4cm*4cm*16cm (set No. 2)

Composition No.	Lime paste, gr.	Slate powder, gr.	Dolomitic sand fr. 0-2.5, gr.	Opus signinum, gr.	Water, gr.	Casein glue gr.	Powder lime, gr.	Wood ash, gr.	Egg mass, gr.	Curd degrease, gr.	Fibre-optic gr.	Curdled milk gr.
1	-	180	310	40	-	80	25	7	-	-	-	-
2	-	180	310	35	-	78	24	8	-	-	-	-
3	**84**	**165**	**280**	**35**	**7**	**-**	**-**	**7**	**-**	**17**	**-**	**-**
4	84	165	280	31	8	-	-	14	15	13	-	-
5	-	235	150	25	-	96	35	9	-	-	-	-
6	90	230	150	26	30	-	-	-	-	9	-	-
7	84	168	299	26	47	2	-	14	-	-	-	-
8	85	220	143	25	35	2	-	-	-	-	1	-
9	**84**	**161**	**300**	**19**	**24**	**5**	**-**	**13**	**-**	**-**	**-**	**-**
10	70	272	86	18	22	7		10				
11	70	272	86	25	22	7		10				
12	**-**	**330**					**60**	**40**	**43**			**30**
13		360					103	30				55
14		350						30	88			
15		350					60	32				37
16		360						30				58
17		365							52			
18		365					39	10	55			40

5. Conclusion

Based on the analysis of the physical and mechanical properties of the solution for stone masonry walls of ancient tower and necropolis buildings in the North Caucasus, the following compositions are recommended:

Composition A (consumption per 1.0 m^3)
1. Slate powder – 628 kg.
2. Dolomitic sand – 1172 kg.
3. Lime paste-328 kg.
4. Wood ash – 50 kg.
5. Opus signinum - 75 kg.
6. Casein glue - 20 kg.
7. Water-l.

Composition B (consumption per 1.0 m3)
1. Slate powder – 644 kg.
2. Dolomitic sand - 1094 kg.
3. Lime paste-328 kg.
4. Wood ash – 27 kg.
5. Opus signinum - 137 kg.
6. Curd degrease-20 kg.
7. Water - 27 l.

Composition C (consumption per 1.0 m^3)
1. Slate powder – 1289 kg.
2. Powder lime - 234 kg.
3. Chicken egg-168 kg.
4. Wood ash-156 kg.
5. Curdled milk - 117 kg.

References

[1] Akhmadov Sh B and Kidirniyazov D S 2016 From the history of the material culture of the Chechens in the XVIII century: monuments of architecture *Bull. of the Kalmyk Institute for humanitarian research of the RAS* **2** 74-80

[2] Akhmadov Sh B 2016 From the history of the development of the culture of the peoples of the North Caucasus during the Great Patriotic War (1941-1945) *Collection of mater. of the all-Russian sci. conf. dedicated to the 70th anniversary of Victory in the Great Patriotic War of 1941-1945,* ed Sh A Gapurov (Grozny: Publishing and Polygraphic center "Groznenskij rabochij") pp 59-66

[3] Akhmadov Ya Z, Gumba G D, Kurumov D S and Hasmagomadov E H 2019 *The history of the Nakhs of West Asia, the Caucasus and Chechnya from the earliest times to the end of the XV century* (Moscow: Litera)

IOP Conf. Series: Materials Science and Engineering 1103 (2021) 012002 doi:10.1088/1757-899X/1103/1/012002

[4] Akhmadov Ya Z and Akaev V Kh 2016 Gumba G. D. Nakhi: questions of ethnocultural history (I millennium b.c.) *Proceedings of higher educational institutions. The North Caucasus. Social Sciences* **4** 95-98

[5] Aliroev I Yu et al 2012 *Chechens* (Moscow: Academic research and publishing, production and printing, and book distribution center of the RAS Nauka)

[6] Bataev D K S, Uzayeva A A and Murtazayeva T S A 2019 Repair modified compositions on the basis of expanding additives *Bulletin of Grozny State Oil Technical University. Technical Sciences* **15** 7-14

[7] Bataev D K S, Apkarov Sh I and Edilsultanova M V 2017 Improved compositions of repair concrete based on technogenic raw materials *Collection "Ecology, health and education in the XXI century. Global integration of modern research and technology. Mater. of the III Caucasian ecological forum"* (Grozny: Chechen State University) pp 115-118

[8] Gumba G D 2007 Alans, aces, and digors on "Ashkharatsuytsu" *Bulletin of the Academy of Sciences of Abkhazia* **2** 225

[9] Mazaeva T A 2019 Chechen Medieval Towers-obelisks To The Issue of Architectural Form Interpretation *Rilem Bookseries* **18** 225-233

[10] Mazaeva T A 2014 Stone towers of Chechnya. On the issue of interpretation of architectural form *Proc. of the Int. sci. and historical conf. named after academic L. Blumentrost* (Berlin, Wissenschaftliche Welt) pp 57-82

IOP Conf. Series: Materials Science and Engineering 1103 (2021) 012003 doi:10.1088/1757-899X/1103/1/012003

Tendencies of non-assur mechanisms use in the structure of technological machines

V S Isakov, V S Deriglazov, O V Isakova and A V Yereyskiy

Platov South-Russian State Polytechnic University (NPI), 132, Prosveshcheniya str., Novocherkassk, 346428, Russia

E-mail: ddp_prof@npi-tu.ru

Abstract. The article gives the scientific research analysis results in the field of non-assured mechanisms on the basis of which the conclusion is made about the need for further development of mechanisms and methodological approaches to the synthesis of machines with special properties of the kinematic structure. This makes it possible to reduce energy consumption and improve technical performance, in particular, grinding machines and presses with gear and jagged actuating devices. The developed general scheme for the formation of the adaptable parameters of the technological machine establishes the dependence of these parameters on the conditions determined by the external environment, the properties of the initial and final product, and also implements the design of a materials grinder with different physical and mechanical properties, a press for molding fuel granules based on gear and jagged actuating devices with tense closed kinematic contours. The results of experimental studies have confirmed the existence of a relationship between the magnitude of the closed kinematic loop prestressing and the energy of the grinding and pressing process. The revealed regularities make it possible to minimize energy consumption in accordance with the strength of the mill charge or the size of the deformed sample by choosing rational parameters of the circuit stress, i.e. to adapt the technological machine to the properties of the processed material. Experimental studies of a gear mill on a sample of sandstone, quartz, marble have confirmed that, with the same energy consumption, the maximum grinding efficiency (an increase in the proportion of fine particles) is achieved only in a certain range of closed kinematic circuit loading corresponding to the strength of the mill charge. The data obtained make it possible to use the effect of a tense closed kinematic contour of gear and jagged actuating devices in technological machines for various purposes.

1. Introduction

The variety of processed materials properties, the variability of technological processes and operations determine the variety of technological machines and equipment: mankind responded to each new technological need by creating a new device.

If we consider the path of machines development, then initially they were represented as something single and integral (a mill, a ship, etc.), intended for some operation or type of work. Over time, it became clear that machines, used in different areas of life, have many similarities, the same functional blocks. For example, the drive, transmission mechanism, working body. The division of the machine into parts according to their functional characteristics was very hard for the humanity and took a long time. Functional and structural diagrams used in modern design practice are still not unambiguous.

The division of the machine into individual elementary mechanisms, kinematic pairs, links, the classification of mechanisms by the function of transforming motion are usually associated with the names of Gaspard Monge and Lazar Carnot. Perhaps this was one of the first attempts to systematically present the theory of the simplest machines and their elements. This division made it possible to make more accurate calculations in static and then dynamic modes. The foundations for the development of the of mechanisms structure and classification theory belong to F. Reuleaux and L.V. Assuru [1].

F. Reuleaux introduced the concepts of a kinematic pair and a kinematic chain, and L.V. Assur developed the classification and principles for constructing mechanisms. However, mechanisms formed by the layering of normal type chains (the so-called normal kinematic chains that do not have excess or missing leads) were taken into consideration. This approach introduced certain restrictions. The classic task of the mechanisms analysis is to identify "extra" kinematic connections and their elimination, which simplified the determination of the number of the degrees of freedom, the calculation of the kinematics and dynamics of the mechanism.

However, the kinematic chains of positive and negative orders, which were later called non-assure kinematic chains, continued to be used in the practice of designing complex mechanisms, giving them special properties and confirming the generally accepted pattern of technical systems development - the pattern of structure complication, the number of elements and connections. So, any technological machine based on a certain physical principle (effect) in its development becomes more complex due to quantitative and qualitative changes, first of all, kinematic chains. Structural formations theory development was not included in the Assur classification.

Particular interest in non-assur mechanisms manifested itself in the second half of the 20th century, which is confirmed by numerous studies in the field of structural theory: adaptive and indifferent mechanisms [1, 2], mechanisms with anomalous structure [3], mechanisms with a negative degree of freedom (static) [4], mechanisms with a tense closed loop [5], mechanisms with a variable structure [6, 7], etc.

In the above theories, structures of mechanisms that were not included in the classical theory of L.V. Assur are considered. The existence and possibility of using such mechanisms in technology was also pointed out by L.V. Assur in his doctoral thesis [1]: “In addition to the mechanisms formed by layering of a normal type chains, there can and indeed be mechanisms formed by chains with excess and missing leashes. These mechanisms drop out of the study, since the study of such chains would complicate the classification and introduce complete uncertainty into it. "Later V.I. Tainov [8], considering anomalous structures of mechanisms, called them "isotopic".

2. Materials and methods of non-assured mechanisms research in the structure of grinding machines with toothed working bodies

The variety of terminology indicates the need for the structural formations’ theory development, methodological approaches to their study and their classification. Today the above studies have resulted in creation of specific machines that use the special properties of mechanisms to improve technological or technical performance.

So, for example, in [6, 7] mechanisms that can change the number of links of kinematic pairs and their mobility, inertial and elastic parameters, variety and reunification of kinematic chains are considered. So, “mechanisms based on the constancy of the structure can be divided into two groups: mechanisms of constant and variable structure” [8].

For the first time mechanisms with variable structure were considered in the works of E.Ya. Antonyuk and S.N. Kozhevnikov [9, 10]. They were further developed in the studies of S. Abdraimov, T.O. Nevenchanaya [6, 7, 11] and others. Numerous designs of impact machines used in construction and mining, technological machines in light industry, etc. were developed on the basis of the created classifications and mechanisms synthesis methods. Further classification showed the presence of two types of basic links: constantly attached to other links and periodically connected to other links.

Sets of basic elementary solutions of mechanisms, built according to strict formal criteria are given in work [4] in a systematic form - the form of a catalog. However, the limitations in the classical theory of L.V. Assur do not allow taking into account all the variety of possible design solutions included in the catalogs. In this regard, K. Roth notes [4]: "In order to be able to describe any (including static) structures, we will also apply the concept of "chain" to single-link pairs ...". He also concludes: “chains with the number of links of the degrees of freedom f≤0 are called here" static ", and kinematic - with the number of degrees of freedom f≥0. That is, in the catalog with f = -1, static circuits can be suitable for use as fasteners and fixed connections. The “statically stressed cycle” is a closed chain of links f = -1, and the force effect is carried out not only through the working surfaces, but also through the working volumes, i.e. processing objects.

Kutzbach [12], Hein [13] and many other authors in their works pointed to the expediency of using "stressed cycles" in "paired connections". The effective use of tense closed kinematic circuits (contours) is realized in wear test stands, in sampling gaps mechanisms, etc. The works [5,14] present the results of a systemic machine mechanisms consideration using tense closed circuits. The mechanisms proposed in [5, 14] were developed and implemented in machines for grinding and cutting various materials, pressing equipment, energy-saving and energy storage drives of construction, lifting and mining machines, etc. According to the proposed classification, mechanisms were investigated with preliminary loading, with variable loading of kinematic variable structures.

Most of the studies [2, 15, 16, 19, 20–23], developing the theory of mechanisms systematization, consider the basic classification of L.V. Assur, which is the main, central core of the general classification. So, in work [16] V.V. Dobrovolsky calls the mechanisms formed by chains with excess and missing leashes, "non-assur chains of positive and negative orders," depending on whether they increase or decrease the number of degrees of freedom and by how many units. In the works of L.T. Dvornikov [17] the proposed by him general (universal) structural classification was complemented and analyzed with non-assur mechanisms.

Substantiated approaches to the synthesis of adaptive and indifferent mechanisms [1, 2] made it possible to obtain a number of fundamentally new technical solutions in various industries, to confirm the advantages and prove the effectiveness of non-assur mechanisms of both positive and negative orders.

From the point of view of the structural theory, adaptive or non-assur chains of positive orders are chains that have “a greater number of mobility degrees than the number of leading links per the number of adaptive links of a given mechanism that connect its driven links in a certain way” [1]. The introduced concept - an adaptive connection - is a condition for the interaction of two moving links on their relative displacement, in which their movement is not determined by either kinematic or dynamic parameters of the mechanism, but depends only on the parameters of the technological (working) process [1]. This makes it possible to automatically regulate the operation of the machine as a whole or its units, depending on external conditions, to redistribute the installed power between the executive bodies in a given way. Research questions and the introduction of adaptive mechanisms were carried out by G.M. Vodyanik, E.V. Rylev, A.N. Drovnikov and others.

The presence of non-assur groups of negative orders in the structure of the mechanism causes the appearance of opposite properties: immunity, indifference to the magnitude of the closing connection. By analogy with adaptive mechanisms, the concept of an indifferent connection is introduced, i.e. closing connection, which reduces the mechanism degree of freedom, but does not reduce its mobility. If the adaptive structural group has an “extra” link that serves the development of a closed contour of the kinematic chain, then in an indifferent mechanism this link forms a new closed loop, serves to increase the number of contours. In this case, the tensile or compressive force applied to the "extra" link cannot be brought to the output link, which makes it possible to create an interference fit in it.

A number of presses and other equipment mills designs developed on this basis by A.N. Drovnikov, S.A. Kuznetsov, V.S. Isakov, V.B. Balashov et al. are given in [18].

3. The results of non-assur mechanisms studies in the structure of grinding machines with gear working bodies

Grinding of various natural and technological materials is a mandatory process in almost all industries: mining, coal, chemical, etc. In agriculture and woodworking industry, about 100 million tons of biological waste are left annually, which are also grinded during subsequent processing. Thus, physical properties, chemical compositions of materials are characterized by a variety and a wide range of values. At the same time, some physical properties, for example, the strength of particles change with an increase in the thickness of grinding. The structure of new composite materials is characterized by the heterogeneity of the individual inclusions' properties: strong, brittle, viscous, etc.

Therefore, the ideal shredder structure should be one that adapts to the physical properties of the starting material. An adaptation in this case means adjustment to the properties of the destroyed material by introducing certain changes in the design (reconfiguration, replacement of individual elements, structural changes or in another way) in order to increase the efficiency and reduce the energy consumption of the process. Such changes can be made through the use of adaptive, indifferent and other mechanisms structures discussed in section 1 of this article. Existing mechanical shredders with gear working bodies, in our opinion, are the most suitable for these requirements.

The general diagram of the adaptable parameters is shown in Figure 1.

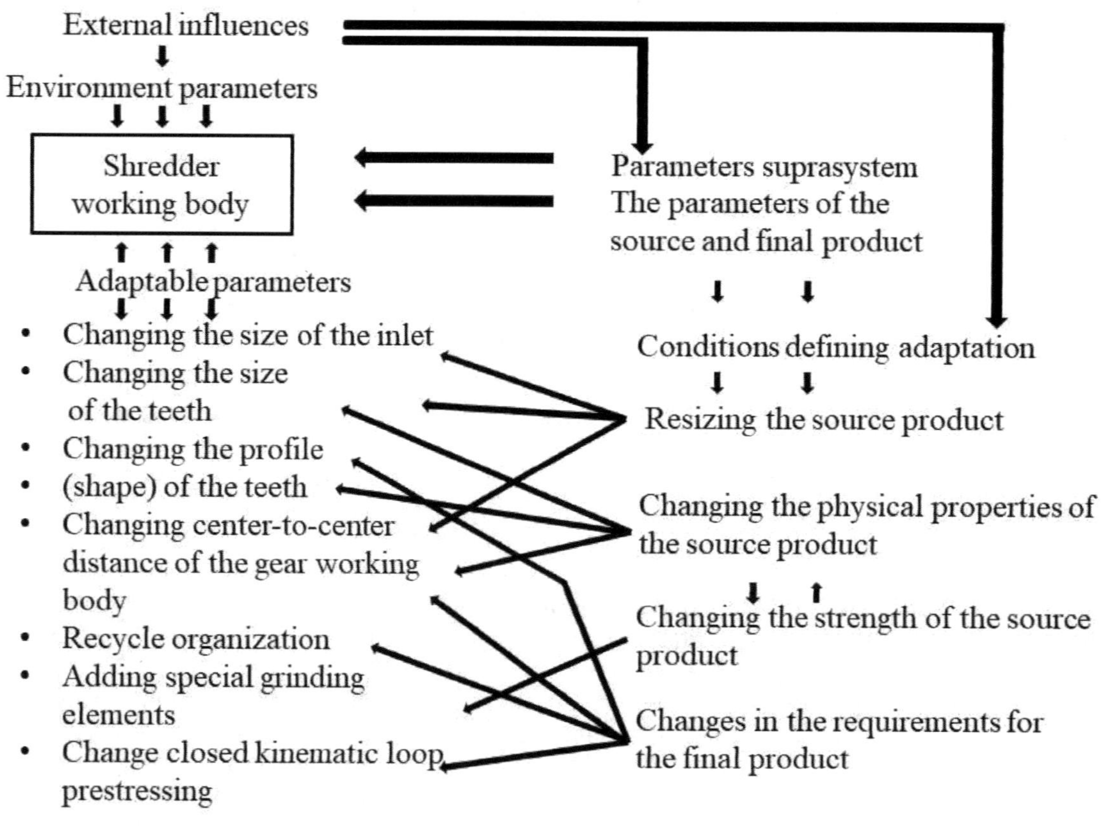

Figure 1. Formation of adaptable parameters

As it can be seen from the presented diagram, the parameters of the initial and final product determine the design features of the grinder working body, determine the need for changes (adaptation), and also directly affect the adaptable parameters. Generally, the conditions that determine non-adaptation can be represented by three groups: changes in the size of the original product, changes in requirements to the final product. Each group requires adequate changes in one or a set of adaptable parameters.

The increase in the size of the initial product determines the size of the mill inlet, the size of the teeth and possibly the center-to-center distance in order to arrange the pre-grinding degree. The requirements for the final product determine the profile (shape) of the teeth, the center-to-center distance, the organization of the recycle, and for mills with closed kinematic loop [5] and the value of prestressing. Changes in the physical properties of the initial product correspond to changes in the size and profile (shape) of teeth, center-to-center distance, the addition of special (for example, cutting) elements, etc. Each change presupposes the presence of dominant methods of destruction. Shredders with a toothed (toothed) working body are able to implement various methods of destruction: crushing, breaking, impact, abrasion, cutting, etc. It is advisable to choose the dominant methods depending on the requirements for the final product and the type of material.

Singling out (Figure 1) the strength of the destroyed material from the whole variety of physical properties, it can be assumed that the main adaptable parameter will be the amount of preliminary compression of the stressed closed kinematic contour. By determining the correct stress for different materials, you can maximize mill efficiency.

4. The results of experimental studies on mechanisms with a tense constantly closed kinematic circuit.

The purpose of experimental studies is to confirm the influence of design factors on the operating properties of the actuator, the presence of a relationship between the closed circuit voltage and the efficiency of the mechanism, determining the amount of prototype shrinkage depending on the load of the closed stressed circuit and the load on the drive, as well as determining the amount of work expended on deformation of the prototype, depending on varying factors.

The general view of the laboratory stand with a stressed permanently closed kinematic circuit is shown in Figure 2.

Figure 2. External view of the laboratory stand with a tense closed kinematic loop

Two blocks are installed on fixed supports. Involute teeth are installed respectively on the handles rigidly fixed to the blocks. The involute engagement allows rolling with minimal slip.

The distance between the supports of the blocks is chosen equal to two radiuses of the pitch circle.

To create the load that simulates the voltage of a closed kinematic loop, the blocks are interconnected by means of a chain, and the load value can be changed stepwise from 25 to 120 kg.

Lead shot with a diameter of 3; 3.9; 4.7; 4.9 mm is chosen as processing objects. Before the start of the experiment, the shot underwent a visual control for spherical shape and instrumental control for

IOP Conf. Series: Materials Science and Engineering 1103 (2021) 012003 doi:10.1088/1757-899X/1103/1/012003

diameter deviation. The admissible value of the diameter deviation is no more than 1%. The material of the object is chosen in accordance with the fact that it deforms plastically, elastic deformations are minimal.

The process of crushing the prototype is carried out by installing it between the teeth, the installation is carried out in such a way that the pressure on it is produced at the point of teeth contact. Additional voltage is gradually installed on the platform of the block, changing in steps of 100 g until the system starts to move. The teeth, rolling towards each other, crush the sample and interlock. The amount of the prototype shrinkage was determined with a micrometer. The spread of the reading did not exceed 6%. The test results are summarized in Table 1, describing the magnitude of the loading, the thickness of the sample after deformation, the force required to actuate the mechanism, the linear shrinkage of the material, the magnitude of the relative shrinkage, the volume of the test specimen, the deformed volume and the magnitude of the relative change in volume.

Table 1. Tests results of on a laboratory bench

Sample	Parameter						Sample					
$D_0 = 4.9$ mm; $V_0 = 61.56$ mm^3	m_Γ, kg	25	50	75	100	112.5	$D_0 = 4.7$ mm; $V_0 = 54.33$ mm^3	25	50	75	100	112.5
	H_1, mm	4.05	2.80	1.71	1.60	1.43		3.89	2.80	1.80	1.52	1.40
	Q_H	43.8	96.9	175	188	193.7		43.2	88.8	160.5	182.3	190.3
	Δ, mm	0.85	2.10	3.19	3.30	3.47		0.81	1.90	2.90	3.18	3.30
	$\frac{\Delta}{D_0}$, mm	0.17	0.43	0.65	0.67	0.70		0.17	0.40	0.62	0.68	0.70
	ΔV, mm^3	2.62	25.5	30.6	35.5	35.4		2.28	11.53	24.66	28.91	0.532
	$\frac{\Delta V}{V_0}$, mm^3	0.043	0.236	0.498	0.528	0.575		0.042	0.212	0.454	0.532	0.567
$D_0 = 3.9$ mm; $V_0 = 31.04$ mm^3	m_Γ, kg	25	50	75	100	112.5	$D_0 = 3.0$ mm; $V_0 = 14.13$ mm^3	25	50	75	100	112.5
	H_1, mm	2.90	1.65	1.13	1.01	1.2		2.20	1.32	0.84	0.70	0.59
	Q_H	41.5	79.1	120.5	156.8	190.1		42.7	75.2	108.3	122.9	141.1
	Δ, mm	1.00	2.25	2.77	2.89	2.70		0.8	1.68	2.16	2.30	2.41
	$\frac{\Delta}{D_0}$, mm	0.26	0.58	0.71	0.74	0.70		0.27	0.56	0.72	0.77	0.80
	ΔV, mm^3	2.80	12.53	17.94	19.26	0.54		1.374	5.408	8.354	9.278	10.02
	$\frac{\Delta V}{V_0}$, mm^3	0.090	0.403	0.578	0.621	0.017		0.097	0.383	0.591	0.657	0.71

5. Analysis of the experimental studies results.

By the amount of information, the conducted experimental research gives an idea of use effectiveness of the closed-loop power mechanisms.

The increase in the value of the relative shrinkage $\frac{\Delta}{D_0}$ is accompanied by a corresponding increase in the additional effort (Figure 3).

Points 1, 2, 3, 4, 5 show the value of the initial load mass (for the sample with $D_0 = 4.9$ mm). That is, with an increase in the weight of the load over 150 kg, the work efficiency increases slowly. For the sample with $D_0 = 3.0$ mm with the same weight of the load, the process is carried out more intensively. Thus, for objects of different diameters and different physical nature, the magnitude of the effective stress of the closed loop is different. Point 2 corresponds to a weight of 50 kg; an additional drive force is 96.9 N.

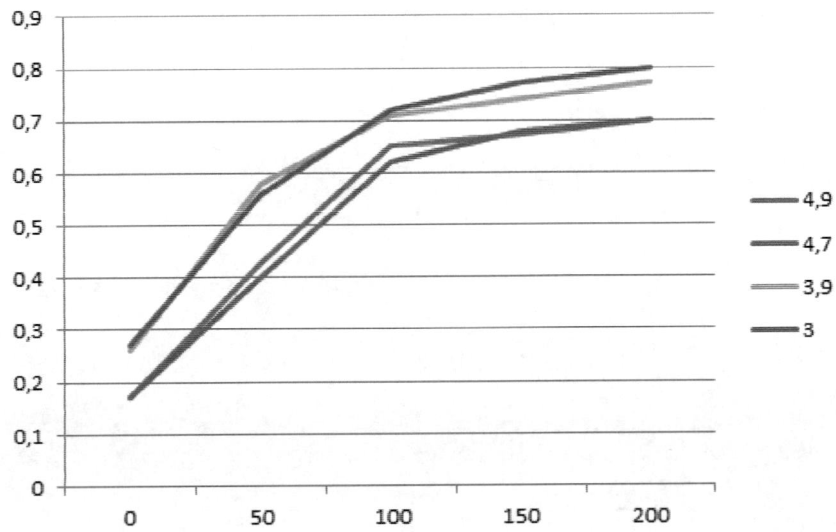

Figure 3. Dependence of the relative shrinkage value on the driving force

According to the obtained data, it is possible to calculate the specific work, that is, the work spent on deformation of 1 mm³.

With the same weight for a sample with a diameter of $D_0 = 3.0$ mm, the additional drive force is 75.2 N.

From a linear shrinkage value, you can go to a more accurate characteristic - the value of the deformed volume ΔV, which can be calculated by the formula:

$$\Delta V = \frac{1}{6}\cdot\pi\cdot\left(\frac{D_0}{2}-\frac{1}{2}\cdot H_1\right)\cdot\left[3\cdot\left(\frac{D_0^2}{4}-\frac{1}{4}\cdot H_1^2\right)+\left(\frac{D_0}{2}-\frac{1}{2}\cdot H_1\right)^2\right].$$

And we calculated the amount of work required to deform a given volume (Figure 3, 4 and Table 2).

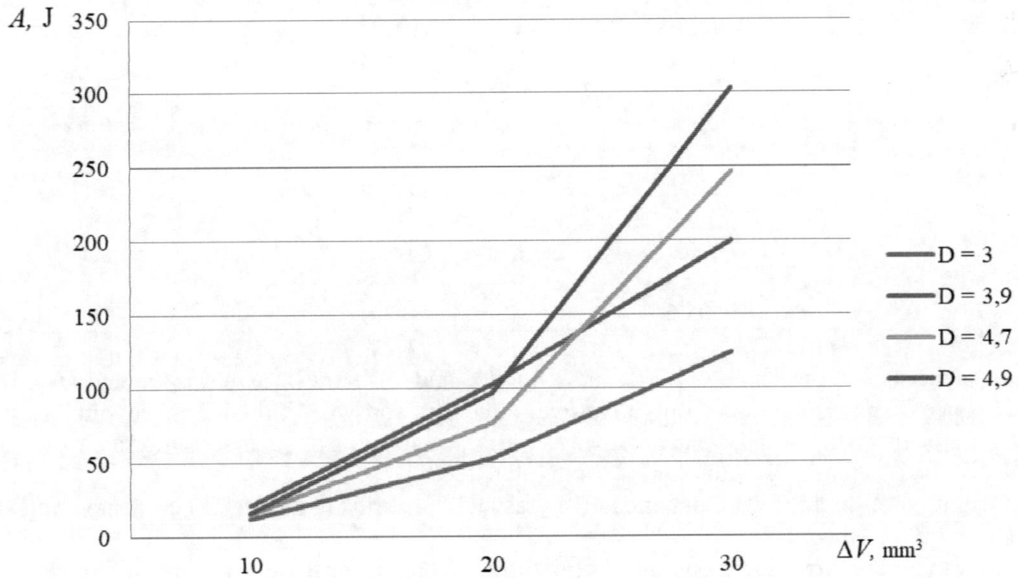

Figure 4. Change in work spent on volume change for samples of different shapes

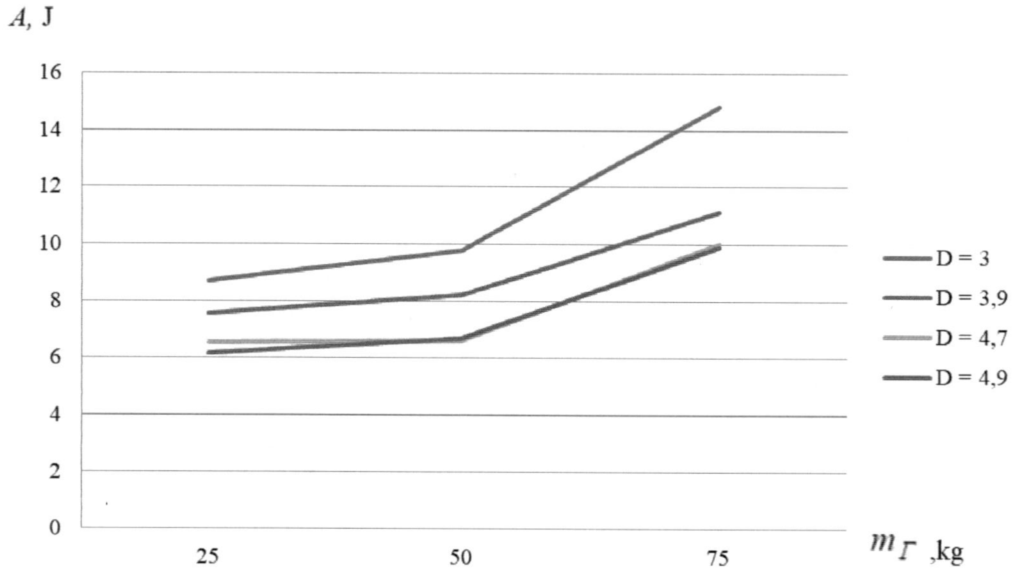

Figure 5. Influence of contour preloading on the amount of work per 1 mm^3 of the deformed volume

Table 2. The results of calculating the work per 1 mm^3 of the deformed volume (for the model of the mechanism)

$m_Г$, kg	ΔV, mm^3	A	$A/\Delta V$
	$D_0 = 3.0$ mm		
25	1.374	12	8.73
50	5.408	52.9	9.78
75	8.354	124.0	14.84
	$D_0 = 3.9$ mm		
25	2.8	21.1	7.55
50	12.53	102.7	8.2
75	17.94	199.8	11.14
	$D_0 = 4.7$ mm		
25	2.28	14.9	6.54
50	11.53	76.5	6.63
75	24.66	246.9	10.01
	$D_0 = 4.9$ mm		
25	2.62	16.2	6.18
50	14.55	97.3	6.69
75	30.66	303.6	9.90

The comparison of pressing processes on a model and in a press in terms of energy costs is incorrect, since the press stress has only a vertical direction, and when pressing on involute teeth the tangent component will also take place. However, it is possible for comparison in terms of technical efficiency of the two pressing technologies, on involute teeth and in a press.

For this purpose, samples were obtained on a press with a shrinkage value (Δ) corresponding to the samples pressed on a model.

The growth rate of work spent on 1 mm^3 of deformed volume for the press is much higher than for the model. The ratio of the amount of work spent on the model for the same deformations is shown in Figure ure 6.

- for $D_0 = 3.0$ mm
$\Delta = 0.80; 1.68; 2.16$ mm;
- for $D_0 = 4.9$ mm
$\Delta = 0.85; 2.10; 3.19$ mm

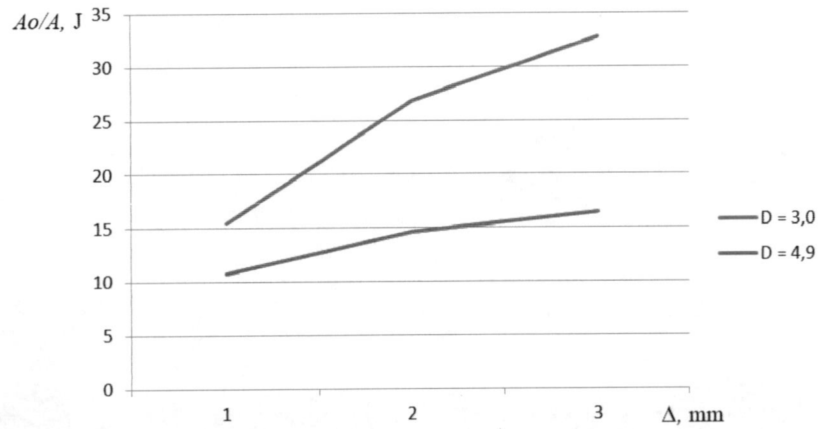

Figure 6. The ratio of the amount of work spent in the press to the amount of work spent on the laboratory bench

The work when pressing the sample $D_0 = 3.0$ mm up to the shrinkage $\Delta = 2.16$ mm using the experimental stand is 32 times less than the work expended by the press. In general, the ratio of works ranges from 10 to 32 times. Thus, the proposed technology of pressing (or grinding) on machines with a permanently closed stressed structure formed by gear drives has a significant economic and technical effect. The presence of tangential stresses on the deformed sample, rolling of the working surfaces can significantly reduce the energy consumption of the process. The value of the circuit rational stress depends on a number of factors: the geometric dimensions of the processing object, the physical properties of the material, etc. Therefore, for each specific case, the magnitude of the circuit voltage, the engagement profile and the parameters of the mechanism must be determined individually, which confirms the validity of the theoretical development of the generalized method for calculating the mechanisms given in Chapter 2.

Based on the results of the research, the design of the mill was developed, shown in the diagram (Figure 7).

The graphs (Figure 8) show the change in the granulometric composition of the material crushed in a mill without loading a closed kinematic loop and with preliminary loading. The volume of particles of larger fractions decreases, and the volume of particles > 0.15 mm increases with the same energy consumption for grinding. In addition, the grinding time is sharply reduced. The presented data confirm the efficiency of using a mill with a stressed closed kinematic loop.

Comparative bar diagrams presented in Figure 9 serve as a more visual form of the results presentation. The increase in the mass of fine fractions is highlighted in the graphs. The maximum increase in the percentage of particles with a size of 0.15-0.315 mm is 7.8% for sandstone. In crushed quartz and marble, the main increase in the mass of small particles occurs in the range > 0.15: 8% and 4.52%, respectively.

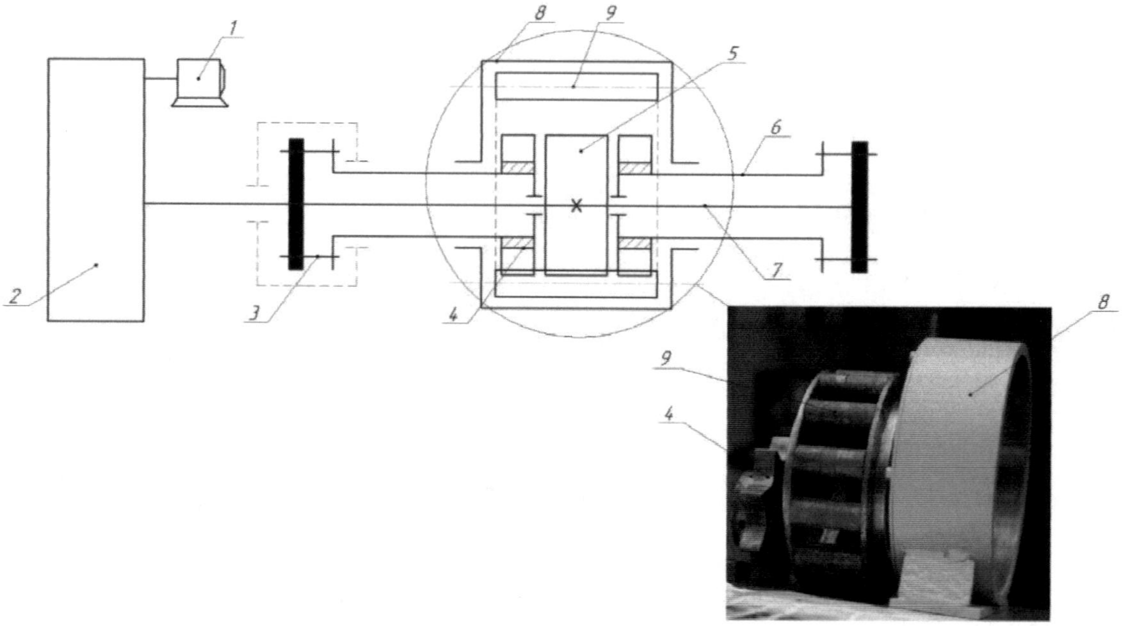

Figure 7. Gear-pin mill with a tense closed kinematic circuit: 1 – motor; 2 – gearbox; 3 - differential couplings; 4 - working body fixed on the hollow shaft 6; 5 - working body fixed on the inner shaft 7; 8 - body mills; 9 - lantern wheel

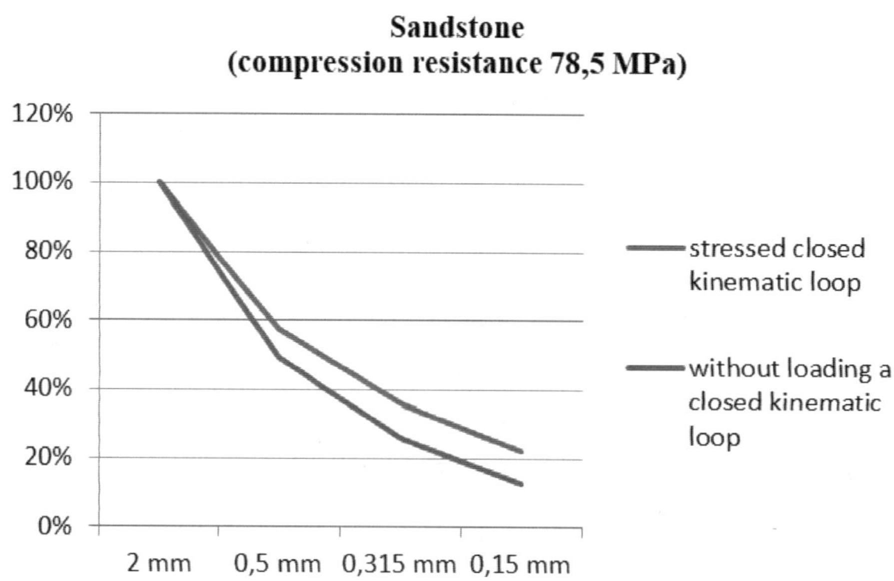

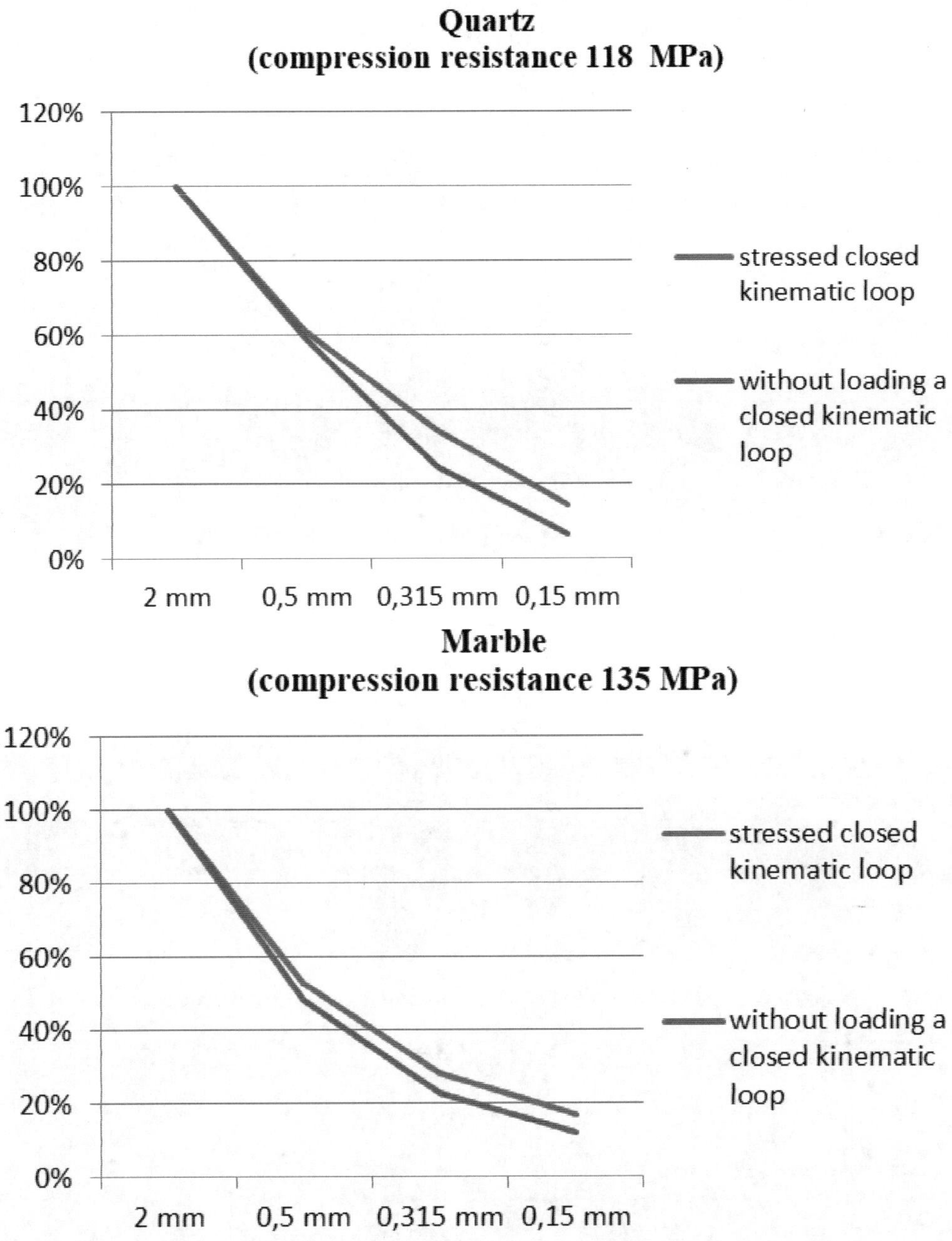

Figure 8. Diagrams of the material granulometric composition crushed in the mill

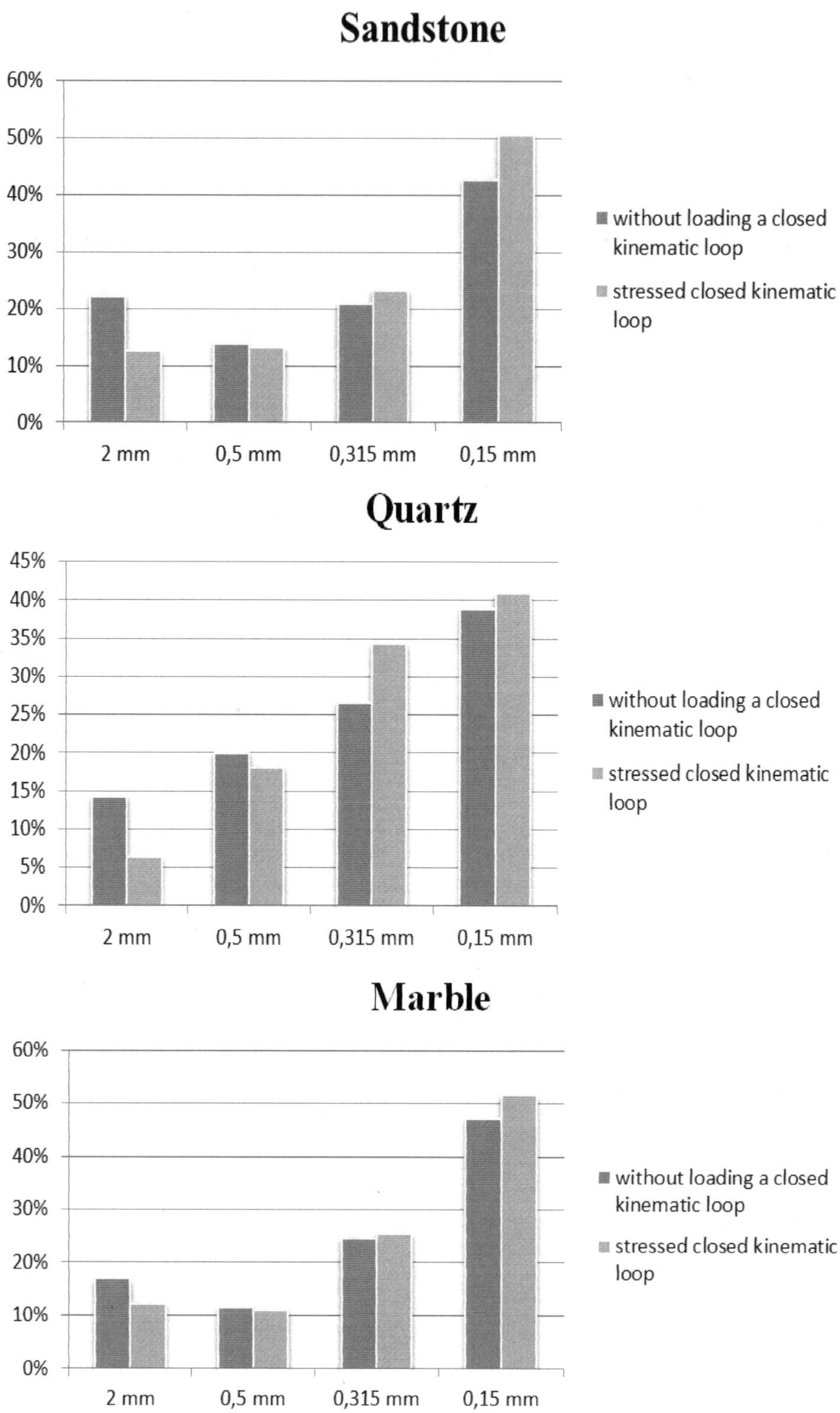

Figure 9. Granulometric distribution diagrams after grinding in a mill

6. Conclusion

The results obtained, firstly, confirmed the operability and efficiency of strained closed kinematic circuits' application in mills with a gear (toothed) working body. Secondly, they showed that the efficiency of crushing and pressing is determined by the magnitude of the closed kinematic loop prestressing, set in accordance with the physical and mechanical properties of the material being processed.

Having carried out experimental studies on the laboratory bench and the developed gear-pin mill, it was possible to obtain the rational loading forces of the closed kinematic loop in relation to different ranges of the crushed materials strength (Figure10) in mills equipped with a gear (toothed) working body.

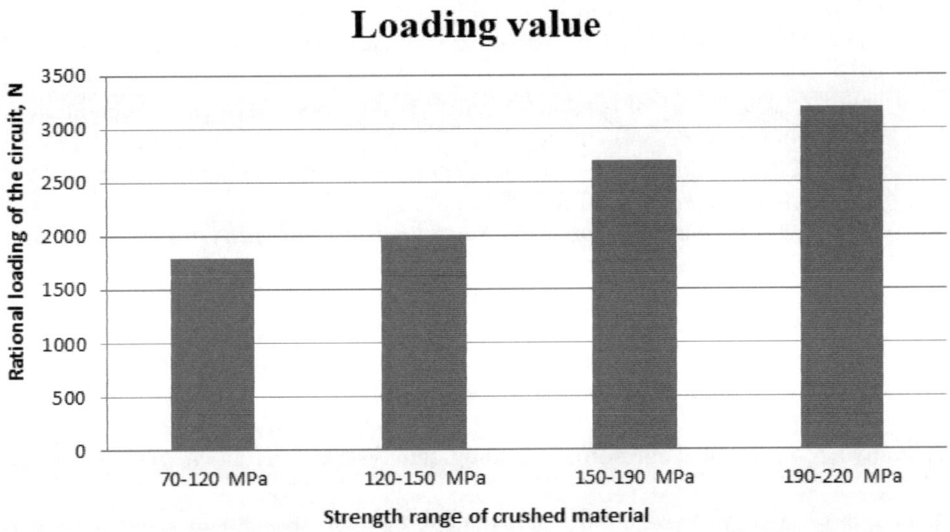

Figure 10. Rational forces of loading a closed kinematic loop

Based on the results obtained in the process of grinding materials, confirming the operability and determining the rational loading range of a closed kinematic loop, using gear (toothed) working bodies, it becomes possible, in addition to grinding or cutting various kinds of materials, to use this technical solution to obtain fuel granules. As in the case of grinding, the manufacture of fuel pellets is a very energy-intensive process that requires an optimal squeezing force of the starting material for the best molding, and the use of the proposed system will significantly reduce the energy consumption in the process of obtaining fuel pellets with the possibility of implementing a sufficient force during molding.

References

[1] Drovnikov A N 1984 *Adaptive structures of mechanisms and machines* (Rostov-on-Don)
[2] Drovnikov A N 2000 *Non-assur structures of mechanisms and machines*: Monograph (Rostov-on-Don: Pegasus) 136 p
[3] Kuznetsov S A and Drovnikov A N 1999 *Integral mechanisms of indifferent structure. Analysis and synthesis* (Novocherkassk: Publishing house of SRSTU (NPI))
[4] Roth K 1982 *Design with Construction Catalogs*
[5] Drovnikov A N and Isakov V S 2006 *Mechanisms of construction and materials-handling vehicle with the use of strained closed circuits*
[6] Abdraimov E S 2001 *Structural synthesis of flat mechanisms of variable structure* (Bishkek: Ilim)
[7] Nevenchannaya T O and Khokhlova O A 2005 Structural and design features of variable structure mechanisms kinematic chains *Bull. of the Astrakhan State Technical University* **2**

[8] Tainov A I 1958 *Foundations of the theory of mechanisms structure* (Department of BPI)
[9] Antonyuk E Ya 1988 *Dynamics of variable structure mechanisms* (Kiev: Naukova dumka)
[10] Kozhevnikov S N and Antonyuk E Ya 1983 Systematization of dynamic models of mechanical units *Theory of mechanisms and machines* **35**
[11] Abdraimov S and Nevenchannaya T O 1990 *Construction of mechanisms of variable structure and study of their dynamics* (Frunze: Ilim)
[12] Kutzbach K 1930 *Basics of fixed pair connections* (VDI-Z) vol 74(45) pp 1541-1545
[13] Hain K 1959 Die Entwicklung von Spannvorrichtungen mit mehreren Spannstellen aus kinematischen Ketten *Das Industrieblatt* **59(11)** 559-564
[14] Isakov V S 2006 On the formation of structures of closed stressed hydromechanical contours *J. of the North Caucasus region Universities* 81-88
[15] Kozhevnikov S N 1979 *Foundations of structural synthesis of mechanisms* (Kiev: Naukova dumka)
[16] Dobrovolsky V V 1939 Theory of mechanisms with two or more degrees of freedom *Proceedings of the Machine Tool Institute* **4** 41-76
[17] Dvornikov L T 2011 Fundamentals of the general (universal) classification of mechanisms *Theory of mechanisms and machines* **9(2)** 18-29
[18] Drovnikov A N 1999 *Indifferent structures of mechanisms and machines:* Monograph (Rostov-on-Don: Pegasus) 108 p
[19] Jinkui C and Weiqing C 1998 Systemics of Assur groups with multiple joints *Mechanism and machine theory* **33(8)** 1127-1133
[20] Li Shujun, Hongguang Wang and Jian S Dai 2015 Assur-Group inferred structural synthesis for planar mechanisms *J. of Mechanisms and Robotics* **7.4**
[21] Isakov V S, Erejskiy A V and Kvetkin A A 2020 Structural analysis of Braking Energy-Conservative Systems of Transport Technological System *IOP Conf. Ser.: Mater. Sci. Eng.* **786(1)**
[22] Sun Yuanxi et al 2016 Solving the kinematics of the planar mechanism using data structures of assur groups *J. of Mechanisms and Robotics* **8.6**
[23] Li S and Dai J S 2012 Structure synthesis of single-driven metamorphic mechanisms based on the augmented assur groups *J. of Mechanisms and Robotics* **4(3)**

IOP Conf. Series: Materials Science and Engineering 1103 (2021) 012004 doi:10.1088/1757-899X/1103/1/012004

Evaluation of significance of control directives for software errors detection in functional programming of on-board computers

V A Drabov, N V Romanova, T S Novikova, A V Vernigor, A V Rekovets and N V Ptitsyna

Smolensk State Agricultural Academy, 10/2, Bolshaia Sovetskaia street, Smolensk, 214000, Russian Federation

E-mail: romanovanv0403@yandex.ru, sgsha@sgsha.ru

Abstract. Modern farm machinery is equipped with on-board computers that execute complicated tasks. The main efforts are aimed at selecting and justifying methods for adequate hardware-software translation of functional programming from one computer system to another, while increasing its stability to failures of computer hardware. The development of approaches to the selection of procedures for detecting errors determines the degree of their relative information value when analyzing the programs progress. As it follows from the analysis of error types in functional programming, it is established that errors occur at the stages of forming the technical task, when developing algorithms, when programming, and when configuring programs. Errors have different frequency and different weight in terms of the impact on the functioning of on-board computers used in agricultural machinery. The article contains a detailed list of errors that must be detected when debugging working programs of on-board computers. Based on expert assessments, the errors are divided into 33 different types and formed into 6 groups. By combining errors, it is possible to minimize the time spent on performing static debugging of the functional programming of on-board computers. Using the hierarchy analysis method, matrices of paired comparisons of control directives with error types in each group are formed. Thus, the relationship between groups of errors, control directives, and values of error types is established. The first thing, programs debugging is performed using the control directives that have the highest degree of significance, since they can be used to identify and eliminate the most dangerous errors.

1. Introduction

Computer-based system as a main element in agricultural machinery, impliments the operation algorithm and completes tasks of functional programming in real time mode. Moreover with a serial upgrading of any unit of agricultural machinery where on-board computers are used, the existing functional programming is widely used along with a newly written one, as it is practically impossible to repeate the whole volume of its in-process tests [1].

The so-called simulators, special program models that simulate the funtioning of computer systems are used for debugging the functional software. Nowadays this issue is becoming more and more intensified as far as the volume of processed data in on-board computers of agricultural machinery is increased and the topology of information flow is becoming more complicated [2; 3].

The main stages of validity testing of functional programming are the so-called static and dynamic debugging. The main purpose of a static debugging is the identification and elimination of errors in program bodies.

2. Materials and methods

The research material is functional programming of on-board computers used in agricultural machinery. In the process of developing functional software errors inevitably occur in the programs. They are caused by inaccuracies in the technical tasks, imperfection of the used algorithm and programming errors. [4]

The assessment of the degree of experts' opinions was made on rank correlation coefficients:

$$P_{\alpha\beta} = 1 - \frac{6}{N(N^2-N)}\sum_{i=1}^{N}(P_\alpha - P_\beta)^2 \quad (1)$$

Where N – the initial number of experts, P_α and P_β – evaluation of ranking of some parameters of text tasks α and β by the experts.

To calculate the expert evaluation coefficient (EEC) the expert is not to specify a certain number but a possible interval $[q1, q2]$,q2>q1, then for the j-type error and m number of experts a Qj range of interval expert assessments is formed,

$$Q_j = \left\{\left[q_{j1}^{(1)}, q_{j2}^{(1)}\right], \left[q_{j2}^{(2)}\right], \ldots., \left[q_{j1}^{(m)}, q_{j2}^{(m)}\right]\right\} \quad (2)$$

To choose a generalized interval of rates of EEC the minimum and maximum intervals are used:

$$q_{j1} = min\left\{q_{j1}^{(1)}, q_{j1}^{(2)}, \ldots.., q_{j1}^{(m)}\right\}$$

$$q_{j1} = max\left\{q_{j2}^{(1)}, q_{j2}^{(2)}, \ldots.., q_{j2}^{(m)}\right\}$$

The arithmetical average of EEC is taken as shared experts' opinion

$$Q_j = \frac{1}{2}\left(q_{j2} + q_{j1}\right) \quad (3)$$

3. Evaluation of software error detection

The organization of the functional programming involves the development of supporting means of the static debugging process based on the applicationt of special patches in the general program body, the so-called control directives (CD) for the search for possible crashes or errors in main program commands. It is obvious that their quantity should not be large as they demand some amount of memory and time resources consumption of the computer system.

The evaluation of control directives for software error detection of functional programming is a complicated, time-consuming and considerably creative process. Its peculiarity results in impossibility of its formalization by means of common mathematical methods as they do not allow working out a solution in conditions of fuzzy initial data presented not in numeral but in qualitative form. That is why a method of expert evaluation is applied for the offered methodology for solving this hard-to-formalize task. [5]

The first stage of a methodology for evaluating of control directives for software error detection of functional programming is the forming of the technical expert advisory group.

The final choice of expert group is carried out according to a maximum value of the rank correlation coefficient based on the Formula 1. This figure is m = 6.

At the second stage of a methodology for each error type of the fuctional software EEC within the interval [0,1] is calculated. It reflects the influence rate of errors on the on-board computer system operation.

At the same time the one-dimensional or qualimetry condition is met. Due to this condition EEC for each error types is calculated by one number. The error classification is shown in the Figure 1.

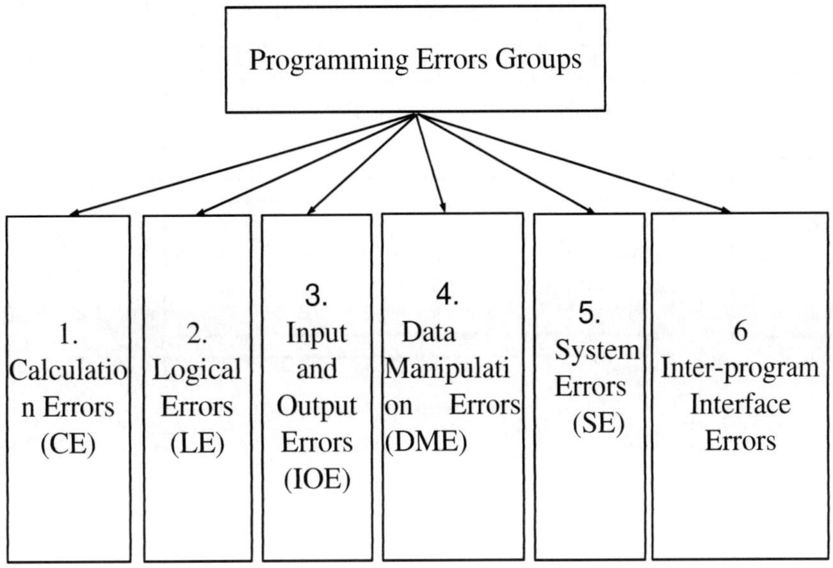

Figure 1. Error Classification

At the third stage a choice of control directives for software error detection in functional programming is carried out. A control directive is a special technological program as a component of the software development automatization system. It is designed for support a user's checkout method of the selected part of the program with address (memory and register), operand values and results visualization.

It is a natural programmers' attempt to compose a separate control directive to detect each type of errors. However, this approach makes the task of debuggingthe programs more complicated. It is important to minimize the quantity of directives. One and the same control directive should be preferably used to detect several error types in different groups. It can be actualized on the assumption of the possibility of joint application of several control directives for error detection in each group.

As a result of this approach the experts calculated and formed five universal control directives for the error detection in six selectad groups. (Table 1.).

Table 1. Control Directive's (CD) Name and Purpose

Code	CD Name	CD Purpose
CD1	Recording the program progress by the control counter when passing the address interval $Ai - Aj$	Reading a functional programming fragment from the address Ai to Aj for further analysis and checkout
CD2	The timing of the progress of the program counter between addresses $Ai - Aj$	Changing of the program execution time and comparing it with the allocated time resource
CD3	Circularity Measurement	Calculation of the number of cycles and comparison of it with the required value
CD4	Recording the program progress at the RAM Executive address	Checking operands and results accuracy in RAM
CD5	Stop recording the program progress at the RAM executive address	Control of the fact that a given operand is written to a specific RAM executive address

IOP Conf. Series: Materials Science and Engineering 1103 (2021) 012004 doi:10.1088/1757-899X/1103/1/012004

At the fourth stage of the methodology, models of relationships between control directives and error types in each group are built to assess the significance of each control directive.

In Figure 2 for example, a model of relationship between control directives and error types in inter-program interfaces. The control directives values for each type of error are shown in circles.

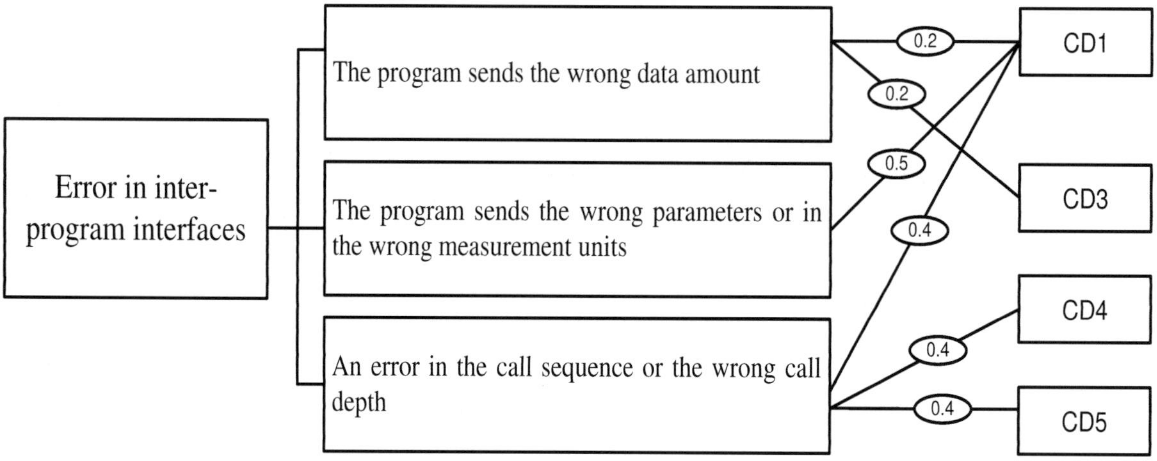

Figure 2. Model for informative value of control directives for detecting errors in inter-program interfaces

The hierarchy analysis method is used to process expert information when determining the significance values of control directives [5; 6].

The value of each control directive for a group is defined as the weighted average sum of the values of the control directives in relation to the types of errors in the group (Table 2).

Table 2. Degree of Control Directives Value for Groups of Errors

	CE	LE	IOE	DME	IIE	SE
CD1	0,57	0,77	0,37	0,42	0,56	0
CD2	0	0	0	0	0	0,54
CD3	0,15	0,23	0	0	0,1	0
CD4	0,08	0	0,315	0,29	0,17	0,23
CD5	0,2	0	0,315	0,29	0,17	0,23

The degree of control directives value is essential for the organization of the debugging procedure for functional software, since they may be able to identify and eliminate the most dangerous errors. Software components are developed by separate groups of programmers.

The procedures for the detection of functional software errors were taken into account when forming the initial prerequisites for the development of a static debugging systems.

A feature of this logical organization of the system for static debugging of functional software is the modular multi-level construction of its interconnected hardware and software components.

4. Conclusion

Thus, to find and localize errors while analysing the programs progress, it is proposed to use special control directives designed to control the process of debugging functional software. The control directives are divided into two main types. The first type is implemented by placing stop addresses or interrupting the program's progress. The second type is implemented by specifying the operands names (physical addresses), whose values are registered each time the program accesses them, either only when reading or only when writing their values.

IOP Conf. Series: Materials Science and Engineering 1103 (2021) 012004 doi:10.1088/1757-899X/1103/1/012004

References

[1] Morozov D S 2014 Topicality of the principles of test driven development within the scope of assesing results of practical activity regarding software engineering *European Applied Sciences* **10** 69-72

[2] Bazhenov A S and Itsykson V M 2013 Forecasting software development project characteristics using meta-modeling *ACM Int. Conf. Proceeding Series*

[3] Furrer F J 2019 Future-proof software-systems: a sustainable evoution strategy *Future-Proof Software-Systems: A Sustainable Evolution Strategy* pp 1-376

[4] Kalikhman D M, Kalikhman L Y, Sadomtsev Y V, Deputatova E A, Turkin V A, Nakhov S F, Nikiforov V M and Tchaikovsky M M 2013 Principles for designing digital robust controllers for measurement channels of angular elocity and proper acceleration in modern strapdown inertial navigation systems and software for their monitoring *20th Saint Petersburg Int. Conf. on Integrated Navigation Systems, ICINS 2013 - Proceedings* pp 343-350

[5] Godlevsky M D, Shekhovtsov V A and Braginsky I L 2012 Principles of modeling, assessment and quality management of the software development process *East European J. of Advanced Technologies* **5(3(59))** 45-49

[6] Kulakov A V, Golovanev V E, Gorbacheva L A, Yusupov S K, Semishkin V P and Shary N V 2016 Development of principles for restoring software and hardware complexes as a means of increasing reliability and maintainability in complex automated process control systems *Heavy mechanical engineering* **5** 22-26

Comparative analysis of experimental methods to assess the resistance of milling road asphalt concrete

D V Furmanov

Yaroslavl State Technical University, 88, Moskovskiy prospekt, Yaroslavl, 150023, Russia

E-mail: furmanovdv@ystu.ru

Abstract. The paper is devoted to the comparative analysis of experimental evaluation of cutting resistance of road asphalt concrete to cutting elements with cylindrical carbide tips, which are widely used in rotors of road milling machines. The paper shows the comparison of the results obtained from the study of cutting of typical asphalt concrete with a single cutter on a pendulum stand and on a stand with a mechanical low-speed drive of the cutting element. The advantages and disadvantages of both methods are determined and provided in the paper. The conclusions obtained allow us to recommend reasonably the methods of research of strength indicators of asphalt concrete from the point of view of cutting resistance to cutting with working bodies of road machinery.

1. Introduction

It is difficult to choose and justify the rational parameters of new milling machines, as well as choose the rational technological parameters of existing machines without detailed studying of interaction of a cutting element with a material. With regard to cutting resistance, asphalt concrete of various grades shows a wide spectrum of the properties depending on the changes in external conditions. Such properties essentially make studying of this process more complex [1]. Being a dispersedly filled thermocomposite [2, 3], asphalt concrete has a set of complex strength properties, which depend on the strength of the filler, fractional composition and structure, properties of the binder, and temperature [4]. The structure of the asphalt concrete is formed as early as in the mixing chamber and changes during paving and operation [5].

Currently, these circumstances do not allow simulating the cutting of this material using only a set of analytical methods, and the correct choice of the experimental method will influence not only the volume and accuracy of information received, but also the degree of compliance of the method with the actual process.

2. Study methods

The first method that can be potentially used in the study of cutting is the method based on measuring the cutting energy of a single cutting element when using a freely dropped pendulum (Fig. 1). This method is widely used to determine the energy of ductile failure of standard samples of metals and polymer materials [6, 7]. With regard to determination of energy intensity of cutting, this method was effectively used in research of working processes of peat deposit milling [8].

The operation of the test bench is based on measuring the cutting forces based on changes in kinetic and potential energy during the cutting:

$$A = m_p \cdot g \cdot (h - h_0) \tag{1}$$

where h is pendulum drop height; h_0 is pendulum deflection height after impact; g is the gravity acceleration; m_p is the specified mass of the moving parts of the test bench at the impact point of the cutting force.

Figure 1. Pendulum test bench for studying inertial cutting of materials.

The following ratio can be used to calculate the last variable:

$$m_p = \frac{T^2 \cdot M}{4 \cdot \pi^2 \cdot L^2} \tag{2}$$

where T is experimentally determined pendulum period; M is the first moment of the pendulum, which is also determined experimentally by lifting the pendulum; L is the distance between the cutting zone and the rotation axis.

The following shows how the cutting resistance forces are calculated:

$$A = \int_0^{l_{max}} F(l)dl \tag{3}$$

where l_{max} is the sample length; F(l) is the horizontal cutting force that change functions.

Since the last value is unknown, the average of the horizontal cutting resistance force can be estimated:

$$F_{av} = \frac{A}{l_{max}}. \tag{4}$$

This value allows us to calculate the energy values of the working process, determine the drive power and the average torque of the milling cutter rotor. However, the following values remain unknown: the values of the vertical cutting resistance forces, the peak values of cutting forces, the pulse rate of cutting forces due to the uneven structure of the material.

Nevertheless, the undisputed advantage of this method is that it most accurately reproduces the real process: since a real cutter is installed on the pendulum, the required cutting speed is achieved by selecting the desired drop height, and it is also possible to install thermostatted samples with different temperatures.

The second method is based on the direct measurement of the cutting resistance forces by power sensors (Fig. 2). This method is most conveniently implemented when using a metalworking (milling) machine equipped with special fixtures for sensors, cutting elements, and material. Such design provides high rigidity of mountings of main elements of the measuring complex, accuracy of positioning of the cutting tool relative to the workpiece, and uniformity of feed. The design allows obtaining oscillograms of vertical and horizontal components of cutting resistance forces (Fig. 3). Thus, the figure shows an oscillogram of the horizontal resistance force to cutting of asphalt concrete of grade B2 [9] with 2-mm thick chips to be cut. The data were obtained at a sample temperature of 22 °C.

Let us consider the basic advantages of this type of equipment. The undoubted advantage is that this equipment allows us to determine not only horizontal but also vertical cutting resistance forces, which provides us with a way to fully analyze the loads on the working body and the cutting element. In addition, the analysis of the oscillogram of the components of cutting resistance forces allows us to determine not only the average values, but also the amplitude of the components and the full cutting forces arising from the interaction of the cutting element with asphalt concrete components.

Figure 2. A mechanically driven test bench for cutting asphalt concrete.

The only disadvantage of this method is that it is not possible to achieve cutting speeds that correspond to actual speeds when milling asphalt concrete. Taking into account that the cutting speed of road milling machines reaches 3.5 - 4.5 m/s [10], and the feeds of existing universal milling machines reach 1 - 1.5 m/min, the components of cutting resistance forces measured by this method will be somewhat underestimated.

3. Comparative analysis of the methods

The experience gained with both methods allows us to compare the obtained values of cutting resistance forces and draw conclusions. In the paper published earlier [11], we obtained the following regression equation describing the average value of the cutting resistance force component using a pendulum test bench on asphalt concrete, grade B2:

$$F = 1971 \cdot \Delta h^{0.45} \cdot t^{-0.445} \tag{5}$$

where h is the chip depth, mm; t is the temperature, °C.

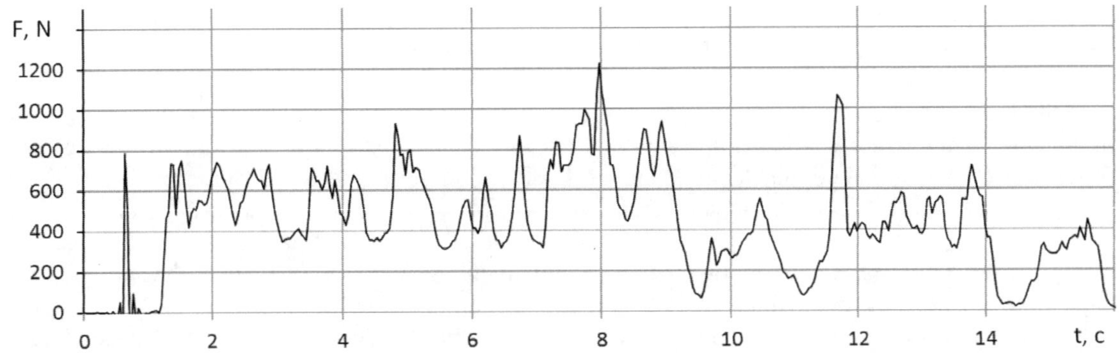

Figure 3. Oscillogram of the horizontal component of cutting resistance forces obtained by cutting asphalt concrete, grade B2 at a chip depth of 2 mm.

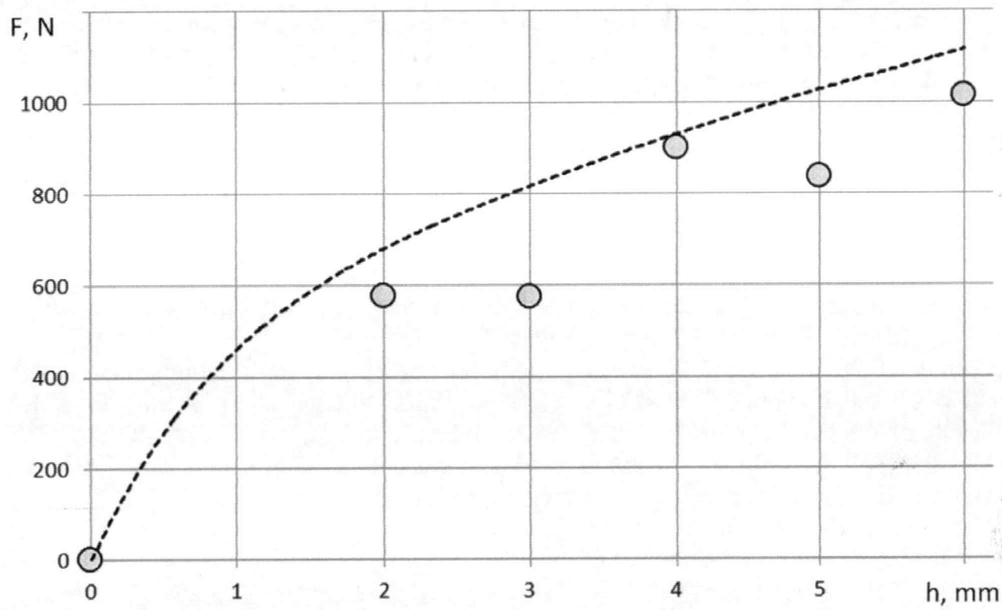

Figure 4. The dependencies of the horizontal component of the cutting resistance force of asphalt concrete of grade B2 obtained on the pendulum bench (indicated by the plot) and on the mechanical bench (indicated by points).

The cutting resistance force component obtained using this formula is 680.4 N. The oscillogram analysis in Fig. 3 shows that the average value of the horizontal component of cutting resistance force is 578.6 N.

Let us also consider the rest of the plot points (Fig. 4). The curve on the graph reflects the dependence on the chip thickness of the tangent component of the cutting force of asphalt concrete, obtained on the pendulum test bench and reflected in the paper mentioned previously. The curve represents a section of the approximating surface of the empirical function (5) in the temperature plane of t = 22 °C. Points on the plot show the results of the experiment, obtained on a mechanical test bench when cutting a sample of asphalt concrete of the same grade with the same cutting element.

Comparative analysis of the two plots shows the average divergence not exceeding 15%, which can be caused by factors related to the measurement error and spread of the strength values of structural components of asphalt concrete and their bonds in two different samples. However, the most likely reason for the underperforming mechanical bench is the difference in cutting speeds.

4. Conclusion

Both methods demonstrate a satisfactory relative error and, consequently, the possibility of using both methods in practice. Both methods demonstrate workability and good mutual convergence.

The pendulum test bench is simple and cheap; the cost of experiments on such equipment is much lower than the use of more complex mechanical methods. In addition, when it comes to cutting asphalt concrete, the pendulum bench makes it possible to implement the working process at the same speeds as those implemented in real machines. Unfortunately, the pendulum bench is not fit to study dynamic phenomena arising in the contact of the material with the cutting element. The use of a pendulum bench should be recommended to study the energy capacity of milling various road construction materials (not only asphalt concrete).

The method based on the use of mechanically driven test benches can be recommended for detailed study of cutting of road-building materials, including determining peak values of cutting forces, the horizontal component, as well as the full cutting resistance force, with dynamic phenomena taken into account. It is to be noted, however, that not all workflow parameters can be implemented on such machines in the same manner as they are implemented on real machines. Cutting speeds on such equipment will be significantly lower, which may lead to lower cutting resistance forces.

The question of the impact of cutting element speed on cutting resistance for modern asphalt concrete is also subject to study.

References

[1] Baratashvili M P 2019 Determination of factors influencing the operation modes of a machine and their importance for destruction of the surface asphalt layers *Scientific digital archive*. Retrieved from: http://econf.rae.ru/article/6606
[2] Bazhenov S L 2014 *Mechanics and technology of composite materials* (Dolgoprudny: Intellect)
[3] Matthews F and Rawlings E 2004 *Composite materials. Mechanics and technology* (Moscow: Tehnosfera) 408 p
[4] Ryb'ev I A 1969 *Asphalt concrete* (Moscow: Vysshaya shkola) 399 p
[5] Gezenzvey L B 1976 *Road asphalt* (Moscow: Transport) 336 p
[6] 2017 GOST 4647-2015 *Plastics. Method for determination of Charpy impact strength. Active since 20.11.2015* (Moscow: Standardinform)
[7] 1994 GOST 9454-78 *Metals. Method for testing the impact strength at the low, room and high temperature. Active since 17.04.78* (Moscow: Standards Publ. House)
[8] Samsonov L N 1985 *Peat deposit milling* (Moscow: Nedra) 211 p
[9] 2010 GOST 9128-2009 *Asphaltic concrete mixtures for roads and aerodromes and asphaltic concrete. Specifications. Introduced since 1.01.2009* (Moscow: Standartinform)
[10] Agarkov A M, Tikhonov A A and Prokhorov D S 2016 Analysis of asphalt concrete milling equipment *Innovative development* **5(5)** 7-8
[11] Furmanov D V, Chizhov V S and Lysakov N E 2020 Experimental determination of cutting resistance strength at asphalt concrete destruction with a single cutting element *The Russian Automobile and Highway Industry J.* **17(2)** 196-207. Retrieved from: https://doi.org/10.26518/2071-7296-2020-17-2-196-207.

Vibrodiagnostics of hoisting crane mechanisms using simulation modeling

A Yu Ganshkevich[1] and Ph O Fadeev[2]

[1] Academy of Water Transport, Russian University of Transport, 22, building 3, Novosuschevskaya street, Moscow, 127055, Russian Federation
[2] Original Soft LLC, Original Group JSC, 20, building 27, Ogorodny passage, Moscow, 127322, Russian Federation

E-mail: ganshkevich.a.yu@edu.rut-miit.ru

Abstract. The subject of the research is an opportunity to increase the efficiency of vibration diagnostics of hoisting cranes through the use of simulation of the vibration signal. In the article the basic approaches developed by the authors to construction of the type diagram of a mechanism and modeling of a vibrating signal of the crane mechanism with operational defects are resulted. The basic design dependences for the description of a vibrating signal of the mechanism with defects at presence of amplitude and frequency modulation are obtained, approaches to modelling of stages of acceleration and braking of the mechanism are formulated, the technique of modelling of a noise component of a signal is identified. The possible areas of application of simulation modeling during vibration diagnostics are shown and a brief description of the results obtained is given. Pre-imitation modeling will improve the detection of defects during the vibration diagnostics of cranes and reduce the time spent on the analysis of the obtained results.

1. Introduction

Nowadays, one of the most pressing problems in the operation of hoisting machines serving large transport terminals is sudden failures, which lead to unplanned downtime of equipment and serious losses. Statistical analysis of accidents and incidents in Russian ports showed that about 50% of failures arise from technical reasons: wear and tear, fatigue, etc. Decrease in the number of sudden failures can be achieved by introducing early fault diagnosis methods.

Vibrodiagnostics is one of the most effective methods of diagnosing mechanical equipment that allows one to detect defects at the stages of its origin and early development without disassembling and dismantling the units [1, 2]. The interpretation and analysis of crane mechanism vibration signals cause specific problems [3, 4], some of which can be successfully solved by using frequency-time analysis and adaptive filtering [4]. However, when analyzing mechanisms with a large number of vibration sources, the authors have faced a number of problems:

- for effective application of bandpass filters, it is necessary to know a priori the frequencies at which certain defects are most likely to be detected;
- to identify defects by frequency peaks when signals from several defects are superimposed, you need to know the values of combination frequencies

- when checking hypotheses about the relationship of individual frequency peaks with specific defects, it is necessary to be able to investigate in detail the contribution of individual defects to the overall vibration spectrum.

To solve these problems, the authors developed a technology of simulation of the mechanism with operational damage, which allowed one to get a complete and more detailed picture of the composition of the vibration signal.

2. Task Setting

In any crane mechanism there are many components that are sources of vibration: clutches, shafts, bearings, gears, etc., each of which is characterized by its forced frequency, depending on the frequency of the excitation force at the input link of the mechanism. This frequency is commonly called the fundamental frequency of the unit. When defects occur, the vibration signals of the units are modulated by frequency, phase or amplitude, which leads to the appearance in the spectrum of harmonics and subharmonics of fundamental frequencies. In addition, some units demonstrate a vibro-response at its natural frequency, which does not depend on the fundamental frequency of the unit [1, 3].

The interaction of different sources of excitation leads to the appearance of a large number of combination frequencies. In case of non-stationary operation of mechanisms, which is caused by changes in the load and repeated short-term operation of the cranes, these frequencies change significantly during the cycle.

The simulation model of the mechanism should consider these factors for correct work.

3. Imitation model of mechanism

3.1. Model schematization

At the first stage of modeling, the structural scheme of the mechanism was constructed, the components of which are units and defects. A unit is a structural element that generates a vibration signal during operation, such as gearing, clutch, bearing ring, etc. Each unit is characterized by its fundamental frequency. A defect is a structural element that modulates the fundamental frequency of a unit, such as static or dynamic coupling unbalance, pitting in a gearing pair, etc. When constructing the structural diagram, the elements were in the same sequence as in the mechanism. In fig. 1, as an example, a fragment of the structural scheme of the winch of the crane hoisting mechanism "Albatros" 10/20-32/16 is given.

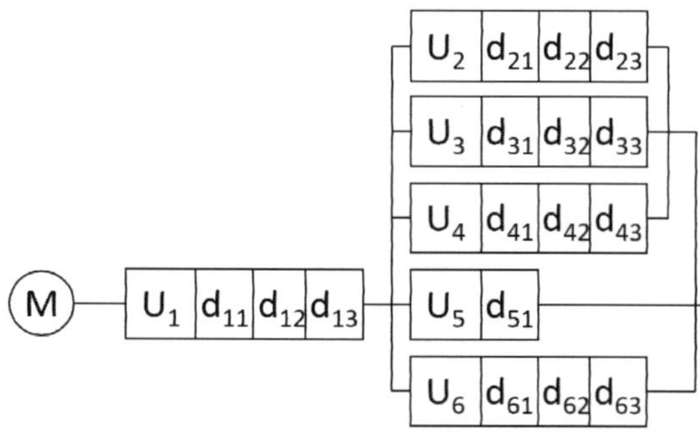

Figure 1. Fragment of the structural diagram of the hoisting mechanism winch.
M – Primary excitation source: electric motor;
$U_1 \dots U_6$ – units: clutch, cage, inner and outer bearing rings, high speed shaft and gearing, respectively;
$d_{11} \dots d_{63}$ – possible defects of the respective units: unbalance, displacement, deformation, wear, pitting, etc.

3.2. Excitation effect simulation

Since most of the crane mechanisms mainly operate at variable speed, the vibration signal at the input was simulated to consider the change in the frequency of the excitation force. The variation of the frequency occurs during start and brake of the mechanism, as well as due to the instability of the drive load. The law of excitation frequency change was simulated by function (1):

$$f_0(t) = f_{0n} \cdot k_{ab}(t) \cdot k_f(t), \qquad (1)$$

f_{0n} – rated frequency of the excitation force during steady motion; $k_{ab}(t)$ – coefficient that takes into account the acceleration or braking of the mechanism and is determined by the formula (2); $k_f(t)$ – coefficient that takes into account changes in motor speed due to variable load.

The law of motor speed change at start and braking modes was set based on experimental studies. The experiment was conducted on portal slewing cranes "Albatros" 10/20-32/16, "Albrecht" 10-32, "Condor", belonging to "North Port" JSC, Moscow. The mentioned cranes have asynchronous motors with a phase rotor and relay and a contactor control circuit as a part of the electric drive. Estimation of motor rotation speed was carried out based on measurement of the motor supports vibration frequency measured by accelerometers. Accelerometers VS 201 (no. 49619-12 in the Register of Measuring Instruments of the Russian Federation) and analog-to-digital converter ZET 220 produced by CJSC "Electronic Technologies and Metrology Systems" ("ETMS" CJSC) were used for measurements. Measurements were made by the "blinded" method, i.e., the crane operator was not aware of the measurement and worked in the normal mode.

As shown by the analysis of the dynamics of acceleration and braking modes of portal slewing cranes, the dependence of the drive speed on time during the cycle is well described by the piecemeal function of the type (2):

$$k_{ab}(t) = \begin{cases} \sin\left(\frac{\pi \cdot t}{2 \cdot t_a}\right), & t < t_a \\ 1, & t_a \le t \le (t_c - t_b), \\ \sin\left[\frac{\pi \cdot (t - t_c - 2 \cdot t_b)}{2 \cdot t_b}\right], & t > (t_c - t_b) \end{cases} \tag{2}$$

t_a – acceleration time of the mechanism to the nominal speed; t_c – operation cycle time of the mechanism; t_b – braking time of the mechanism.

The law of coefficient change $k_f(t)$ depends on the mechanism that is modeled. For the hoisting mechanism, its value depends only on the load on the ropes and the direction of movement of the load, and is within the range of $k_f \in [0.97; 1.05]$, and for the derricking mechanism changes smoothly when the load moment varies, which is well described by the dependence (3):

$$k_f(t) = \frac{1}{p+1} \cdot [sin(\varepsilon \cdot t + \vartheta) + p], \tag{3}$$

$p, \varepsilon, \vartheta$ – parameters determined empirically for each crane model.

Simulation of vibration signals was performed with the sampling frequency f_s equal to the sampling frequency of the measuring complex, which was used for measurements. The excitation signal was described by a vector $\overrightarrow{a_0}$ whose elements were calculated using formula (4):

$$a_{0i} = \sin\left(2 \cdot \pi \cdot f_{0n} \cdot \int_{i-0.5 \cdot \Delta t}^{i+0.5 \cdot \Delta t} k_{ab}(t_i) \cdot k_f(t_i) dt\right) \cdot k_{ab}(t_i) \cdot k_f(t_i), \tag{4}$$

$\Delta t = f_s^{-1}$ – sampling interval.

3.3. Defect modeling

The dependence between the excitatory frequency f_0 and the stimulated frequencies of individual units f_{ui}, is in most cases described by the linear dependence of species (5):

$$f_{ui} = u_i \cdot f_0 \tag{5}$$

u_i – gear ratio for unit *i*.

In the presence of defects, the forced frequency of the unit is subjected to frequency or amplitude modulation, which manifests itself on the spectrum in the form of side harmonics with frequencies $(f_{ui} \pm k \cdot f_{di})$, where f_{di} – the frequency of defect manifestation, $k = 1, 2, ...$ In addition, shock pulses occurring when some defects get into the contact zone, cause a response of the defective elements in the form of fast damping oscillations. Such response appears on the spectrum in the form of combination frequencies $(f_{ci} \pm k \cdot f_{di})$, where f_{di} – shock pulse repetition frequencies, $k = 1, 2, ...$

The most common defects in crane mechanisms are: coupling and shaft misalignment, radial and angular misalignment of the coupling halves, bearing defects - wear or pitting of the rolling elements and raceways, the destruction of cages, cracks or chipped rings, gear defects - pitting, abrasive wear, cracks, chipped teeth, sticking in the pair, etc. Vibration patterns of such defects are well studied and described in detail in the literature [1, 5-9] and others. Thus, for example, surface tooth decoration manifests itself in the spectrum in the form of amplitude modulation of the tooth-fixing signal, while jam - in the form of frequency modulation of the same signal [9]. In the presence of amplitude and frequency modulation of the mathematical model of the signal defective unit was described by the dependence (6):

$$a_u(t) = \sum_d \left[(1 + M_d \cdot \cos(\Omega_{d1} \cdot t + \Phi_d)) \cdot \cos\left((2 \cdot \pi \cdot f_u \cdot t + \varphi_u) + m_d \cdot \sin(\Omega_{d2} \cdot t) \right) \right] \quad (6)$$

M_d and m_d – indices, and Ω_{d1} and Ω_{d2} – angular frequencies of amplitude and frequency modulation generated by the defect *d*; Φ_u and φ_u – initial phases of amplitude modulation signals and fundamental unit vibrations, respectively.

When modeling the mechanism as a whole, the signals of individual units were summed up (7):

$$a(t) = \sum_j \left[\xi_j \cdot a_{ui}(t) \right], \quad (7)$$

j – serial number of the unit; ξ_j – coefficient that takes into account the signal attenuation on the way from the source of excitation to the sensor installation point.

Based on the analytical dependence (7), a_d binary signal with a given sampling rate f_s by formula (8) was generated:

$$a_{di} = a\left(\frac{i}{f_s}\right). \quad (8)$$

An additional factor complicating the interpretation of the results of vibration diagnostics is the presence of the noise component of the signal, which was modeled by adding to the members of the series $\overrightarrow{a_d}$, random value, distributed evenly in the interval [-1,1] and multiplied by the noise factor.

4. Modeling results

The developed simulation model was used for diagnostics of the hoisting mechanism with defects of the first and second stage gears.

The modulating frequency values Ω_{u1} and Ω_{u2} for dependence (6) were taken on the basis of known reference data [1, 7, 8] depending on the type and nature of possible defects. The initial phases Φ_d and φ_u were assumed to be zero

When planning a diagnosis, a simulation model was used to determine the required sampling rate. To avoid loss of diagnostic information, condition (9) must be fulfilled for the sampling rate:

$$f_s \geq 2 \cdot f_{max}, \quad (9)$$

f_{max} – maximum frequency in the signal spectrum that can carry information about defects. Manipulation of M_d and m_d, modulation indexes made it possible to determine f_{max} taking into account combinational frequencies that occur when multiple vibration processes occur simultaneously.

When analyzing the diagnosis results, the simulation model was used to specify the bandpass filter cut rates when searching for certain defects. For this purpose, all foreign sources of excitation with corresponding defects were "switched off" in the model signal, which could mask the frequencies of the required defect.

The comparative analysis of the model and experimental signals performed by the same methods allowed to reveal some uncertainties observed in the experimental signal. For example, a mathematical experiment with a simulation model of the hoisting mechanism of the «Albatros» portal slewing crane allowed one to determine that unidentified peaks in the Fourier spectrum, based on the results of a field experiment, correspond to the frequencies of gear defects shifted to the low frequencies region as a result of aliasing.

5. Conclusion

1. The pre-simulation modeling allows one to choose the right characteristics of the measuring system.
2. Pre-simulation with "deactivation" of individual defects allows selecting the frequency bands to search for specific defects.
3. Comparative analysis of the model behavior and the diagnosed unit allows one to reveal uncertainties arising from the analysis of the vibration diagnosis results.

So, the simulation modeling allows increasing the efficiency of application of vibration diagnostic methods.

References

[1] Gavrilin A N and Moises B B 2014 *Technology systems diagnostics: a training manual, Part 2* (Tomsk: Tomsk Polytechnic University Press)

[2] Mustafin R S and Makarcheva E V 2015 Experience of diagnostics of gearboxes of the main hoisting mechanism of oxygen-converter shop load-lifting machines by the example of KG-1830 *J. of the Eurasian Union of Scientists (EUS)* **6(15)** 132-136

[3] Omez D V and Dragan A V 2015 Problems of Vibroacoustic Diagnostics of Gear Damages of Multiple Drives *Scientific Notes* (Lutsk) **49** 106-110

[4] Ganshkevich A Yu, Alexandrova O A and Stoyantsov N M 2020 Topical issues of vibration-based diagnostics of the mechanical equipment of portal slewing cranes *IOP Conf. Ser.: Mater. Sci. Eng.* **747** 012112. DOI: 10.1088/1757-899X/747/1/012112

[5] Chernyshenko A V, Melnichenko A A and Pavlova A A 2008 Features of diagnostics of the low-speed rolling bearings in tows of the overhead traveling crane travel mechanism *Mechanical engineering: collection of scientific* **2** 36-48

[6] Balitsky F Ya, Sokolova A G, Dolaberidze G V and Ivanova M A 2013 Diagnostics of machine equipment defects in nonlinear vibration diagnostics by cascade vibration demodulation method *J. of scientific and technical development* **12(76)** 3-12

[7] Barkova N A and Borisov A A 2009 *Vibration diagnostics of machines and equipment. Calculation of basic vibration frequencies of machine units, parameters of measuring equipment and practical expertise: a training manual* (St. Petersburg: St. Petersburg State Maritime Technical University Press)

[8] Rusov V A 2012 *Diagnostics of rotating equipment defects by vibration signals* (Electronic Materials)

[9] Sokolova A G, Balitsky F Ya, Dolaberidze G V and Ivanova M A 2014 Analysis of parameters of the modulated vibration processes and their envelopes based on the results of simulation modeling *J. of scientific and technical development* **8(84)** 39-53

IOP Conf. Series: Materials Science and Engineering 1103 (2021) 012007 doi:10.1088/1757-899X/1103/1/012007

Simulation of internal forces of a cantilevered beam with a moving load

E Kudryavtsev and M Jafari

Moscow State University of Civil Engineering (National Research University), 26, Yaroslavl highway Moscow, 129337, Russia

E-mail: sdm@mgsu.ru

Abstract. The simulation of cantilevered beam internal forces in any beam section with a moving load is presented. The processes of analytical and computer modeling of the structural analysis are provided depending on the location of the moving load. The cantilevered beam in this article can be a schematic for a gantry crane or an overhead crane with a moving trolley. These calculations are not only very useful for cranes, but also all kinds of bridges and trusses with a moving load. The mathematical model is provided and numerically solved in Mathcad software. As a result of the study, the program effectively determines the internal forces in any beam section depending on the location of the moving load. The proposed procedures will reduce the time and cost of conceptual and detailed design very easily when the beam parameters change, which is useful for optimization studies.

1. Introduction

Structural analysis is a very important part of a beam design and should be done before any dynamic modeling. Static and dynamic analysis of a beam with a moving load is an interesting topic [1, 2].

The article describes the internal forces computer modeling process of a cantilevered beam under the action of a moving load at any point (cross-section) for a freely supported beam in the Mathcad system [3].

The analysis can be presented in analytical, numerical, and graphical forms (in the form of influence lines). Influence lines are used to calculate many structures under the moving loads (beams and trusses of a bridge and gantry cranes; tower cranes booms; mobile and railway bridges, etc.). The internal forces modeling makes it possible to determine any load positions, at which the value of the calculated forces is more than permissible.

The purpose of the study is to create a set of analytical expressions, and programs for the internal forces automated calculation in any cross-section of the beam. The process of influence lines computer modeling is considered by a simple beam example.

2. Definitions

The general scheme of the cantilevered beam on two supports with the load on it in the form of a single mobile force $P = 1$ N is shown in Figure 1.

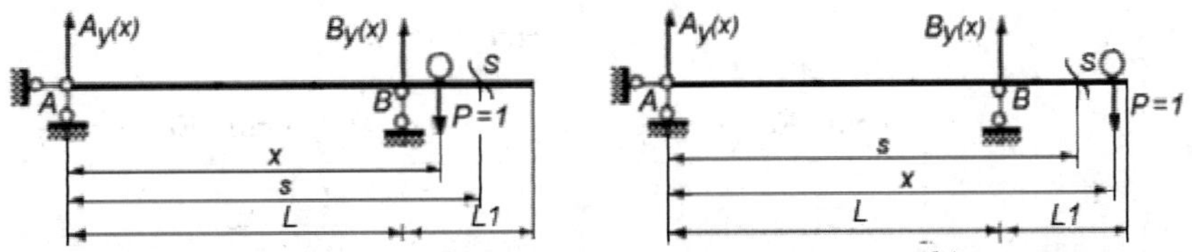

Figure 1. General scheme of the cantilevered beam on two supports with a mobile single load: the considered section *a)* before the moving load on the console and *b)* after the moving load.

3. Basics of the theory

We consider the process of determining internal forces on the sample flat-cantilevered beam on two supports with moving a single load of $P = 1\ N$. So we have:

3.1. The influence lines of the support reactions:

- support reaction lines in the reference points *A* and *B* are being determined from the moments relative to the support point, depending on the location of the single load.

$$\sum M_A(x) = P\cdot x - B_Y(x)\cdot L = 0, \text{ where } B_Y(x) = \frac{P\cdot x}{L}. \tag{1}$$

$$\sum M_B(x) = P\cdot(L-x) - A_Y(x)\cdot L = 0, \text{ where } A_Y(x) = \frac{P\cdot(L-x)}{L}. \tag{2}$$

$$\sum P_i = 0, \text{ where } A_x = 0.\ B_Y(x) = \frac{P\cdot x}{L}. \tag{3}$$

3.2. The influence line of the bending moment:

Two variants should be considered to determine the influence line of internal forces on the console (the s section):

when the single load is on the left and right of the section *s*. If the single load is located to the left of the s section on the console ($x \le s$):

$$\begin{aligned} M_S^L(x) &= -A_y(x)\cdot s - B_y(x)\cdot(s-L) + P\cdot(s-x) = \\ &= -\frac{P\cdot(L-x)\cdot s}{L} - \frac{P\cdot x}{L}\cdot(s-L) + P\cdot(s-x) = \\ &= -P\cdot s + P\cdot\frac{x\cdot s}{L} - P\cdot\frac{x\cdot s}{L} + P\cdot x + P\cdot s - P\cdot x = 0. \end{aligned} \tag{4}$$

If the single load is located to the right of the s section on the console ($x > s$):

$$\begin{aligned} M_S^R(x) &= -A_y(x)\cdot s - B_y(x)\cdot(s-L) = -\frac{P\cdot(L-x)\cdot s}{L} - \frac{P\cdot x}{L}\cdot(s-L) = \\ &= -P\cdot s + P\cdot\frac{x\cdot s}{L} - P\cdot\frac{x\cdot s}{L} + P\cdot x = P\cdot(x-s), \end{aligned} \tag{5}$$

when, $x = s$ $\quad M_S^R(x) = P\cdot(x-s) = P\cdot(s-s) = 0.$

So, for the console:

$$M_S(x) = \begin{Bmatrix} 0,\ for\ x \le s \\ P\cdot(x-s),\ for\ x > s \end{Bmatrix} \tag{6}$$

3.3. The shear forces:

If a single load is located to the left of the s section on the console ($x \leq s$):

$$\sum P_{yi}^{L} = A_y + B_y - P = \frac{P \cdot (L-x)}{L} + \frac{P \cdot x}{L} - P = 0, \tag{7}$$

and for ($x > s$):

$$\sum P_{yi}^{R} = A_y + B_y = \frac{P \cdot (L-x)}{L} + \frac{P \cdot x}{L} = P,$$

$$Q_S(x) = \begin{Bmatrix} 0, \; for \; x \leq s \\ P, \; for \; x > s \end{Bmatrix}.$$

4. Computer simulation

4.1. Source data:

- amount of the mobile load, *kN* $P: = 1$
- distance between the supports, the length of the console, *m* $a_0 := 6$ $a_1 := 2$
- length of the beam (m) and the number of sections under consideration: $l: = a_0 + a_1 = 8$ $NS: = l \cdot 10 = 80.$

4.2. Algorithm of calculation:

1. Determining the reactions forces of the beam at different coordinates for the single force *P*: By (x) – in the reference point *B* and A_y (x) – in the reference point *A*:

- analytical $By(x) := \frac{P \cdot x}{a_0}$ $Ay(x) := \frac{P \cdot (a_0 - x)}{a_0}$

$\Delta x := \frac{l}{NS}$ $t := 0..NS$ $x_i := i \cdot \Delta x$ $By_i := \frac{P \cdot x_i}{a_0}$ $Ay_i := \frac{P \cdot (a_0 - x_i)}{a_0}$ $x0_i := 0$

- graphical (Figure 2).

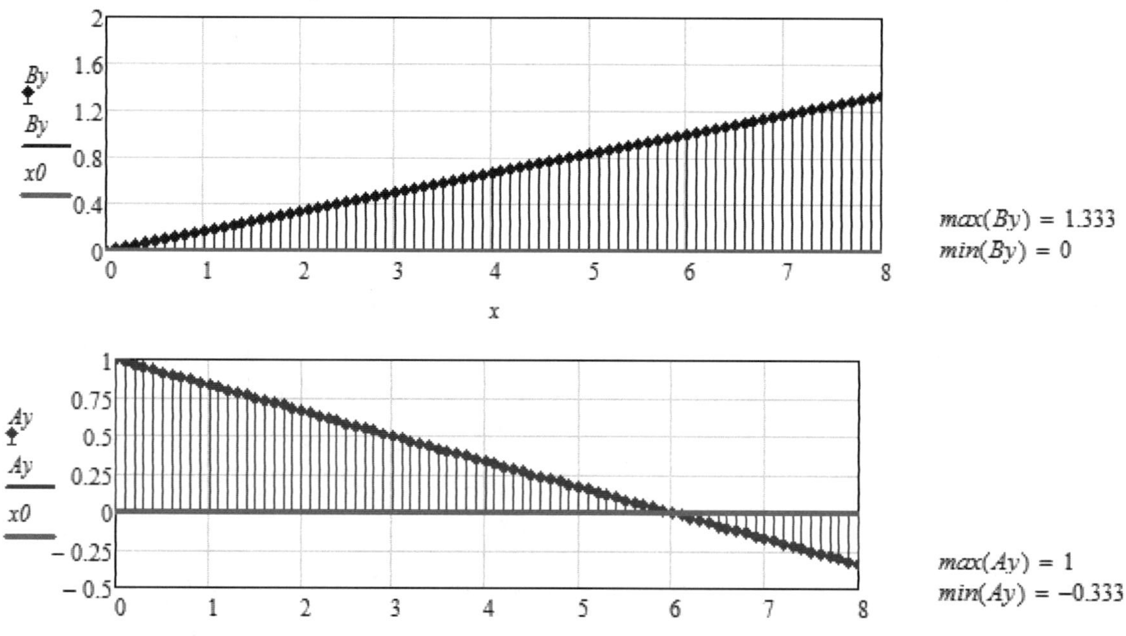

Figure 2. Reaction forces at different coordinates for the single force in points *B* and *A*

2. Determining the influence line of the bending moment in any beam section s: Program 1 (Figure 3).

$$M(s) := \left| \begin{array}{l} \text{for } i \in 1..NS \\ \quad \left| \begin{array}{l} Y_i \leftarrow -P \cdot x_i \cdot \left(1 - \frac{s}{a_0}\right) \text{ if } x_i \le s \le a_0 \\ Y_i \leftarrow -P \cdot s \cdot \left(1 - \frac{x_i}{a_0}\right) \text{ if } x_i > s \le a_0 \\ Y_i \leftarrow 0 \text{ if } x_i \le s \ge a_0 \\ Y_i \leftarrow P \cdot (x_i - s) \text{ if } x_i \ge s \ge a_0 \end{array} \right. \\ Y \end{array} \right.$$

Program 1. Determining the influence line of the bending moment in any section

- An influence line for section s = 4.4 in graphical kind $s := 4.4$

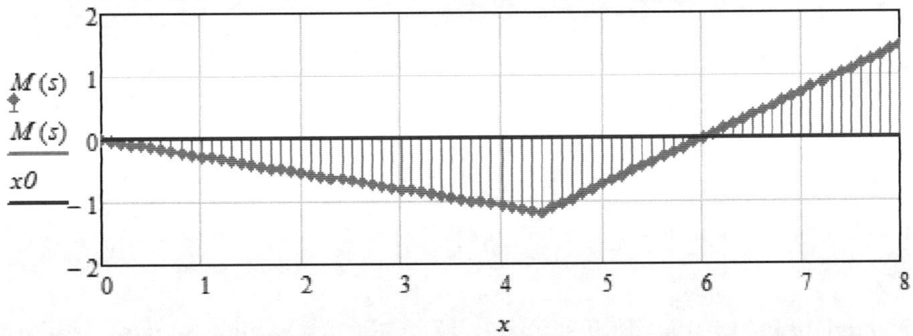

$max(M(s)) = 1.467$
$min(M(s)) = -1.173$

Figure 3. Determining the influence line of the bending moment for section s = 4.4

3. Determining the influence line of the shear force in the section – s: Program 2; Program 3; Program 4 (Figure 4):

$$Q1(s) := \left| \begin{array}{l} \text{for } i \in 0..NS \\ \quad \left| \begin{array}{l} Q_i \leftarrow -\frac{P \cdot (a_0 - x_i)}{a_0} + P \text{ if } 0 \le x_i \le s \le a_1 + a_0 \\ Q_i \leftarrow 0 \text{ if } s \ge a_0 \end{array} \right. \\ Q \end{array} \right.$$

Program 2. Determining the influence line of the shear force between the supports at the left of the section:

$$Q2(s) := \left| \begin{array}{l} \text{for } i \in 0..NS \\ \quad \left| \begin{array}{l} Q_i \leftarrow -P + \frac{P \cdot x_i}{a_0} \text{ if } 0 \le x_i \ge s \le a_1 + a_0 \\ Q_i \leftarrow 0 \text{ if } s \ge a_0 \end{array} \right. \\ Q \end{array} \right.$$

Program 3. Determining the influence line of the shear force between the supports at the right of the section

$$Q3(s) := \left| \begin{array}{l} \text{for } i \in 0..NS \\ \quad \left| \begin{array}{l} Q_i \leftarrow 0 \ \text{ if } x_i \leq s \leq a_0 \\ Q_i \leftarrow P \ \text{ if } x_i \geq s \geq a_0 \end{array} \right. \\ Q \end{array} \right.$$

Program 4. Determining the influence line of the shear force on the console at the right and left of the section

For example, s:= 3.5

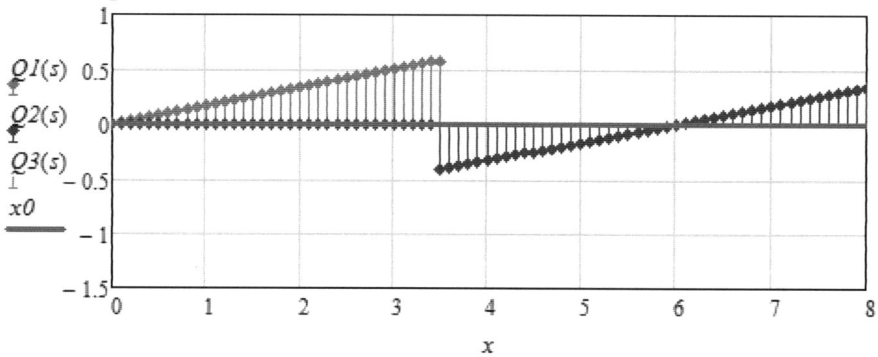

$max(Q1(s)) = 0.583$
$max(Q2(s)) = 0.333$
$min(Q2(s)) = -0.417$
$max(Q3(s)) = 0$

Figure 4. Determining the influence line of the shear force, for example, s = 3.5

5. Conclusion

The simulation of cantilevered beam internal forces in any beam section with a moving load is presented. The processes of analytical and computer modeling of the structural analysis were provided depending on the location of the moving load. The mathematical model was provided and numerically solved in Mathcad software. The proposed procedures of modeling of internal efforts in any cross-section of a cantilevered beam from the moving load allow one to reduce time and cost for the performance of such calculations at least ten times and to carry out effectively the research when the beam parameters, moving load parameters, and the supports position is changing.

References

[1] Xiaogang Q, Gening X, Xiaoning F and Xiaoheng B 2014 Intelligent Optimization Methods for the Design of an Overhead Travelling Crane *Chinese J. of Mechanical Engineering* **28** DOI: 10.3901/CJME.2014.1008.157.

[2] Scalera L, Gallina P, Seriani S and Gasparetto A 2018 Cable-Based Robotic Crane (CBRC): Design and Implementation of Overhead Traveling Cranes Based on Variable Radius Drums *IEEE Transactions on Robotics* **3** 474-485

[3] Kudryavtsev E 2018 Computer modeling influence lines of farm elements *MATEC Web of Conf.*

[4] Bear F and Jonston E 2011 *Vector Mechanics for Engineers I – Static, II – Dynamics,* 9th Ed (Tata McGraw Hill)

[5] Shames 2002 *Engineering Mechanics: Statics and Dynamics*, 9th Ed (PHI)

IOP Conf. Series: Materials Science and Engineering 1103 (2021) 012008 doi:10.1088/1757-899X/1103/1/012008

Monitoring of water quality in ponds using a miniature unmanned underwater vehicle

D I Safarov, S V Efimov, S I Knyazev and S F Jatsun

Department of mechanics, mechatronics and robotics, South-West State University, 94, 50 Let Oktyabrya, Kursk, 305040, Russia

E-mail: teormeh@inbox.ru

Abstract. The proposed article is based on mathematical models of study and development setting up computational experiments of a miniature unmanned underwater vehicles (MUUV) The development and study of systems for automated monitoring of water quality in reservoirs (SAWM) allows the prompt real-time assessment of the environmental situation at various points in the monitoring area. The research objectives are the development of mathematical models describing the process of MUUV control, as well as the development of a mathematical model of the MUUV system. The development of a motion control strategy in the study of water quality is carried out to organize the sampling process. A mathematical model which describes perturbed motion is proposed and a method for the synthesis of the parameters of regulator of the control system based on the decomposition of the equations of motion of the MUUV and transient process is developed. Mathematical models of the force factors acting on the MUUV were developed in view of external pulse disturbances. Mathematical models of control factors were developed which ensured the motion of the MUUV with a variable thrust vector.

1. Introduction

Water objects and hydraulic structures are an integral part of the infrastructure of modern cities. Overhead and underwater structures of bridges, supports, tunnels, collectors need periodic monitoring of their technical condition in order to extend their service life. At the same time, there is an urgent need to control the quality of water resources and the hydrosphere objects pollution degree in order to improve the quality of human life [1]. For such purposes, stationary sampling sources are ineffective. The above tasks can be solved by mobile robotic platforms for underwater monitoring of water objects [2, 3].

The development of systems for automated monitoring of water quality in reservoirs (SAWM) allows the prompt real-time assessment of the environmental situation at various points in the monitoring area [4]. One of the main steps preceding the study of the water quality is sampling. Under current conditions, water sampling is carried out by an operator at stationary monitoring posts fitted out with required equipment.

However, the stationary character of such pollution analyzers makes it practically impossible to provide control at various sites of the monitoring object. Therefore, there is a problem of prompt and safe information acquisition relevant and sufficient for the assessment of the quality of the reservoir and the appropriate response.

2. Materials and methods

In recent years, miniature unmanned underwater vehicles (MUUV) have been emerging. Such devices feature a robotic mobile platform (RMP) with equipment installed on it for carrying out water samples. In view of the fact that the motion of the RMP occurs in an autonomous mode, there arises a relevant objective of the control algorithms synthesis which ensure the motion of the MUUV with a given accuracy along the desired trajectory, with respect to the impact of uncontrolled external disturbances [5-8].

To solve the problem of the control algorithms synthesis, let us consider the motion scheme of the MUUV, shown in Figure 1. We assume that the object under consideration is a system of three rigid bodies: the body and two screw propellers, which are connected to the body by cylindrical hinges. MUUV moves relative to the fixed coordinate system *OXYZ* (not shown on the diagram). The moving coordinate system $CX_2Y_2Z_2$ is related to the body, the rotation of which, relative to the fixed coordinate system, is given by the matrix T_{20}. The center of mass is denoted by *C*. The position and spatial orientation of the MUUV are determined by six generalized coordinates [9-11].

The coordinates *X,Y,Z* set the position of the point *C*. The body orientation is determined by Krylov angles θ, ψ, φ. Let us consider mathematical force models, acting on the MUUV:

1) Weight

$$\bar{G} = -\sum m_i g \bar{j}; \; i=1\text{-}3, \tag{1}$$

where $\bar{j}$ - unit vector of axe *OY*;

2) The power of Archimedes

$$\bar{P} = \sum V_i \rho \bar{j}, \tag{2}$$

where V_i - the volume element MUUV; ρ - the density of water.

3) The resistance force $\bar{H}$, acting in the point C,

$$\bar{H} = (1/2)\varsigma S \rho \bar{V} \,|\, \bar{V} \,|. \tag{3}$$

where ς - the dimensionless drag coefficient, *S* - the effective area of hull; $\bar{V}$ - the speed of movement relative to the water flow.

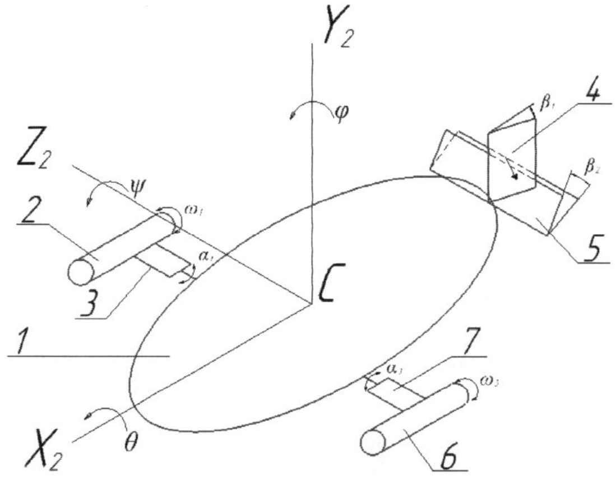

Figure 1. MUUV design diagram. 1 - MUUV body; 2 - right-handed screw propeller; 3 - right-handed electric drive of the screw propeller rotatory motion; 4 - steering rudder; 5 - depth rudder; 6 - left-handed screw propeller; 7 - left-handed electric drive of the screw propeller rotatory motion; β_1 - steering rudder tilt angle; β_2 – depth rudder tilt angle; α_1- angle of rotation of the right screw propeller, set by the matrix T_{12}; α_3 - angle of rotation of the left screw propeller, set by the matrix T_{32}; ω_1 - angular velocity of the right screw propeller; ω_3- angular velocity of the left screw propeller.

The motion of the MUUV is controlled by changing the following forces:

$$\overline{F}_1^{(0)} = T_{12}T_{20} \cdot \overline{F}_1^{(1)}, \tag{4}$$

$$\overline{F}_3^{(0)} = T_{32}T_{20} \cdot \overline{F}_3^{(3)}. \tag{5}$$

$\overline{F}_1^{(1)}, \overline{F}_3^{(3)}$ - control forces generated by screw propellers in local coordinate systems. $\overline{F}_2^{(0)} = (R_x, R_y, R_z)^T$ - control force arising from the rotation of the rudders and the depth rudder which in a fixed coordinate system is determined by the formula:

$$\overline{F}_2^{(0)} = T_{20} F_2^{(2)} \tag{6}$$

$F_2^{(2)}$ the control force generated by turning the steering rudder and the depth rudder in the second coordinate system $CX_2Y_2Z_2$. . Vector of control forces is presented in the form:
$\overline{F} = (\overline{F}_1, \overline{F}_2, \overline{F}_3)^T$.

Thus

$$\overline{\Phi} = \overline{G} + \overline{P} + \overline{H} + \overline{F} \tag{7}$$

$\overline{\Phi}$ - the main vector of external forces applied to the system. $\overline{\Phi} = (\Phi_x, \Phi_y, \Phi_z)^T$.

The obtained system of first-order differential equations allows determining the value of the projections of the center of mass velocity of the MBPC under various types of control actions:

$$\begin{cases} \dfrac{d\upsilon_c^X}{dt} = (\Phi_x) \cdot \dfrac{1}{m} \\ \dfrac{d\upsilon_c^Y}{dt} = (\Phi_y) \cdot \dfrac{1}{m} \\ \dfrac{d\upsilon_c^Z}{dt} = (\Phi_z) \cdot \dfrac{1}{m} \end{cases} \tag{8}$$

Differential equations describing the rotational motion of a system relative to the center of mass:

$$\begin{cases} J^x \dot{\omega}_x + \omega_y \omega_z \left(J^z - J^y\right) = M_{x2} \\ J^y \dot{\omega}_y + \omega_x \omega_z \left(J^x - J^z\right) = M_{y2} \\ J^z \dot{\omega}_z + \omega_x \omega_y \left(J^y - J^x\right) = M_{z2} \end{cases} \tag{9}$$

With use of these equations we can get control forces providing desired motion of MUUV.

Decomposition-based method of the controller synthesis

In reality different disturbances are acting on the MUUV. Therefore, it is needed to take into consideration the perturbed motion of MUUV for one point to another.

The points of water sampling are denoted by M_j. The coordinates of these points are known. The desired MUUV trajectory is a straight line segment $M_j M_{j+1}$ (Fig.2)

To simulate the motion of the MUUV, we introduce the coordinate system $M_j X_j Z_j$ the origin of which, we associate with the point M_j. We consider the motion of the MUUV along this straight line. The desired position of the MUUV in a fixed coordinate system is determined by the radius-vector:

$$\overline{r}_M = (x_M, z_M, \varphi_M)^T. \tag{10}$$

We assume that the desired motion of M occurs along the axis $M_j X_j$ according to the law:

$$x_M(t) = \sum_{i=0}^{n} d_i \cdot t^i \qquad 0 < x_M < L \tag{11}$$

$$z_M = 0\ ; \tag{12}$$

$$\varphi_M = 0\ . \tag{13}$$

To determine the fixed values d_i we write down the conditions which determine the position of the MUUV at the initial and final moments of time at the points M_j and M_{j+1}:

$$t = 0:\ z = z_1 = 0, x = x_1 = 0;\ \dot{z} = 0, \dot{x} = 0;\ \varphi = \varphi_1, \dot{\varphi} = 0, \tag{14}$$

$$\mathrm{t} = \mathrm{t}_1:\ z = z_2 = 0, x = x_2 = L;\ \dot{z} = 0, \dot{x} = 0;\ \varphi = \varphi_1, \dot{\varphi} = 0. \tag{15}$$

L - the length of the line segment $M_j M_{j+1}$.

t_1 - the time which MUUV needs for movement from point M_j to point M_{j+1}.

Using boundary conditions (14), (15), we obtain the coefficients of the polynomial (11).

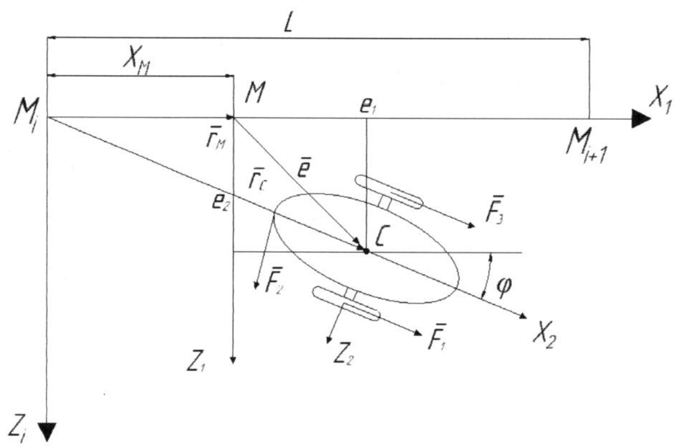

Figure 2. MUUV motion pattern. $M_j X_j Z_j$ – the fixed coordinate system; MX_1Z_1- the moving coordinate system associated with the given point M; CX_2Z_2- the moving coordinate system associated with MUUV.

The real motion of the MUUV determines the radius -vector

$$\bar{r}_C = (x_C, z_C, \varphi)^T \tag{16}$$

The deviation of the real position from the given one determines the vector

$$\bar{e} = (e_1, e_2, e_3)^T, \tag{17}$$

where e_1, e_2, e_3 - errors in coordinates and the angle of rotation.

Then the vector $\bar{r}_C$ is defined by the formula:

$$\bar{r}_C = \bar{r}_M + \bar{e} \tag{18}$$

We represent the system of linearized differential equations describing the perturbed motion of the MUUV in the form:

$$M \frac{d\bar{\xi}}{dt} = A\bar{\xi} + B\overline{U} + \bar{f}(t), \tag{19}$$

where $\bar{\xi} = \frac{d\bar{e}}{dt}$ -the velocity error vector; A, B, M - MUUV parameter matrices; $\bar{U}$ - vector of control actions; $f(t)$ - the disturbance vector.

In the studied case

$$\bar{\xi} = (\nu_1, \nu_2, \nu_3)^T, \tag{20}$$

where

$$\nu_i = \frac{\mathrm{de_i}}{\mathrm{dt}} \tag{21}$$

The initial state is determined by the vectors $\bar{\mathrm{e}}(0)^T, \bar{\xi}(0)^T$. We assume that all the state variables are measurable and formulate the problem to find the state feedback control law.

We represent the vector of control actions *U*, which ensures the decomposition of the considered system in the form:

$$\bar{U} = \begin{vmatrix} F^{(0)} \\ F_2^{(0)} \\ M_C(F) \end{vmatrix} \tag{22}$$

Let us assume that

$$\bar{U} = \bar{a}\bar{\xi} + \bar{b}\bar{e} = -\begin{vmatrix} a_{11} & 0 & 0 \\ 0 & a_{22} & 0 \\ 0 & 0 & a_{33} \end{vmatrix} \begin{vmatrix} \nu_1 \\ \nu_2 \\ \nu_3 \end{vmatrix} - \begin{vmatrix} b_{11} & 0 & 0 \\ 0 & b_{22} & 0 \\ 0 & 0 & b_{33} \end{vmatrix} \begin{vmatrix} e_1 \\ e_2 \\ e_3 \end{vmatrix} \tag{23}$$

where $\bar{a}, \bar{b}$ - are matrices of the regulator parameters. Then:

$$F^{(0)} = F_1^{(0)} + F_3^{(0)} = a_{11}\nu_1 + b_{11}e_1; \tag{24}$$

$$F_2^{(0)} = a_{22}\nu_2 + b_{22}e_2 \tag{25}$$

$$M_C(F) = F_1^{(0)} l_1 + F_2^{(0)} l_2 - F_3^{(0)} l_1 = a_{33}\nu_3 + b_{33}e_3 \tag{26}$$

After rearrangements of the equation (20), we obtain three independent differential equations describing the perturbed position of the center of mass and the MUUV orientation:

$$\ddot{e}_1 + (n_x + a_{11})\dot{e}_1 + b_{11}e_1 = -n_x\dot{x}_M - \ddot{x}_M, \tag{27}$$

$$\ddot{e}_2 + (n_z + a_{22})\dot{e}_2 + b_{22}e_2 = 0, \tag{28}$$

$$\ddot{e}_3 + (n_\varphi + a_{33})\dot{e}_3 + b_{33}e_3 = 0 \tag{29}$$

where $\dot{x}_M, \ddot{x}_M$ - the desired speed and the MUUV center of mass acceleration. n_x, n_z, n_φ - reduced drag coefficients for the corresponding generalized coordinates.

Definition of the coefficients of regulator is very important problem because matrix $\bar{a}$ and $\bar{b}$ seriously influence on the quality of control system of MUUV. Precision, quick actions and stability strongly depends on those parameters. The equation (13) is linear and it allows considering transient process separately independent form different motions of MUUV.

In this paper we will use an approach bases on the analysis of the transient process. The equation (13) is linear and it allows considering transient process separately independent form different motions of MUUV.

IOP Conf. Series: Materials Science and Engineering 1103 (2021) 012008 doi:10.1088/1757-899X/1103/1/012008

Let us consider the solution of the equations of the system using the example of the first equation. The solution can be represented as the sum of the general solution of the homogeneous equation e_1^* and the particular solution e_1^{**} .

$$e_1 = e_1^* + e_1^{**} . \tag{30}$$

From the equation (21) excluding the right side and equations (22), (23) we can get general solution in the next form:

$$e_1^* = k_1 e^{\lambda_1 t} + k_2 e^{\lambda_2 t} \tag{31}$$

k_1, k_2 - unknown coefficients, which can be defined by use of the initial conditions of transient process.

This is the value of the perturbed movement acquired by the center of mass of the MUUV as a result of initial pulse disturbance.

$$\lambda_{1,2} = -\frac{n_x + a_{11}}{2} \pm \sqrt{\frac{(n_x + a_{11})^2}{4} - b_{11}} \tag{32}$$

To determine the coefficients of the matrix $\bar{a}$ and $\bar{b}$ let us consider the transient process described by (25).

The view of the transient process depends on the coefficients $\lambda_{1,2}$

For realization of aperiodic process without vibration we need to get coefficients $\lambda_{1,2}$ less zero and they should be valid values.

Those conditions we will use for definition of the coefficients $\lambda_{1,2}$. Let the parameters of transient process imposing restrictions on the time of the transient process and coefficient of overregulation which determine the quick speed and precision of the control system.

The time of the transient process will be taken equal to T. Natural, we consider the case than $T << t_1$. Further calculations are performed for T=1c

We set for initial condition of transient process $t = 0$

$$e_1^*(0) = e_{10} , \dot{e}_1^*(0) = 0 \tag{33}$$

and final condition at $t = T$, respectively

$$e_1^*(T) = \varepsilon e_{10} , \dot{e}_1^*(T) = 0 \tag{34}$$

ε - small parameter.

Further on, we assume that $\varepsilon = 0{,}001$.

$$e^*{}_1 = 0{,}001 e_{10} . \tag{35}$$

The conditions (33), (34), (35) allow to get equations for definition of the parameters k_1, k_2 , $\lambda_{1,2}$.

From initial conditions:

$$k_1 = -\frac{\lambda_2 e_{10}}{\lambda_1 (1 - \frac{\lambda_2}{\lambda_1})} = \frac{\lambda_2 e_{10}}{\lambda_1 - \lambda_2} , \tag{36}$$

$$k_2 = \frac{e_{10}}{1-\frac{\lambda_2}{\lambda_1}} = \frac{\lambda_1 e_{10}}{\lambda_1 - \lambda_2} \tag{37}$$

From final conditions with reference (36), (37):

$$\lambda_1 = \frac{\ln \varepsilon e_{10} - k_2 \lambda_2 T}{T k_1} \tag{38}$$

$$\lambda_2^{1,2} = A \pm \sqrt{A^2 - B} \tag{39}$$

A, B -parameters, depending on k_1, k_2.

These formulas are transcendental and they solution requires numerical method. In this paper it is offered to convert the formula (25) in to the following form.

$$\lambda_{1,2} = -\frac{n_x + a_{11}}{2} \pm \sqrt{\frac{(n_x + a_{11})^2}{4}(1-\beta)}, \tag{40}$$

where $\beta = \frac{4b_{11}}{(n_x + a_{11})^2}$.

β - is parameter determining quality of transient process. If $\beta > 1$ we have oscillatory process and in case when $\beta < 1$ we get aperiodical process

After that with use (35) we define parameters a_{11} b_{11}.

$$a_{11} = \frac{(\lambda_1 + \lambda_2)^2}{2} - (\lambda_1 - \lambda_2)^2 \tag{41}$$

$$b_{11} = \lambda_1 + \lambda_2 + n_x \tag{42}$$

Parameters $a_{11}, a_{22}, a_{33}, b_{11}, b_{22}, b_{33}$ also can be defined from consideration of the equations (30), (31).

3. Results and discussion

We represent the particular solution in the form:

$$e_1^{**} = \sum_{i=0}^{3} c_i t^i \tag{43}$$

The time derivatives take the form:

$$\dot{e}_1^{**} = \sum_{i=2}^{3} c_i i(i-1)t^{i-2} + \sum_{i=1}^{3} c_i i t^{i-1} \tag{44}$$

$$\ddot{e}_1^{**} = \sum_{i=3}^{3} c_i i(i-1)(i-2)t^{i-3} + \sum_{i=2}^{3} c_i i(i-1)t^{i-2} \tag{45}$$

Substituting (43), (44), (45) into (27) we obtain:

$$(n_x + a_{11})\sum_{i=2}^{3} c_i i(i-1)t^{i-2} + \sum_{i=3}^{3} c_i i(i-1)(i-2)t^{i-3} + \sum_{i=2}^{3} c_i i(i-1)t^{i-2} +$$

$$+ (n_x + a_{11})\sum_{i=1}^{3} c_i i t^{i-1} + b_{11}\sum_{i=0}^{3} c_i t^i = -n_x \sum_{i=1}^{3} d_i i t^{i-1} - \sum_{i=2}^{3} d_i i(i-1)t^{i-2} \quad (46)$$

Hence, as a result of the corresponding rearrangements, we obtain analytical expressions for the coefficients c_i. Similar transformations we make for equations (28), (29).

4. Numerical simulation (experiment)

Now we can calculate the perturbed position of the MUUV relative to the desired.

Setting the parameters:

$n_x = 0{,}1$; $n_z = 0{,}2$; $n_\varphi = 0{,}5$; $d_1 = d_2 = d_3 = a_{11} = 6{,}8$; $a_{22} = 6{,}7$; $a_{33} = 6{,}4$; L = 50м.

The solution of a differential equations (27),(28),(29) obtained by the analytical method allowed calculating parameters of $\bar{e}(t)$-error vector (deviation) and also $\dot{\bar{e}}(t)$ – vector of error velocity. Besides, we can define the diagrams for control forces acting on the MUUV correction of position.

Calculation is made for the order of the polynomial n=5 Diagram of desired function of coordinate Center of mass of MUUV $x(t)$ is presented in Figure 3. This diagram shows that initial and final position of center of mass has zero velocity and acceleration.

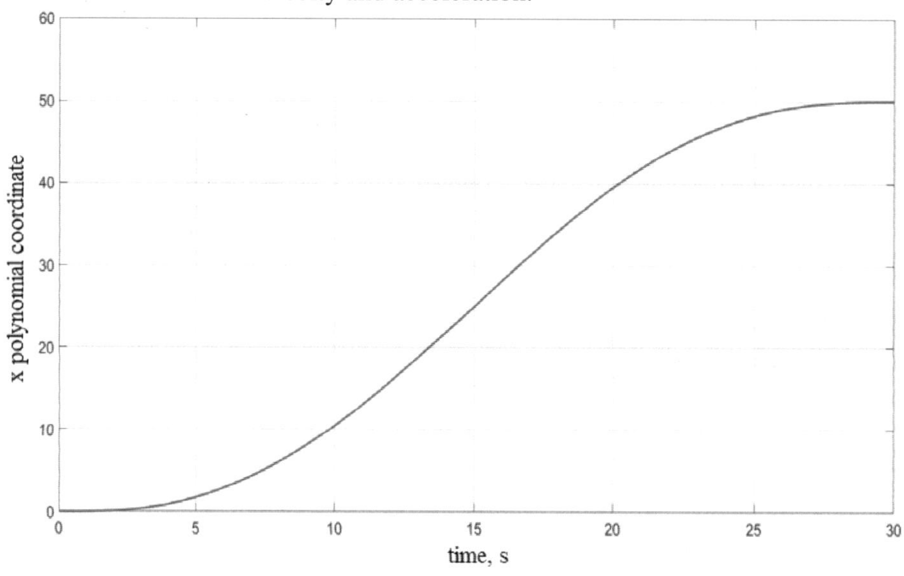

Figure 3. Diagram of desired function of coordinate Center of mass of MUUV $x_M(t)$.

The diagram of deviation real position of the MUUV`s Center of mass relatively desired position during the starting offset is presented on fig.4.In the result of Initial impulse centre of mass has initial velocity $\dot{e}_0(0) = 0{,}1$

In the beginning the case is considered when $\beta > 1$.

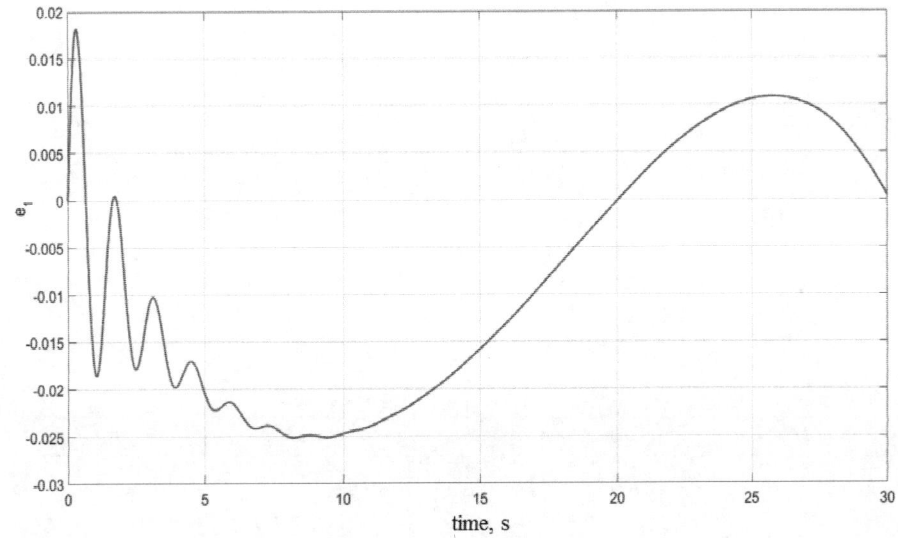

Figure 4. Diagram of deviation real position of the MUUV`s Center of mass relatively desired position during the starting offset $\dot{e}_0(0) = 0{,}1$ $\beta > 1$

The diagram of deviation of real velocity of the MUUV`s Center of mass relatively desired velocity during impulse action is presented on fig.5. In the beginning MUUV has disturbance velocity around 0,1. However, after transient process MUUV has got small additional velocity which do not influence on the general motion process.

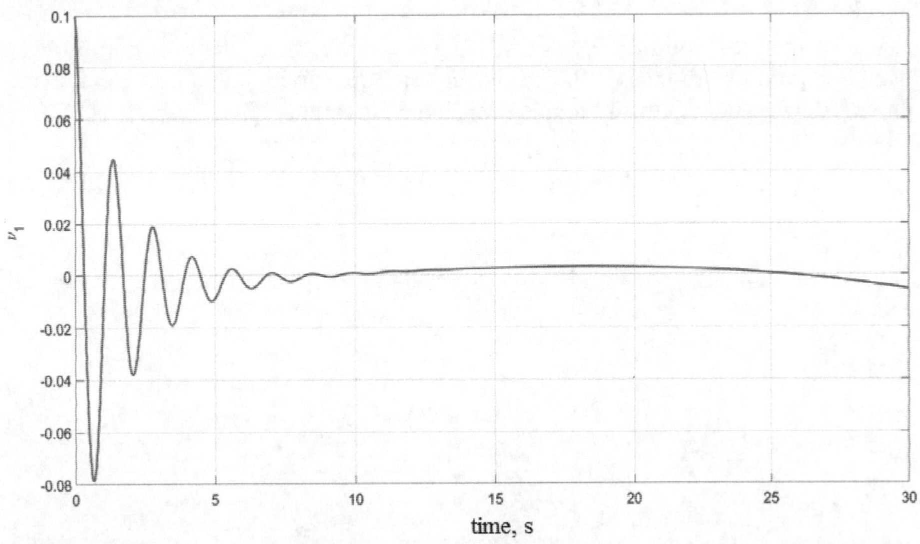

Figure 5. Diagram of deviation real velocity of the MUUV`s Center of mass relatively desired velocity during impulse action $\dot{e}_1 = 0{,}1$ $\beta > 1$

The diagram of control force which allows compensating deviation of real position of the MUUV`s relatively desired during impulse action is presented in Figure 6. It is very oscillatory function, which is difficult to realize by control system.

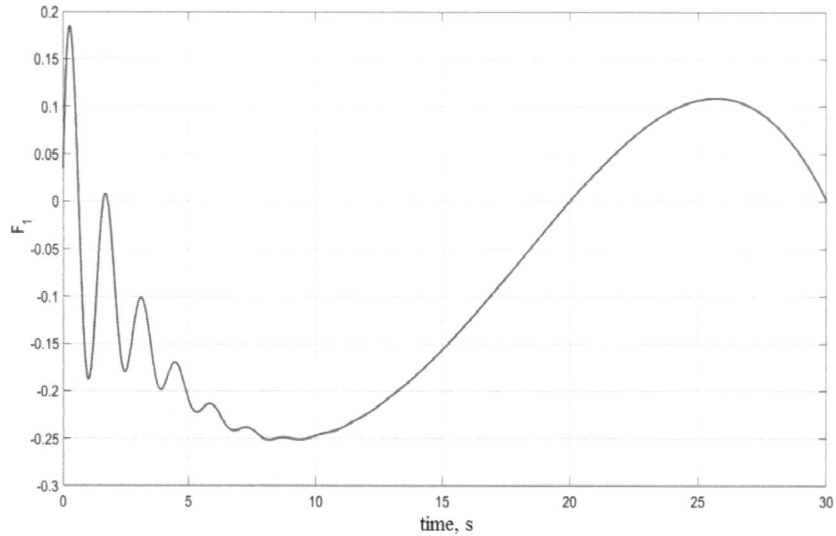

Figure 6. Diagram of control force which allows compensating deviation of real position of the MUUV`s relatively desired during impulse action $\dot{e}_1 = 0{,}1$ $\beta > 1$

Transient process has oscillatory character and this is unacceptable for practical operation of MUUV. Therefore, further the case when $\beta < 1$ is considered.

The diagram of deviation real position of the MUUV`s Center of mass relatively desired position during the starting offset $\dot{e}_0(0) = 0{,}1$ $\beta < 1$ is presented in Figure 7.

The deviation of real position of the MUUV`s is very small on all way of motion of MUUV and changing inside in interval from-0,07 to 0,03. It is very smooth function and it is good result if we take into account that MUUV goes 50 m and relative mistake is less than 1%.

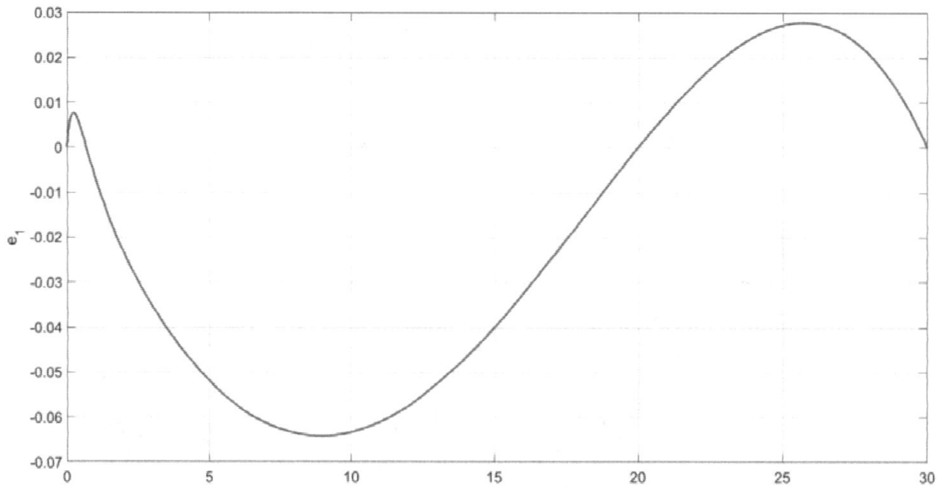

Figure 7. Diagram of deviation real position of the MUUV`s Center of mass relatively desired position during the starting offset $\dot{e}_0(0) = 0{,}1$ $\beta < 1$.

The diagram of deviation of real velocity of the MUUV`s Center of mass relatively desired velocity during impulse action is presented in fig.8.

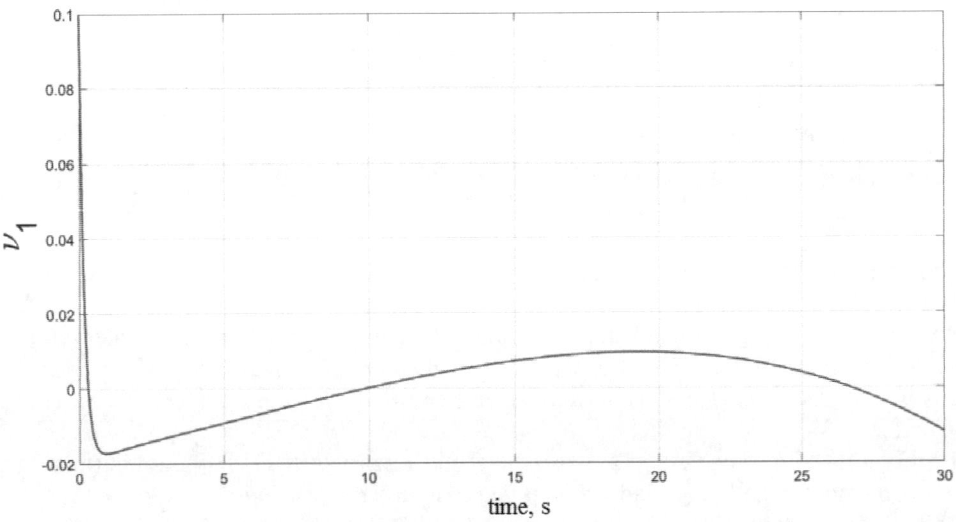

Figure 8. Diagram of deviation real velocity of the MUUV`s Center of mass relatively desired velocity during impulse action $\dot{e}_1 = 0{,}1$ $\beta < 1$.

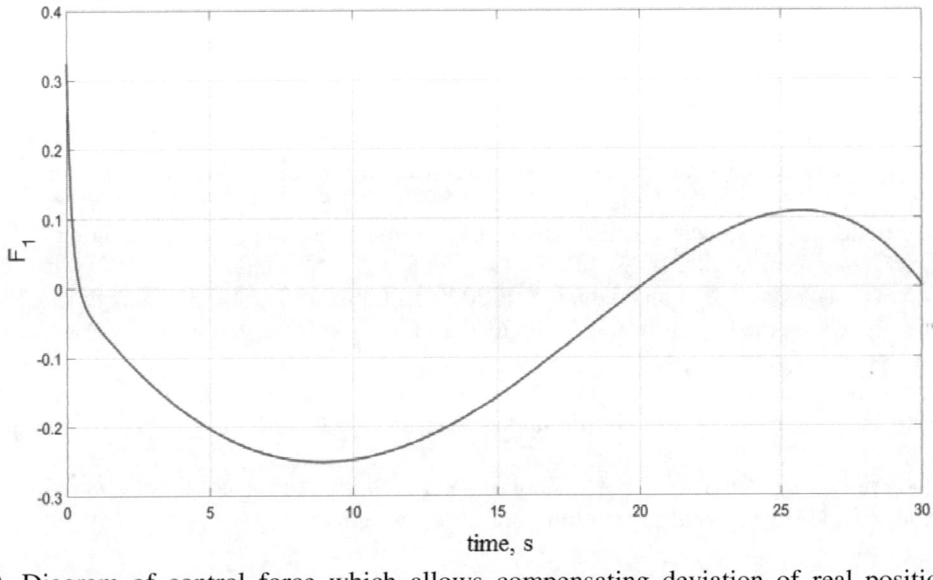

Figure 9. Diagram of control force which allows compensating deviation of real position of the MUUV`s relatively desired during impulse action $\dot{e}_1 = 0{,}1$ $\beta < 1$

The diagram of control force which allows compensating deviation of real position of the MUUV`s relatively desired during impulse action is presented in Figure 9. It is very smooth function, which is quite similar to control system.

5. Conclusion

1. A mathematical model which describes perturbed motion was proposed and a method for the synthesis of the parameters of regulator of the control system based on the decomposition of the equations of motion of the MUUV and transient process was developed.

2. Mathematical models of the force factors acting on the MUUV were developed in view of external pulse disturbances.

3. Mathematical models of control factors were developed which ensure the motion of the MUUV with a variable thrust vector, which allowed carrying out water samples at specified points of the reservoir both on the surface and at a given depth.

Acknowledgments
The reported study was funded by the Russian Fond of basic research 19-08-00440.

References

[1] Robot for diagnosing water pollution. Retrieved from: http://aquavitro.org/2010/12/24/ryba-robot-dlya-diagnostiki-zagryazneniya-vody/
[2] Levin V I 2011 Modeling of optimization problems in conditions of interval uncertainty *Izvestiya of Penza State University. Physical, mathematical and technical sciences* **26** 589–595
[3] Ageyeva M D 2005 *Unmanned underwater vehicles for military purposes* (Vladivostok: Dal'nauka)
[4] Lushnikov B V, Politov Ye N, Tarasova Ye S and Kazaryan K G 2014 Bionic floating robot for monitoring natural and man-made objects in the hydrosphere *Cloud of science* **1(1)** 61-77
[5] Inzartsev A V, Matviyenko Yu V, Pavin A M and Rylov N I 2015 Monitoring of an intelligent system for the application of device data processing technologies aboard an autonomous unmanned underwater vehicle *Underwater Research and Robotics* **2(20)** 20–27
[6] Inzartsev A V, Pavin A M and Bagnitskiy A V 2013 Planning and Implementation of an Underwater Survey Robot Based on Behavioral Methods *Underwater Research and Robotics* **1(15)** 4–16
[7] Inzartsev A V, Pavin A M, Yeliseyenko G D, Rod'kin D N, Sidorenko A V, Lebedko O A and Panin M A 2015 Reconfigurable cross-platform environment for modeling the behavior of an unmanned underwater vehicle *Underwater Research and Robotics* **2(20)** 28–34
[8] Inzartsev A V, Pavin A M, Lebedko O A and Panin M A 2016 Recognition and examination of small underwater objects using autonomous unmanned underwater vehicles *Underwater Research and Robotics* **2(22)** 36–43
[9] Knyazev S I, Yatsun A S and Jatsun S F 2020 Automated Terminal Management System for Miniature Unmanned Underwater Vehicle (MUUV) water surveillance *RusAvtoKon* (Russia, Sochi) pp 205-211
[10] Yefimov S V, Knyazev S I, Yatsun A S and Jatsun S F 2020 Simulation of an automated water monitoring system using a miniature underwater unmanned vehicle *FarEastCon* (Vladivostok) pp 813-817
[11] Bozhenov Yu A, Gavrilov V M, Zhukov Yu I, Ikonnikov I B et al 1986 *Self-propelled unmanned underwater vehicles* (Leningrad: Sudostroyeniye)

Modelling of operator walking with a load of industrial exoskeleton

S F Jatsun, A S Yatsun, E V Saveleva and A V Fedorov

Department of mechanics, mechatronics and robotics, South-West State University, 94, 50-letiya Oktyabrya str., Kursk, 305040, Russia

E-mail: teormeh@inbox.ru

Abstract. The proposed article is based on mathematical models of study and development setting up computational experiments of a person using an exoskeleton walking with a load process. The development and study of anthropomorphic robots' dynamics involves the systematic approach and use of mathematical modelling methods, the automatic control theory apparatus, which makes it possible to study a complex object as a whole and also consideration of the totality of human-machine system elements, united by diverse and multi-level types of connections, as an integral object. The research objectives are the development of mathematical models describing the process of operator- exoskeleton interaction, as well as the development of a three-link mathematical model of the exoskeleton - operator system. The development of a motion control strategy when walking allows obtaining time dependences of state vector parameters changes. Mathematical models describing the process of operator and the exoskeleton interaction are being developed. The research is based on the original three-link mathematical model of the exoskeleton-operator system proposed in the paper. The original method based on interpolation, experimentally obtained data and dynamic analysis of the operator-exoskeleton system behaviour is applied. In the course of numerical modelling, time diagrams of the hinge-mechanical system torques in the process of lifting loads were obtained. Basing on the results of mathematical modelling of an operator's walking with a load in an industrial exoskeleton, the conclusions about the possibilities of an active gravitational compensator in the hip joint are obtained.

1. Introduction

One of the ways of the improvement of the working conditions quality is the exoskeletons usage. They significantly increase the person power capabilities, while the body functional tension level decreases [1-6]. This effect is achieved when the operator and the exoskeleton form an integrated human-machine system (HMS) [9, 10]. The performance of the system is determined by the consistency (synchronicity) degree of the it's elements work [13,14]. The system elements include a person (operator) and the exoskeleton. To ensure this, the system is complemented by another important element - a human-machine interface (HMI), which is a group of technical means that provides the operator-exoskeleton interaction [9].

The development and research of HMS involves the systematic approach, methods of mathematical modelling, and the automatic control theory apparatus usage. That makes it possible to study a complex object as a whole, as well as to consider a set of HMS elements, united by diverse and multi-level types of connections, as an integral object [14]. To construct the HMS design tools, the system elements movement mathematical models are required. They take into account the selected control strategy implemented by the HMI, as well as the operator's physiological characteristics and the exoskeleton technical capabilities. Next, we will consider the development of mathematical models and algorithms for the HMS, the control strategy implemented by the HMI and providing the human functionality expansion.

2. Materials and methods

Figure 1 shows a scheme of a person walking in an exoskeleton with a load. The diagram illustrates forces and moments acting on the operator and the exoskeleton during the process of walking. We will assume that the system consists of three non-deformable bodies: the operator's body -5; the exoskeleton back -4; operator's hips and exoskeleton's joints -3. The interaction of the operator and the exoskeleton is controlled by an elastic element that forms the force $\overline{P}_{54}$. A combined gravitational compensator is installed in the exoskeleton hip joint, which includes an elastic element and a controlled electric drive.

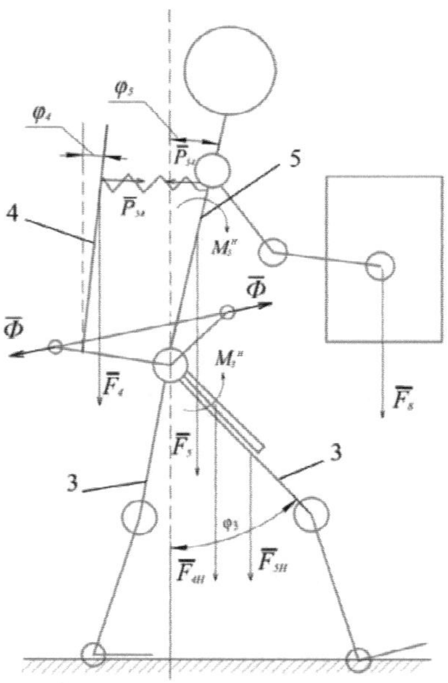

Figure 1. Scheme of a person walking in an exoskeleton with a load

Let us assume that the following forces and moments act on the operator during process of walking: the operator's body gravity force- $\overline{F}_5$, the load gravity force- $\overline{F}_8$, the operator's leg gravity force- $\overline{F}_{5H}$, the force arising in the force meter $\overline{P}_{54}$, the moment of the operator's muscular system M_5^H, the moment the muscular system of the operator's leg M_3^H. The exoskeleton, in turn, is affected by the following set of forces and moments: the exoskeleton back gravity force $\overline{F}_4$, the exoskeleton leg gravity force- $\overline{F}_{4H}$, the force arising in the force meter $\overline{P}_{54}$. $\overline{\Phi} = \overline{\Phi}_1 + \overline{\Phi}_2$ - is the combined gravitational compensator force, where $\overline{\Phi}_1$ - is the elastic element force, $\overline{\Phi}_2$ - is the force, created by the electric drive.

The exoskeleton work is determined to reducing the load on the operator when lifting and moving the load due to the work of an electric drive and an elastic element. For the performance of the high-quality assistance, the exoskeleton links must follow the operator's body and repeat person's movements as accurately as possible. The connecting element between the exoskeleton and the back is a force meter, which forms the $\overline{P}_{54}$ value based on the deviation between the inclination angles of the operator's body and the exoskeleton back. The $\overline{P}_{54}$ value forms the control voltages for the electric drives of the exoskeleton control system.

The system of differential equations describing the movement of the exoskeleton-operator system can be represented as:

$$\frac{d\overline{\xi}}{dt} = A(\overline{\xi}) + B\overline{U} \tag{1}$$

where $\overline{\xi}$ is the speed vector; A , B -matrix of parameters of the HMS; $\overline{U}$ - control actions vector of.

In our case

$$\overline{\xi} = (\varphi_3, \varphi_4, \varphi_5, \dot{\varphi}_3, \dot{\varphi}_4, \dot{\varphi}_5)^T \tag{2}$$

Let us assume that

$$\overline{U} = K\overline{\xi} \tag{3}$$

where K is the matrix of the controller coefficients

The numerical solution of equation (1) taking into account (3) can be presented in the form:

$$\overline{\xi}_i = \overline{\xi}_{i-1} + (A + BK)\overline{\xi}_{i-1}\Delta t \tag{4}$$

where $\overline{\xi}_i$, $\overline{\xi}_{i-1}$, are system state vectors at the previous and next time step; Δt -step in time. At the initial moment of time $\overline{\xi}(0) = \overline{\xi}_0$.

Let us consider an example of human-exoskeleton system walking numerical simulation. Let the desired angle of operator's hip rotation φ_3^* be given by a harmonic function:

$$\varphi_3^*(t) = \alpha \sin(\omega t) + \beta \tag{5}$$

From the analysis of the operator's gait, we assume that $\alpha = -0{,}39$; $\beta = 0{,}13$.

The cyclic frequency ω can be calculated as follows:

$$\omega = \frac{V \cdot 2\pi}{0.7} \tag{6}$$

where V is the operator's movement of linear speed of (m/s). 0.7 is the average value of one step length (m).

Next, we will consider the movement of an operator with a frequency ω = 3, which corresponds to a speed of $V = 0{,}33\,\text{m/s}$. For this case:

$$\varphi_3^* = -0.39\sin(3t) + 0.13 \tag{7}$$

It is also assumed that the operator's back desired angle is $\varphi_5^* = 5^0$.

Table 1. Model coefficient values.

Odds	Slow Walking	Fast Walking
ω	3	15
K_p	300	300
K_d	15	15
K_u	200	200
K_{du}	15	15
K_{pL}	1000	1000
K_{dL}	15	15
C_{54}	20000	20000
M_{54}	200	200
C	10000	10000

3. Results and discussion

The simulation results are presented on graphs in the form of dependences of the state vector parameters, external disturbances and control actions.

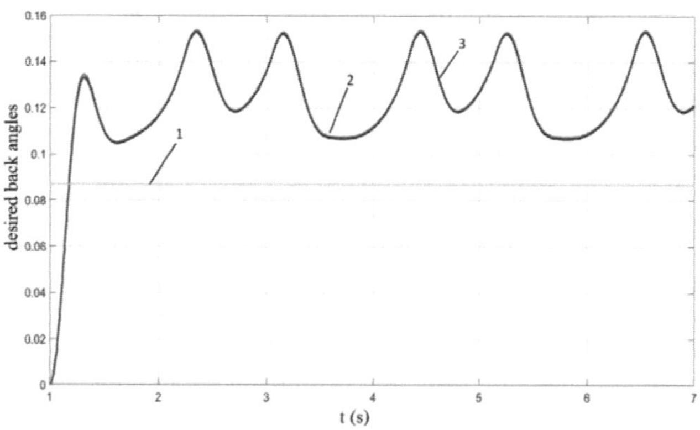

Figure 2. The desired angles of the operator's and the exoskeleton back 1 - φ_5^*, 2 - φ_5, 3 - φ_4

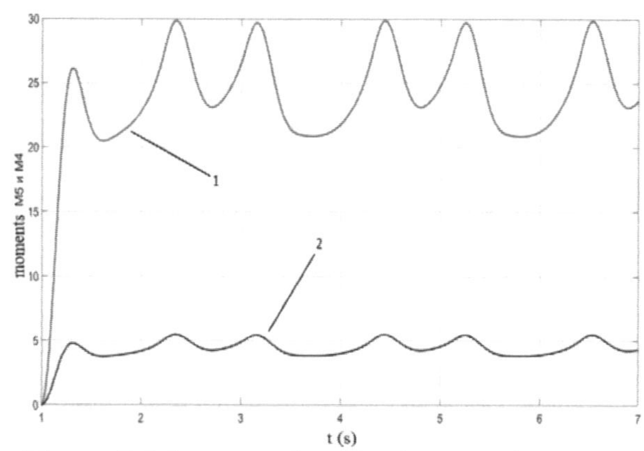

Figure 3. Moments of gravity forces acting on the operator's back - 1 and exoskeleton - 2.

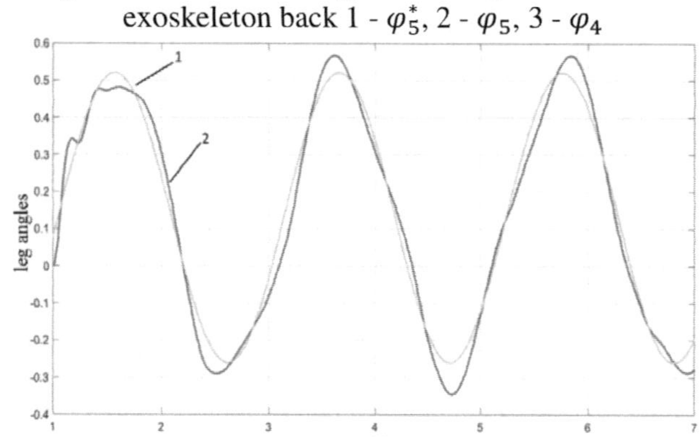

Figure 4. Target -1 and actual -2 operator hip rotation angles. 1 - φ_3^*, 2 - φ_3

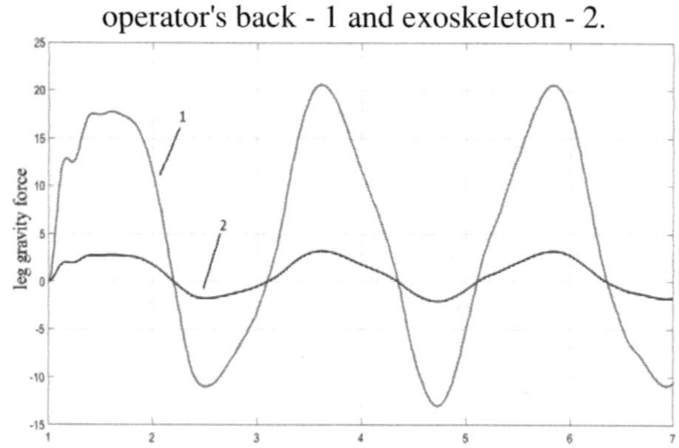

Figure 5. Moments of gravity forces acting on the operator's leg - 1 and exoskeleton – 2

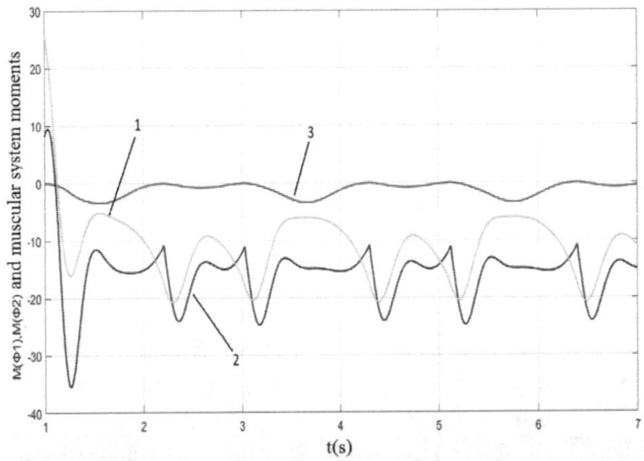

Figure 6. Moments of the operator muscular system - 1, the drive - 2 and the elastic element – 3

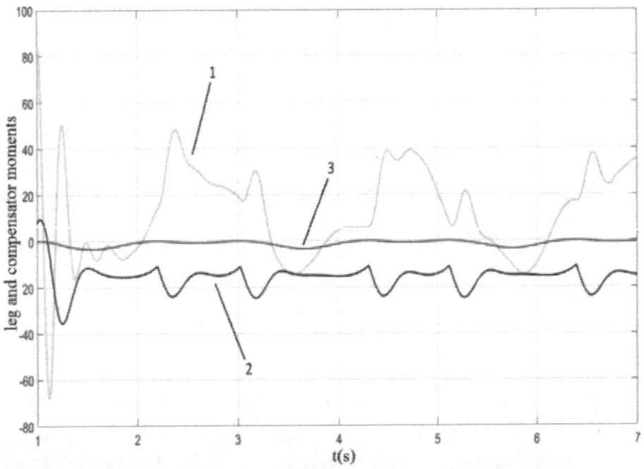

Figure 7. Moments created by the thigh muscles - 1, electric drive - 2 and elastic element – 3

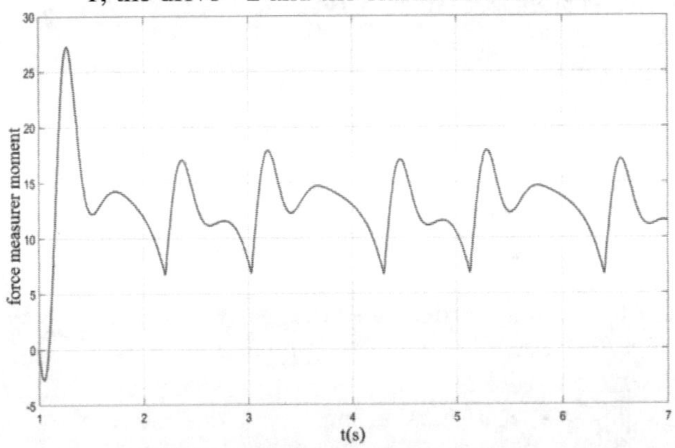

Figure 8. Torque force meter

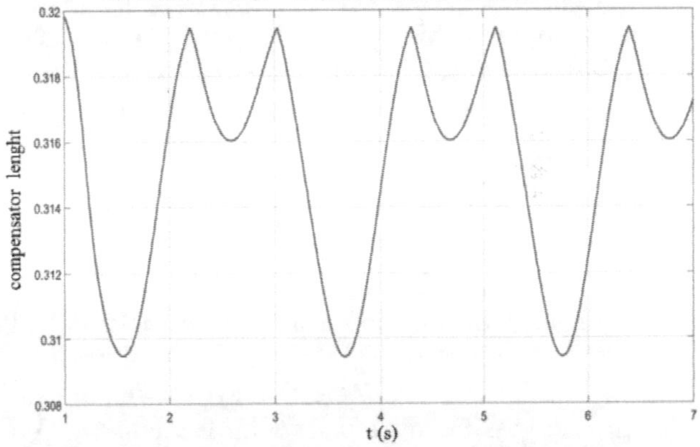

Figure 9. Compensator length

Figure 2 shows that the angles of the operator's and exoskeleton's back are close and oscillatory. There is also a small (up to 9 degrees) deviation of these angles from the given one. Comparing the diagrams in Fig. 2 with the graphs of the moments of gravity acting on the operator's and exoskeleton's back (Fig. 3), it can be seen that the inclination angles change occurs synchronously with the change in the gravity force moments.

Figure 4 shows the dependences of the specified -1 and actual -2 angles of the operator's hip rotation. Here, there is a slight deviation of the real angle from the specified one; the error is associated with the action of the moments of gravity, which are shown in Fig. 5. Figure 6 shows the graphs of the operator muscular system moments - 1, the drive - 2 and the elastic element - 3. The analysis of these dependencies shows that the elastic element moment during walking does not exceed 5 Nm.

The main contribution is made by the operator muscular system moments and the electric drive, and the electric drive torque magnitude reaches 25 Nm, and the muscular system 20 Nm. Fig. 8 shows the dependence of the moment of the force meter on the $\overline{P}_{54}$ time. In the steady state, a periodic process takes place, which is associated with a change in the length of the gravitational compensator (Fig. 9) and the adopted control strategy [3].

4. Conclusion

As a result of the research, a design scheme for the operator-exoskeleton system was developed, and the force factors acting on the system during process of walking with a load were determined [7, 8, 11]. A mathematical model of the system motion and an algorithm for the gravitational compensator electric drive controlling was developed [9, 12]. A numerical experiment was carried out, which showed that the proposed mathematical model reflected the basic laws of the human - exoskeleton system motion and allowed establishing quantitative values of the state vector main parameters.

Acknowledgments
The reported study was funded by the Russian Federation President Grant for young candidates of sciences MK-901.2020.8

References

[1] Gritsenko G P, Yakobson Ya S and Belen'kiy V Ye 1978 Analysis of the temporal structure of human walking, in: *Prosthetics and prosthetic engineering* (Moscow: TsNIIPP) vol 46 pp 25-34
[2] Gurevich R A 1978 Stability of the balance of the human body depending on the localization of muscular control moments *Orthopedics, traumatology and prosthetics* **5** 18-22
[3] Yefimov V A, Kudryavtsev M V and Proskuryakov V B 1975 Study of the dynamics of movements of a biotechnical system such as exoskeleton, in: *Biomechanics* (Riga) pp 604-609
[4] Fantoni I and Lozano R 2012 *Nonlinear control of mechanical systems with a deficit of control actions* (Moscow – Izhevsk: Computer Dynamics LLC) 312 p
[5] Po-Yang Lin Win-Bin Shieh D Y Chen 2013 *A theoretical study of weight-balanced mechanisms for design of spring assistive mobile arm support (MAS)*
[6] Zhang J, Cheah C C and Collins S H 2015 *Evaluation of Force Tracking Controller with Soft Exosuit for Hip Extension Assistance ICRA* pp 5584–5589
[7] Okamoto T, Shiratori T, Glisson M, Yamane K, Kudoh S and Ikeuchi K 2014 Extraction of person-specific motion style based on a task model and imitation by humanoid robot *Intelligent Robots and Systems (IROS 2014) IEEE/RSJ Int. Conf.* pp 1347-1354
[8] Dmitriyev V A and Karlov A Ye 2019 Calculation of a complex criterion of qualitative indicators of an industrial exoskeleton, in: *Science and education: domestic and foreign experience. Coll. of conf. proc. of the Twenty-first Int. Sci. and Pract. Conf.* pp 32-36
[9] Jatsun S F, Al Maji Kh Kh M, Pavlovsky V E, Yatsun A S, Karlov A E and Saveleva E V 2019 Analysis of the effect of the exoskeleton geometrical dimensions on the nature of a linear compensator operation, in: *Developments in eSystems Engineering 2019. Robotics, Sensors and Industry 4.0.* (Kazan)
[10] Yatsun S F, Yatsun A S, Savel'eva E V and Karlov A E 2020 Simulation of Interaction between an Operator and an Exoskeleton *Machinery Manufacture and Reliability* **49(6)** 490–499
[11] Poyezzhayeva Ye V, Babaricheva Ye V, Korlyakova K A and Permyakova T Yu 2018 The use of new classes of exoskeletons in road construction works *Construction and road machines* **1** 59-61
[12] Statnikov A A and Kolokol'nikova A I 2018 Cybernation and the use of exoskeletons in the medical and military fields *Contemporary Research* **1** 53-55
[13] Chigarev A V and Borisov A V 2015 Recursive method for obtaining differential equations of motion of deformable planar anthropomorphic systems in unsupported and two-bearing phases of motion *Theoretical and Applied Mechanics: Int. Sci. and Tech. Coll.* (Minsk: BNTU) **30** 70-81
[14] Borisov A V and Chigarev A V 2019 *Application of generalization of matrices to construct various models of three-dimensional models of endo- and exoskeleton according to the equations of plane mechanisms*

Methodology to calculate the required number of spare parts in order to ensure the required operability of equipment parks

B G Kim[1] and Z N Shakir[2]

[1] Department of construction production, Vladimir State University, 87, Gorky st., Vladimir, 600000, Russia

[2] Department of construction production, Vladimir State University, 34, Karrada st., Baghdad, Iraq

E-mail: kimbg@vlsu.ru

Abstract. Comprehensive mechanization of work is accompanied by large expenditures for ensuring the serviceability and operability of the machines fleet. There is an almost constant need for spare elements (machines, units, parts) to maintain the required use level of machines and equipment, as well as to perform work on schedule. An algorithm of actions allowed finding a rational level of redundancy for machines and their parts under various operating conditions and constructive improvement of technology. The article showed that the rate of failure is influenced by the natural and climatic factor, the "age" of machines, and other conditions. Optimization of the companies' repair policy in terms of ensuring the serviceability of equipment is a complex multi-vector problem. The described methodology makes it possible to determine the necessary range and volumes of spare elements for a known fleet of machines at a given load and under the conditions of ensuring the calculated level of reliability and serviceability of the equipment, maintenance, and repair. The application of the proposed method allows optimizing the cost of spare elements while ensuring the established level of reliable operation for machines and technological sets of equipment. Also, it becomes possible to obtain rational solutions for the formation of a warehouses network on different levels to meet the needs of firms with the dispersed operation of machines and mechanisms at a large number of objects.

1. Introduction

The intensification of machines parks usability is possible only under conditions of increasing the reliability of machines [1], reducing the time of repair and maintenance works [2]. One of the factors in achieving a new, increased level of equipment readiness is the presence of a sufficient number of reserve elements for vehicle fleets [3]. The reserve elements here mean spare parts, repair materials, reserve machines and units. The complexity of solving the problem is exacerbated by the following points: a sharp increase in the level of complex mechanization [4], the enlargement of machines parks, a large number of work objects [5], the desire to reduce the time of their construction. These factors have sharply raised the cost of a single machine downtime since the failure of one machine leads to a shutdown or a sharp decrease in the productivity of the entire technological set of machines at the facility [6]. On the other hand, with such a large amount of equipment, the warehouses of spare parts increased sharply, the cost of temporarily dead capital rose. This, in turn, also leads to the increase in the cost of the work performed. The presence of extremely scarce parts further complicates this task [7]. The complexity of solving the problem of spare parts reserve is aggravated by the fact that some spare parts are suitable for several types of machines, and the other part is only suitable for one brand of machines [8]. At the same time, the nomenclature of the fleet is significant, the number of vehicles of the same type in each group varies [9] within wide limits.The problem of determining the rational volumes of redundancy for machines and their parts were solved earlier [10], but for construction machines in the developed methods, simplifications were adopted that did not provide the desired optimization [11].

2. Methodology

Problem statement: there is a fleet of machines with known characteristics of each unit. The operating time of each machine for the considered planned period is set. It is required to determine the need for spare elements to maintain the serviceability and operability of a machines fleet.

The task can be solved sequentially in the P stage:

Stage I. Determination of the number of machine elements failures;

Stage P. Solving the problem of determining the need for spare elements.

To obtain a specific numerical solution, it is necessary to collect the relevant statistics. For a number of construction equipment models, we collected such information, obtained regularities, which were then verified and showed the reliability of theoretical results.

3. Results

Problem statement: there is a fleet of machines with well-known characteristics of each unit. The operating time of each machine for the considered planned period is set. It is required to determine the need for spare parts in order to maintain the serviceability and operability of a machinery fleet.

The problem can be solved sequentially in II stages.

Stage I. Determining the number of failures for machine elements.

The pattern of machine failures during its operation is established.

$$I_{1i} = (X_{1i}, X_{2i} \dots\dots\dots.. X_{ni}) \quad (1)$$

Where (I_{1i}) is the number of failures, (i) is the machine, (I) = {1,2, ... I, ... m} is the set of indexes for a machine of the same kind, X_{1i}, X_{2i}, ... Xni is the factors affecting the flow of failures of the machine (the total operating time of the machine, the temperature of the environment, the qualification of the driver, soil conditions - for earthmoving machines, etc.).

The probability of a major repair in the period in question is determined.When capital repair is possible; the following condition must be met:

$$\frac{T_{ik}+T_i}{T_k} \quad (2)$$

Where (T_{ik}) is the operating time of the i- machine from the last major repair, mach-hour; (T_i) operating hours of the i- machine in the planned period, mach-hour; (T_k)is the turnaround hours, mach-hour.

The probability of replacing a part during a major repair is revealed.

$$P_{enik} = (1 - P_{eik}) \quad (3)$$

Where (P_{enik}) is the probability of replacing the e- part of the i- machine during a major repair; (P_{eik}) is the probability of the part being usable until the next repair.

If several identical elements are installed on the machine, then the probability of the number of replaced parts (n_{eik}) from the total number can be calculated by the (Eqaution 4).

$$n_{eik} = (1 - P_{eik})n_{ei} + a \quad (4)$$

Where (a) is the integer correction.

Similarly, the probability of replacing the part with a planned current repair is calculated.

Condition (2) is checked if there is no major repair, then the number of current repairs is determined by the dependence:

$$N_T = \frac{T_{iT}+T_i}{T_T} \quad (5)$$

If major repairs are made, then

$$NT = \frac{T_{i\prime T}+T_i}{T_T} - a - 1 \quad (6)$$

Where ($T_{i\prime T}$) is the operating time from the last scheduled repair, mach-hour; (T_i) is standard operating time between two current repairs.

The number of replaced parts n′ during current repairs of the i machine

$$n'_{eiT} = [(1 - P_{eiT})n + a]N \quad (7)$$

Where (P_{eit}) is the probability that the part will not be replaced during the current repair.

A number of parts (for example, filters) are replaced during the maintenance of machines, after reaching the standard operating time.

The number of elements to be replaced in this case will be:

$$Tl'_{eiTo} = (\frac{T_{i3}+T_l}{T_{3ei}} - a-)n_{eiTo} \quad (8)$$

Where (T_{i3}) - time from the last part's replacement, mach-hour; (T_{3ei}) is standard operating time between two replaced parts i of the i - machine.

Besides planned replacements, parts can be replaced as a result of an emergency failure. The probable need for these elements n^{a}_{3c}, for unplanned repairs will be equal to:

$$(9) h'_{eia} = \frac{\lambda'_i}{T_i} \sum P_{eia} + a$$

Where (λ'_i) is the failure rate of the i - machine in the considered period 1 / mach-hour, x_i can vary for different periods of operation; (P_{eia}) is the likelihood that the *e* -part of the *i* - machine will be replaced during an unplanned repair.

If there are several such parts on the *i* - machine, then

$$(10) n'_{eia} = \frac{\lambda'_i}{T_i} \sum P_{eia} + a$$

Where (P_{reia}) is the probability of failure of the r unit

Thus, the total need for *e* - parts for the *i* - machine during its operation will be *e* - part of the *i*-machine.

$$(11) h_{ei} = h'_{eik} + h'_{eiT} + h'^{\,eiTO} + h'_{eia}$$

Since the machines in the considered period T can have different operating times, then for the cars the number of e - parts will be:

$$n_e = \sum_0^T \sum_{i=1}^M n_{ei} \qquad (12)$$

Stage P. Solution of the problem of determining the need for spare parts.

The need for a fleet of machines in spare parts, meanwhile, depends on a number of reasons. These include seasonal uniformity of loading, changes in the intensity of parts replacements both during the year and throughout the entire service life.

We define the need in the e - parts at the T planning period. Since during the planning period T i - machine can have an operating time T equal to or less than this period (if they are given in the same units of measurement), then under the condition

$$T \geq T_i \geq 0$$

There may be peaks of need in individual parts. Using statistical sampling, we can set the amplitude of the oscillations of the part output from the system as a whole for a machinery fleet.

The need for e - part in the most unfavorable way can be expressed through the (Eq.13):

$$n_e^n = k_{en} \frac{n_e^n}{T'} \qquad (13)$$

Where (T') is the planning calendar; (k_{en}) is coefficient taking into account the maximum need for replacing the e – part; n_e^n is the need for the e - part during the period of maximum exit from the structure of this part.

The total number of e -parts taking into account the unevenness of their deliveries to the organization's warehouse is:

$$n_e^c = k_3 n_e \qquad (14)$$

Where (K_3) is the reserve ratio taking into account the unevenness of supplies (k_3 = 1.1 - 1.3).

For repairable revolving parts

$$n_e^c = k_3 \left(n_e + \frac{\sum_{Toei}\, n_e^{\circ}}{I \cdot 365}\right) \qquad (15)$$

Where $(Toei)$ is the turnaround time of the e - repaired part on the n- machine; (n_e) is the number of replaced parts; (n_e°) is the number of revolving parts.

4. Conclusion

This technique allows rationalizing the required amount of spare elements for the entire range of used units, parts, machines. The developed methodology allows solving the problem of the optimization of the formation of a warehouses network on various levels, both large and medium-sized enterprises, as well as factories - machine manufacturers, creating their own warehouses network and support points in the territory where their products are distributed.

References

[1] Kim B, Nasrulloeva R, Shakir Z and Zakrevskaja L 2020 *IOP Conf. Ser.: Mater. Sci. Eng.* **786** 012014

[2] Kim B, Shakir Z and Nasrulloeva R 2020 *IOP Conf. Ser.: Mater. Sci. Eng.* **786** 012015

[3] Kim B 2000 *Serviceability and efficiency of construction equipment parks* (Vladimir: VlSU) p 148

[4] Merna T and Al-Thani F 2005 *Corporate Risk Management* (England: John Wiley & Sons Ltd) p 163
[5] Muna H, Mohamed N and Omar A 2012 *Damascus University J. of Engineering Sciences* **28** 129-150
[6] PMBOK 2004 *A Guide to the Project Management Body of Knowledge* (Pennsylvania: Project Management Institute) p 187
[7] PMBOK 2013 *A Guide to the Project Management Body of Knowledge* (Pennsylvania: Project Management Institute) p 223
[8] Samolazov A and Panadeeva I 2010 *J. Mining equipment and electro mechanics* **2** 2-11
[9] Zbritsky A 2014 *J. Economics of construction* **5** 3–9
[10] Zorin V 2000 *Fundamentals of technical systems performance* (Moscow: Academy) p 287
[11] Drozdov A and Kudryavtsev E 2012 *Construction machines and equipment* (Moscow: Academy) p 188

IOP Conf. Series: Materials Science and Engineering 1103 (2021) 012011 doi:10.1088/1757-899X/1103/1/012011

Optimization of acoustic profiler characteristics in the scope of underwater archeology and monitoring of underwater engineering objects

I A Kirichenko[1], I B Starchenko[2], V Y Vishnevezkiy[1], A I Markolia[3], I I Sizov[3] and T P Strochan[3]

[1] Southern Federal University, 2, Shevchenko St., Taganrog, 347922, Russia
[2] Parametika LCC, 19, Svoboda St., Taganrog, 347902, Russia
[3] Academy of Science of Abkhazia, Sukhum, Abkhazia

E-mail: ikirichenko@sfedu.ru

Abstract. In order to solve the problem of optimizing the characteristics of acoustic profiler with parametric transmitting array in the scope of underwater archeology and monitoring of underwater engineering objects, the results of theoretical studies of the influence of real conditions on the work of parametric hydro-acoustic systems and of the marine soils and bottom sediments features are presented. The necessary amplification range of the receiving path is estimated to compensate for losses during acoustic wave propagation in water and for the most common types of marine sediments: "coarse sand" and "sandy-silty clay". The necessary acoustic power is justified for the given values of the central pump frequency of the parametric profiler and the difference frequency. The most efficient penetration into the marine sediments which was commonly sand and silt occurred at 10-20 kHz of difference frequency. Penetration of frequencies higher than 20 kHz is not effective due to great power losses. The minimal gain of receiving unit must be at least 40 dB. Design and construction of parametric profiler was suggested. Main units were proposed according to the theoretical investigation. In the whole parametric arrays and hydro-acoustical devices with them occurred very effective in subbottom profiling. Small overall dimensions as was shown in construction allow achieving high resolution. So their application is perspective especially in local underwater archeology using small vessels.

1. Introduction

The shallow water environment is one of the most dynamic elements, including coastal zones that are subject to rapid sedimentary flows and are of significant interest to human activities. However, these environments often pose serious technological challenges due to shallow water, strong waves, strong currents, and a large tidal range. In addition, coastal and littoral zones are often marked by the presence of fine gas, which can severely restrict sound penetration [1-5]. As a result, these areas are rarely studied in a structured way, since it is known that such land-sea transition areas are rich in archaeological features [5, 6].

Acoustic profilographs are usually used as a mapping tool for the sea floor and the bottom structure in the upper layers of several meters. Increasing the efficiency and accuracy of profilographers can increase the productivity of research in terms of resolution and increases the potential of acoustic tools for classifying bottom layers and sediments [7].

2. Materials and methods

The most important parameter that characterizes a sonar detection or classification system is the range. This parameter, as it is known [8], significantly depends on the conditions of acoustic energy propagation and therefore the energy range in a homogeneous (boundless) medium and the range in real conditions are distinguished.

The efficiency of the radiating block of the Profiler largely depends on the properties of the medium in which the acoustic wave propagates. The attenuation value is characterized by an absorption coefficient, which shows the degree to which the amplitude of sound waves decreases with distance. To solve the problem of determining the energy loss of an acoustic wave caused by attenuation in water when profiling the upper edge of the bottom, we can use the expression [8], which takes into account these losses. The expression describing the dependence of attenuation K in the logarithmic range takes into account the possibility of using the obtained data for subsequent compensation of losses during propagation of an acoustic wave in the receiving unit of the profiler.

$$K(r) = 20\lg\left(r \cdot 10^{0.1 \cdot \beta_w \cdot r}\right), 0 \le r \le R_{max},$$

where β_W – sound attenuation coefficient in the propagation medium, dB/m, r – distance, m, $R_{max,}$ - maximum distance to the object, m.

The technique that allows predicting the potential capabilities of the profiler, estimate the frequency range required for operation in specific situations, and select the energy potential depending on the parameters of the probing signal, interference conditions, and acoustic properties of the bottom soil is described in detail in the work [9].

To calculate the acoustic power for each of the pump frequencies required to detect a known target at a given distance in the presence of noise interference, you can use the formula [9]:

$$W_a\,(z) = \frac{z \cdot 10^{0.05\alpha z} \cdot \delta \cdot P_{n0} \cdot 10^3 \cdot 8 \cdot c^3}{e^{-(z/L_{att})} \cdot |I(z)| \cdot F^3 \cdot 3.5 \cdot \pi \cdot R_{eqv} \cdot \sqrt{\gamma_{tr} \cdot \tau}}.$$

where W_a – acoustic power, W, z –distance, m, δ - ambient noise level, Pa/Hz$^{1/2}$, P_{n0} - acoustic pressure at center pump frequency, Pa, c - sound speed, m/s, F - difference frequency, Hz, α - attenuation factor, dB/km, L_{att} - attenuation zone length, m, γ_{tr} – transmitting array axial concentration factor, R_{eqv} - equivalent target radius, m, τ - pulse duration, s, $I(z)$ - Bessel function.

To obtain information about the bottom structures in an acoustic parametric profiler, the most promising method is that associated with the movement of the receiving-emitting antenna system of the profiler due to the movement of the carrier [7]. Simultaneous scanning with the directional characteristic in the radiation mode in the traverse plane of the profiler carrier allows getting a strip on the bottom surface.

As it is shown in [9], the high efficiency of converting the energy of high-frequency pump waves into the energy of difference waves in the parametric radiating path is achieved only when a highly directed radiation with a width of the directivity characteristic of no more than 4-6°is formed.

The transmitting array of the profiler is an N-channel antenna array. Each of the N channels is a line of n piezoelectric elements, the distance between the acoustic axes of which is the same. Such antenna systems are called equidistant.

The need to use a separate receiving channel in the antenna of a parametric Profiler is associated with the irreversibility of nonlinear processes that underlie its principle of operation. The geometric dimensions and shape of the receiving antenna are determined based on the type of location scheme, the parameters of the Profiler in the radiation mode, and the manufacturability. The limitations of overall dimensions at difference frequencies, which in practice are 1-20 kHz, do not allow directional reception, which provides optimal spatial filtering when the solid angles of the antenna directivity characteristics are equal in the radiation mode and the reception mode. In this case, the receiving part of the profiler antenna may have a weak orientation and the main requirement for it is to ensure high sensitivity and the possibility of matching with the receiving and amplifying path.

When choosing the values of the width of the directional characteristic in the diametral plane, it is necessary to assume that the width of the main lobe of the transmitting array in the frequency band is 4°-6°, and the minimum distance of the Profiler from the bottom is h_{min}=1-5 m. The area of the bottom voiced by the transmitting array must fall into the solution of the main lobe of the receiving antenna's directional characteristic.

3. Results

Losses during propagation of an acoustic wave in water for a distance corresponding to the maximum range of the acoustic profiler R_{max}=100 m, determine the necessary gain range of the receiving path to compensate for propagation losses, shown in figure 1 (a).

For the maximum distance to the object R_{max}=100 m, the equivalent radius of small objects R_{eqv}=0.2 m, the reduced ambient noise level – 0.05 Pa/Hz 1/2, as well as the values of the central pump frequency of 250 kHz and the difference frequency of 10 kHz, the acoustic power was calculated for the central pump frequency, the results of which are shown in figure 1 (b).

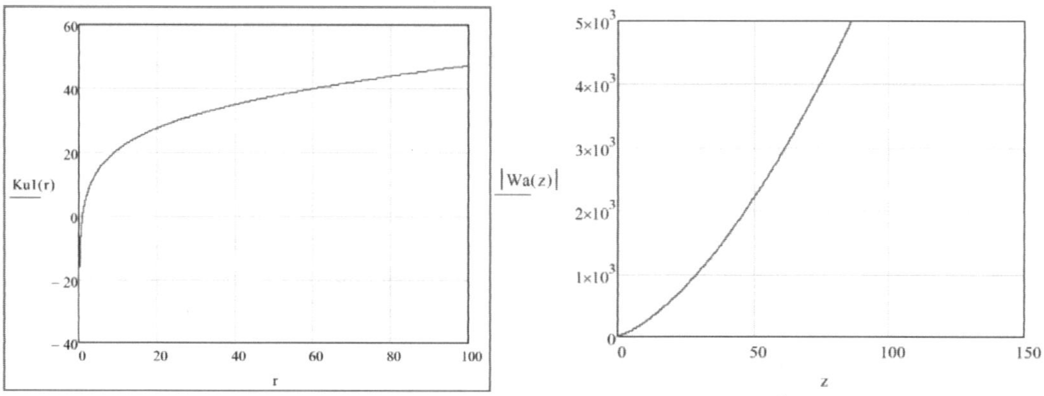

a b

Figure 1. Loss of acoustic wave propagation in water (a) and acoustic power of the profiler (b).

The dependence of losses during propagation of acoustic wave in water is shown in figure 1 (a). It demonstrates that the distance with the maximum one to the object R_{max} does not significantly affect the propagation of acoustic wave. In this case, the dependence of the acoustic power of the pump waves, shown in figure 1 (b), allows determining the required level of acoustic power, which for a distance of 100 m is at least 5 kW.

For bottom rocks, the values of the acoustic wave attenuation coefficient significantly exceed the propagation losses in water [10]. Figure 2 shows the results of calculating losses in a uniform layer of bottom soil 10 m thick for the most common types of marine sediments: "coarse sand" (a) and "sandy-silty clay" (b).

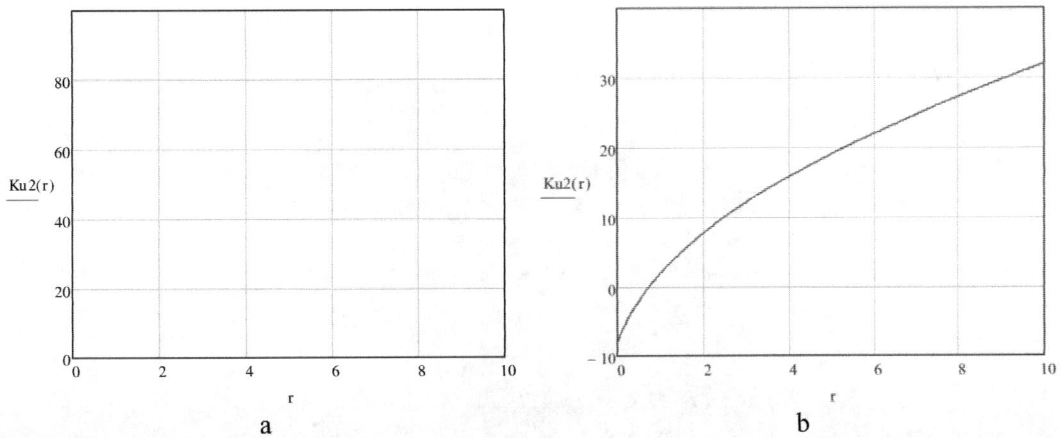

Figure 2. Losses during acoustic wave propagation in the medium: a - "coarse sand"; b - "sandy-silty clay".

The dependences of losses during acoustic wave propagation in the "coarse sand" and "sandy-silty clay", shown in figure 2, demonstrate that for bottom rocks, the values of the acoustic wave attenuation coefficient significantly exceed the losses during propagation in water. The obtained dependences of losses during propagation of an acoustic wave in various media can be used to establish the law of temporary automatic gain control of the receiving path of an acoustic profiler.

4. Construction

The main criteria for the design of the layout of the profiler compartment are: 1) minimal effect governing the input and output power signals to each other; 2) separation as far as possible control lines of transmitting array and the signal lines of the receiving array; 3) the minimum length of the connecting conductors and 4) optimal and uniform heat dissipation by the air convection inside the compartment on the outer wall of the housing.

Parametric profiler is structurally a hollow sealed cylindrical body made of aluminum alloy, inside which electronic equipment is located. The case is covered with protective powder enamel, the color of the coating is determined at the stage of working design documentation. The hydroacoustic receiving and transmitting array of the profiler with a fairing is fixed on the outside of the body. For protection during maintenance and transportation, the working surfaces of the antennas are protected by a removable cover. In the front and back of the profiler compartment, there are flanges for securing on the rod. Electrical connection is provided by connectors located on the side walls of the housing. A crate is located in the cylindrical case, which provides a rigid attachment of electronic units and modules and the ability to remove electronics. The appearance of the profiler is shown in figure 3.

The compartment and antenna housing are shown transparent, the fairing and protective cover are not shown. Figure numerals indicate: 1 – body; 2 – signal module; 3 – power supply module; 4 – receiving arrray; 5 – transmitting array; 6 – cable glands for antenna connection; 7 – electric connectors; 8 – housing o-ring; 9 – power amplifier (6 units); 10 – power cell crate. The antenna body must be protected from incoming water flow by a plastic or metal fairing with a window opposite the active surface of the antenna.

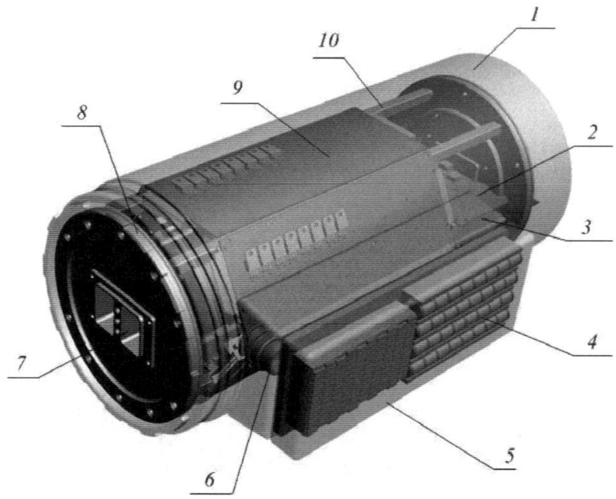

Figure 3. Appearance and layout of the profiler

5. Conclusion

After studying the main characteristics of the acoustic profiler, taking into account the nonlinear process of formation of waves of different frequency and parameters of marine precipitation, we can draw conclusions. The obtained dependence of losses during propagation of an acoustic wave shows that the distance determined by the depth of the place does not significantly affect the propagation of an acoustic wave in water due to the small value of the attenuation coefficient in water and the losses themselves can be easily compensated in the receiving unit with a gain of 40 dB.

The greatest efficiency of penetration into the bottom soil, which is sand and silt deposits, at optimal values of acoustic energy occurs in the frequency range of 10-20 kHz. At frequencies below 10 kHz, penetration is limited by the need to increase the acoustic power of the initial pump waves to a threshold close to cavitation. For frequencies above 20 kHz, penetration into a 10 m thick layer is limited by the attenuation of the acoustic wave in the ground itself.

Taking into account the obtained results make it possible to increase the efficiency of using acoustic devices with parametric antennas when conducting ocean research related to measurements in the interests of underwater archaeology and monitoring of underwater engineering structures.

6. Acknowledgments

The reported study was funded by RFBR and ASA according to the research project №19-52-40005.

References

[1] Missiaen T, Evangelinos D, Claerhout Ch, De Clercq M, Pieters M and Demerre I 2017 Archaeological prospection of the nearshore and intertidal area using ultra-high resolutionmarine acoustic techniques: Results from a test study on the Belgian coast at Ostend-Raversijde *Geoarchaeology* 1–15

[2] Anderson A L and Bryant W R 1990 Gassy sediment occurrence and properties: Northern Gulf of Mexico *Geo-Marine Letters* **10** 209–220

[3] Missiaen T, Murphy S, Loncke L and Henriet J-P 2002 Very high resolution seismicmapping of shallow gas in the Belgian coastal zone *Continental Shelf Research* **22(16)** 2291–2301

[4] Weschenfelder J, Correa I, Aliotta S, Pereira C and De Vasconcellos V 2006 Shallow gas accumulation in sediments of the Patos Lagoon, Southern Brazil *Annals of the Brazilian Academy of Sciences* **78(3)** 607–614

[5] Wilkinson T J and Murphy P 2009 Archaeological survey of an intertidal zone: The submerged landscape of the Essex coast, England *J. of Field Archaeology* **13(2)** 177–194

[6] Bates M, Bates R and Briant R 2007 Bridging the gap: A terrestrial view of shallowmarine sequences and the importance of the transition zone *J. of Archaeological Science* **34** 1537–1551
[7] Kirichenko I and Starchenko I 2019 Directivity of hydroacoustic systems with parametric array in marine conditions *J. of Phys.: Conf. Ser.* **1353(1)**
[8] Matvienko V N and Tarasuk Yu F 1981 *Range of hydroacoustic devices* (Leningrad: Sudostorenie) 206 p
[9] Novikov B K, Rudenko O V and Timoshenko V I 1987 *Nonlinear Underwater Acoustics* (Acoustical Society of Amer.)
[10] Hampton L 1985 Acoustic properties of sediments: An update *Rev. Geophys*. **23(1)** 49–60. doi: 10.1029/RG023i001p00049.

Investigation of return movement process of vibrocleaver blade

E S Kligunov and G G Voskresenskiy

Pacific National University, 136, Tihookeanskaya st., Khabarovsk, 680036, Russia

E-mail: 005196@pnu.edu.ru

Abstract. The lack of the necessary amount of equipment for winter maintenance, significant precipitation leads to a decrease in the efficiency of roads. Inside, courtyards and sidewalks are also covered with a layer of compacted snow due to untimely snow removal. The mechanical method of breaking compacted snow is currently insufficiently studied and so far there is no effective equipment to combat snow and ice formations. A peculiarity of the process of compacted snow destruction in the mode of "vibration cutting" is intermittent contact of the blade with the destroyed medium. Condition of steady mode of vibration cutting is criterion of vibration scraper blade release from contact with medium to be destroyed. The purpose of the study is to establish the time at which the contact of the blade with compacted snow is interrupted. Obtained results of calculations make it possible to set values at which vibration cutting mode is possible and to select parameters of hydraulic accumulator, which provides movement of the vibrocleaver blade of compacted snow.

1. Introduction

The role of roads in the Far East is especially great. Highways - in most regions are the only mode of transport. The lack of the necessary amount of equipment for winter maintenance [1-7], significant precipitation leads to a decrease in the efficiency of roads.

The fallen snow forms a layer of compacted snow, which has reduced coupling qualities, at which the transport speed decreases, the number of accidents and injuries of traffic participants increases.

Inside, courtyards and sidewalks are also covered with a layer of compacted snow due to untimely snow removal.

The main method of breaking a layer of compacted snow is chemical using salts that negatively affect the environment [5, 6]. The mechanical method of breaking compacted snow is currently insufficiently studied and so far there is no effective equipment to combat snow and ice formations.

Finding ways to improve the efficiency and performance of snow removers remains an urgent task. In order to solve this problem, experts of many countries on the winter road maintenance and transport facilities have conducted significant studies [8, 9].

2. Materials and methods

Pacific State University has developed several prototypes of equipment for the destruction of compacted snow on roads and sidewalks using the vibration pulse effect of a blade on a layer of snow and ice [8, 9].

The working process of a compacted snow vibrocleaver, designed to destroy snow-ice formations on sidewalks in a mechanized way, is distinguished by a pulse loading mode. The cutting force depends on the parameters of the layer of snow to be cut: density, depth and width of the cut, angle of cutting, as well as air temperature. In addition, the cutting force is characterized at the beginning of the process of interaction of the blade of the blade with snow by an increase in its value, and then, with chips of snow particles, a sharp decrease in the maximum values characteristic of the destruction of solid brittle media.

The impulse method of interaction of working bodies with the medium finds increasing application. There is a problem of creating simple and reliable vibration exciters. A new type of pump hydropulser is proposed in Russian Patent No. 2162030 (fig. 1).

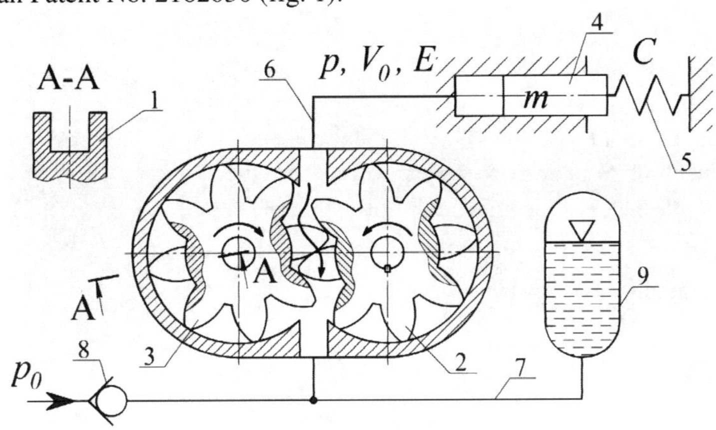

Figure 1. Gear type hydraulic pulser design diagram

Peculiarity of gear-type hydropulser is cutout in contacting teeth of windows 1 in order to preserve kinematic connections between gears 2, 3. During rotation of gears with full teeth, working fluid is sucked in and pumped into working cavity of hydraulic propeller 4. Pressure at output of hydraulic pump-pulsator grows under action of elastic force of spring 5. At the moment of interaction of teeth with openings, pressure chamber 6 of hydraulic propeller 4 communicates with suction chamber 7 of hydraulic pump and under action of compressed spring 5 rod of hydraulic propeller 4 moves to initial position, displacing working fluid into suction line. The process of overflow of the working fluid from the delivery line to the suction line is stopped during the period when the conventional teeth begin to interact, and the working fluid again begins to be pumped into the working line of the hydraulic propeller.

Filling of suction line and makeup of possible leaks in hydraulic pusher 4 is performed through check valve 8 under pressure $p_0 = 0.3 - 0.5$ MPa. Such pressure provides better suction conditions for the working fluid by the pulsator pump. Hydraulic pneumatic accumulator 9 serves as a reservoir and is filled with volume of working fluid supplied from drain line of the main hydraulic system.

A peculiarity of the process of compacted snow destruction in the mode of "vibration cutting" is intermittent contact of the blade with the destroyed medium.

The working process consists at the initial stage of introducing the blade into the compacted snow with movement to the position x_0, and then when the gear windows are opened, the blade returns to the initial position and simultaneously moves forward with the speed of the base machine V_M. The blade cutting edge coordinate will be

$$x = x_0 - x + V_M \cdot t \tag{1}$$

The purpose of the study is to establish the time during which the contact of the blade with compacted snow is interrupted.

The process of returning the blade is presented in the form

$$m \cdot \ddot{x} + k \cdot \dot{x} + c \cdot x = c \cdot x_0 - p \cdot F \tag{2}$$

where m - rod weight; k - coefficient of viscous friction; c - spring stiffness; F - piston area; x_0 - initial position of rod; p - pressure in hydraulic pusher cavity.

Pressure in front of hydraulic push-rod decreases sharply from initial position p_0, and then increases as blade returns to initial position due to overcoming resistance to movement of working fluid [9, 10]

$$p = p_0 \cdot e^{-\alpha \cdot t} + \left(\lambda \cdot \frac{l}{d} + \xi\right) \cdot \left(\frac{2 \cdot F_T}{f_{TR}}\right)^2 \rho_{hf} \cdot \dot{x}^2 \tag{3}$$

where α is the coefficient that takes into account the intensity of the initial pressure drop, $\alpha = 400$ 1/s; λ - is coefficient of resistance to the movement of the working fluid; l, d - the length and diameter of the pipelines; ξ - is the coefficient of local resistance, $\xi = 10$; $F_T = 0.25 \cdot \pi \cdot D^2$ - the area of the hydraulic pushrod; D - the diameter of the hydraulic pushrod; $f_{TR} = 0.25 \cdot \pi \cdot d^2$ - the area of the pipelines, ρ_{hf} - the density of working fluid.

3. Results and Discussion

The solution of equations (2, 3) is performed in the MATLAB program with the accepted values of $m = 0.003$ kg·s²/cm; $k = 0.2$ kg·s/cm; $F_T = 1.13$ cm²; $f_{TR} = 0.5$ cm²; $V_M = 100$ cm/s; $c = 91.8$ kg/cm; $x_0 = 0.6$ cm... 1.6 cm.

Results of transient processes of vibrocleaver blade movement are obtained. It was established that at the moment of opening of gear teeth windows, the blade reaches a value x_0, and the pressure drops sharply from the value p_0 (fig. 2) to values determined by pressure losses on local resistances and along the length of pipelines.

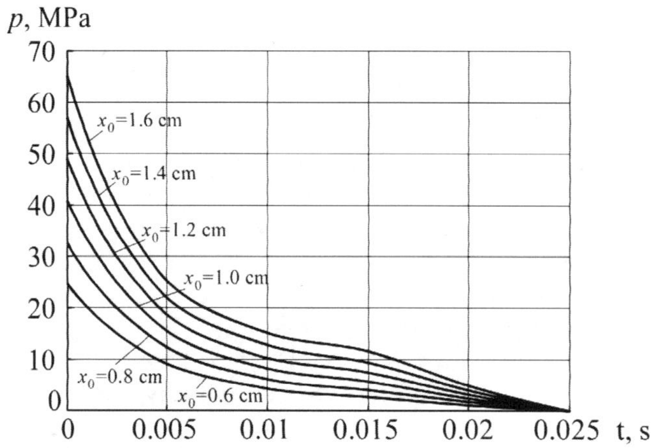

Figure 2. Pipeline pressure change transients

Under the influence of another force, the blade starts a return movement (fig. 3). The speed of the return motion depends on the value x_0 and p_0. At the same time, the vibrocleaver blade continues to move with the base machine at a speed V_M.

The return speed of the blade depends on x_0 and p_0 (fig. 4).

Condition of steady mode of vibration cutting is criterion of vibrocleaver blade release from contact with medium to be destroyed.

$$-V_X > V_M \tag{4}$$

It is found that at the speed of the base machine $V_M = 1.0$ m/s (100 cm/s) only at $x_0 > 0.8$ cm, the speed of return movement can ensure that the blade comes out of contact with the medium to be destroyed.

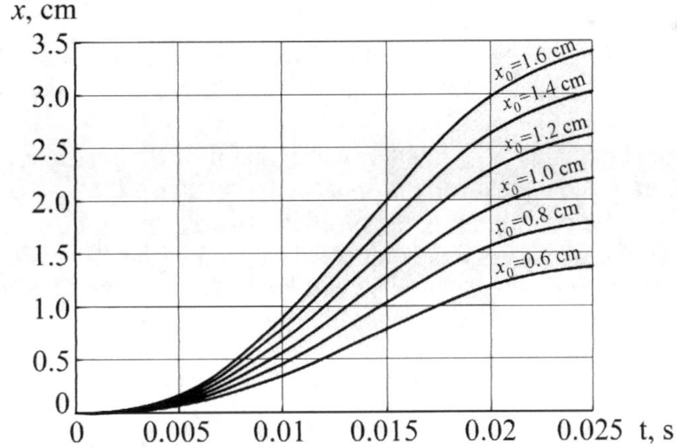

Figure 3. Transient processes of vibrocleaver blade return movement

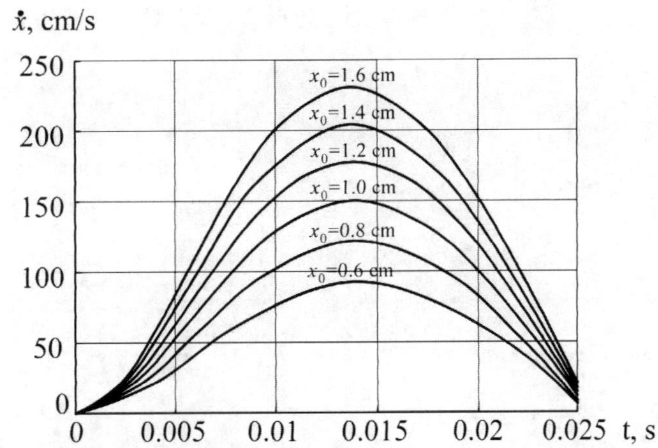

Figure 4. Transients of vibrocleaver blade return speed

4. Conclusion

The obtained calculation results make it possible to set the values at which the vibration cutting mode is possible and to select the parameters of the hydraulic accumulator, which provides movement of the blade of the vibrocleaver of compacted snow up to the values $x_0 > 0.8$ cm.

References

[1] Sajed Sadati S M, Cetin K, Ceylan H, Sassani A and Kim S 2018 Energy and thermal performance evaluation of an automated snow and ice removal system at airports using numerical modeling and field measurements *Sustainable Cities and Society* **43** 238-250

[2] Perrier N, Langevin A and Campbell J F 2007 A survey of models and algorithms for winter road maintenance. Part IV: Vehicle routing and fleet sizing for plowing and snow disposal *Computers & Operations Research* **34** 258-29
[3] Gudac I, Marovic I and Hanak T 2014 Sustainable optimization of winter road maintenance services under real-time information *Procedia Engineering* **85** 183-192
[4] Min J, Fu L, Usman T and Tan Z 2016 Does winter road maintenance help reduce air emissions and fuel consumption? *Transportation Research Part D: Transport and Environment* **48** 85-95
[5] Vignisdottir H R, Ebrahimi B, Booto G K, O'Born R, Brattebo H, Wallbaum H and Bohne R A 2019 A review of environmental impacts of winter road maintenance *Cold Regions Science and Technology* **158** 143-153
[6] Dupuis D, Sprague E, Docherty K M and Koretsky C M 2019 The influence of road salt on seasonal mixing, redox stratification and methane concentrations in urban kettle lakes *Science of The Total Environment* **661** 514-521
[7] Zakirov M F and Baranchik V P 2014 Determination of the resistance to cutting of snow by screw of small rotary snow plow *Interstroimeh – 2014* (Samara: SGASU) pp 25–28
[8] Voskresenskiy G G and Kligunov E S 2020 Distributing Forces and Capacities of the Vibrocleaver Operating Process for Compacted Snow *VIII Int. Sci. Siberian Transport Forum. TransSiberia 2019 (Springer: Advances in Intelligent Systems and Computing)* **1115** 591-599
[9] Voskresensky G G 2008 *Fundamentals of mechanics of compacted snow destruction on roads* (Khabarovsk: Pacific National University) p 250
[10] Korchagin P A, Teterina I A and Rahuba L F 2018 Improvement of human operator vibroprotection system in the utility machine *J. of Phys.: Conf. Ser.* **944(1)** 12–59

Imbalance in the hydrotechnical structure of seaspan lift-transition bridge in sea ferry crossing

E S Kligunov and K P Pozynich

Pacific National University, 136, Tikhookeanskaya str., Khabarovsk, 680035, Russia

E-mail: 005196@pnu.edu.ru

Abstract. The article is devoted to the problem of reliability of functioning of onshore structures of transport systems using the example of railway ferry crossing. A brief assessment of the role ferry crossings at the current stage of the development railway and sea transport was given. The approach to the verification calculation of lifting mechanisms, suspension elements and counterweight devices of the sea span lift-transition road-railway bridge of stage I of the Vanino-Kholmsk ferry complex in the sea port of Kholmsk for the possibility of their use due to an increase in the weight of the span structure after repair was proposed and justified.

1. Introduction

In modern conditions, due to progress in bridge construction, marine (ocean) road-rail ferry lines become the main ones. These lines are a serious competition to lines using universal ships (dry cargo), due to the reduction, and in some cases, the complete elimination of transshipment operations. The effective realization of the maritime potential of our country in the transport component can be given by the synergy of the sea and the railway, which will create the prerequisites for serial, and therefore cheaper production of sea and river ships (ferries), locomotives, rolling stock, containers, digital support of production processes, the development of intermodal transport systems. The lines that provide ferries in the Russian Far East are both a mechanism for solving the problem of transport accessibility and the delivery of socially significant goods to Sakhalin, Kamchatka and Kuril Islands. The railway ferry crossing is a complex engineering structure. In its composition, it has complexes of coastal structures and specialized ships - ferries (Figure 1).

The Vanino-Kholmsk ferry complex (ferry crossing) was built in 1973 is the main transport corridor for the delivery of goods to Sakhalin from the mainland and needs today to carry out an urgent reconstruction of the coastal infrastructure. The global reconstruction of the infrastructure of the Vanino-Kholmsk railway ferry crossing was planned for 2019-2021 years. In accordance with the terms of reference in the sea harbours, berthing facilities are to be upgraded at both ports, and onshore facilities are to be repaired for the reception of new high-capacity sea ferries so that at the time of their reception the port infrastructure meets all the necessary requirements [1-7]. One of the main structural elements of the crossings is lift-transition bridges, which provide connection of sea ferries with the paths of approaches. The reconstruction project provides for the complete replacement of two lift-transition bridges, extension of piers, posts, concreting at a variable level, replacement of baffle frames, replacement of communications.

Figure 1. General view of the ferry complex

2. Methods

The lift-transition road-railway bridge (Figure 2) is a special hydraulic structure of the Vanino-Kholmsk marine ferry crossing.

Figure 2. Ferry loading at Kholmsk seaport

The bridge is constantly in contact with water, which is an aggressive medium in relation to the materials from which the bridge is built. Water has mechanical, physical, chemical and biological effects on them. All these types of impacts lead to premature destruction of hydraulic structures [1-7]. According to published data, at present, according to the results of the inventory, most of the hydraulic structures of Russia (more than 52%) are in a state that requires major repairs and are characterized by a low level of safety. A significant number of hydraulic structures need ongoing repair, or are in emergency and pre-emergency condition.

For more than 40 years of operation of the Vanino-Kholmsk ferry crossing, according to the regulations, technical inspections of the state of the structures and devices of the lift-transition bridge were carried out and technical services were carried out. One of the main tasks at the same time was to maintain the load value established by the project from the own weight of the sea span structure, perceived by lifting mechanisms due to the adjustment of the weight of the main counterweights. At the same time, no preventive and overhaul repairs were carried out to eliminate critical defects in span structures and repairs of lifting mechanisms.

The Vanino-Kholmsk ferry crossing includes two identical two span road-railway bridges. Their design is identical and includes, among others, freely moving counterweights (Figure 3).

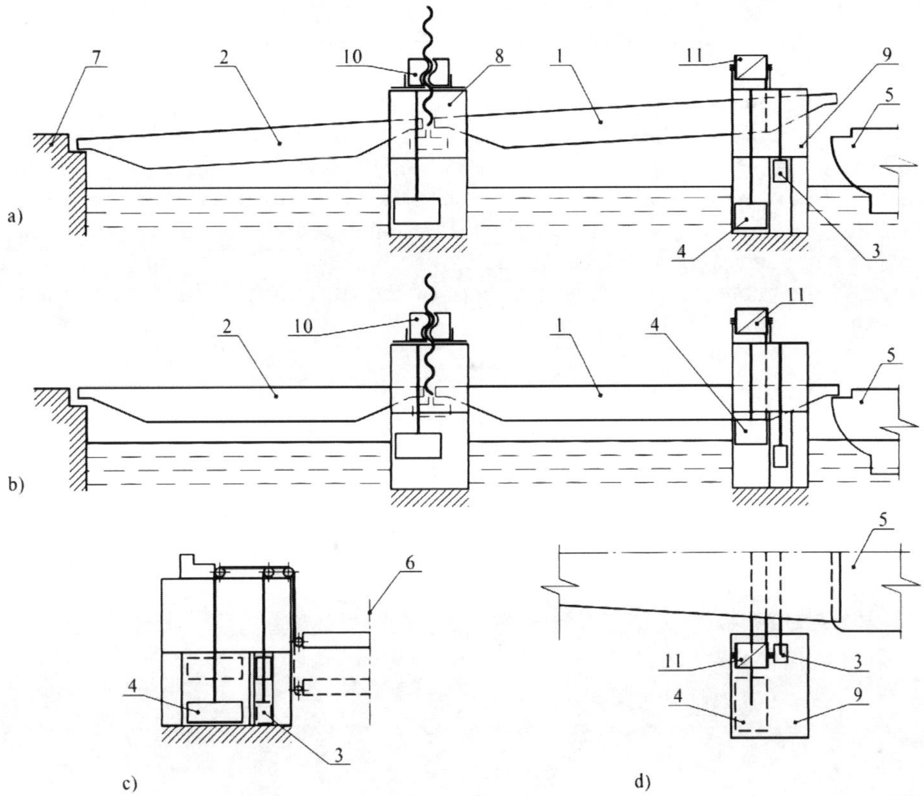

Figure 3. Lifting and transition bridge in the seaport of Kholmsk: longitudinal sections in non-working (a) and working (b) positions, transverse section along the axis of the lifting beam of the sea span (c) and plan (d): 1 - sea span; 2 - coastal span; 3 - counterweight of lifting mechanism; 4 - main counterweight; 5 - steam; 6 - bridge symmetry plane; 7 - coastal support; 8 - intermediate support; 9 - sea support; 10 - screw lifting mechanism on intermediate support; 11 - winch of sea support lifting mechanism

The main purpose of the bridges is to ensure the possibility of transferring railway trains from the shore tracks to the ferry tracks and/or back at a constantly changing depth of the ferry, which depends on the position of the sea level and the different intensity of mobile loads.

The sea end of the sea span during loading and unloading operations, being pre-imposed (with the help of lifting mechanisms) on the stern of the ferry, continuously follows its movements (oscillations) with a special "beak" entering when the bridge is lowered into the "nest" provided in the stern of the ferry.

A two-span lift-transition bridge from the shore to the "bed", formed by two sea lanes to keep the stern of the ship in the correct position, allows locomotive traction to roll up and roll out railway cars from the ferry, as well as load and unload rail-free transport, with the calculated amplitude of vertical movements of the ferry stern up to 4.0 m.

The lifting mechanisms available in the coastal structures are located on the abutments of the intermediate support (vertical drive screws) and on the abutments of the sea support (electric winches). Drawbridge mechanisms belong to the category of mechanisms of lifting machines.

Counterweights are used to balance span structures. Their mass is selected so that the system has a small initial imbalance from the own weight of the span (that is, towards the drawdown span) at any stage of the wiring.

To ensure tight adjoining of the span to the support parts, the value of unbalance in the induced position (initial unbalance) must be at least one percent of the total weight of the span with facilities (full moving mass) and at least 5 tons.

Chambers are arranged in the upper part of counterweights for laying of adjusting ballast from cast-iron or weighted concrete blocks with total weight up to 5% of weight of counterweight. By supplementing or removing the units from the chambers, the balancing of the span structure is adjusted during construction, and if necessary during the operational period. Counterweight devices almost completely balance the own weight of span structures.

Figure 4 shows the cable-unit suspension system of the lift-transition bridge and the location of lifting mechanisms.

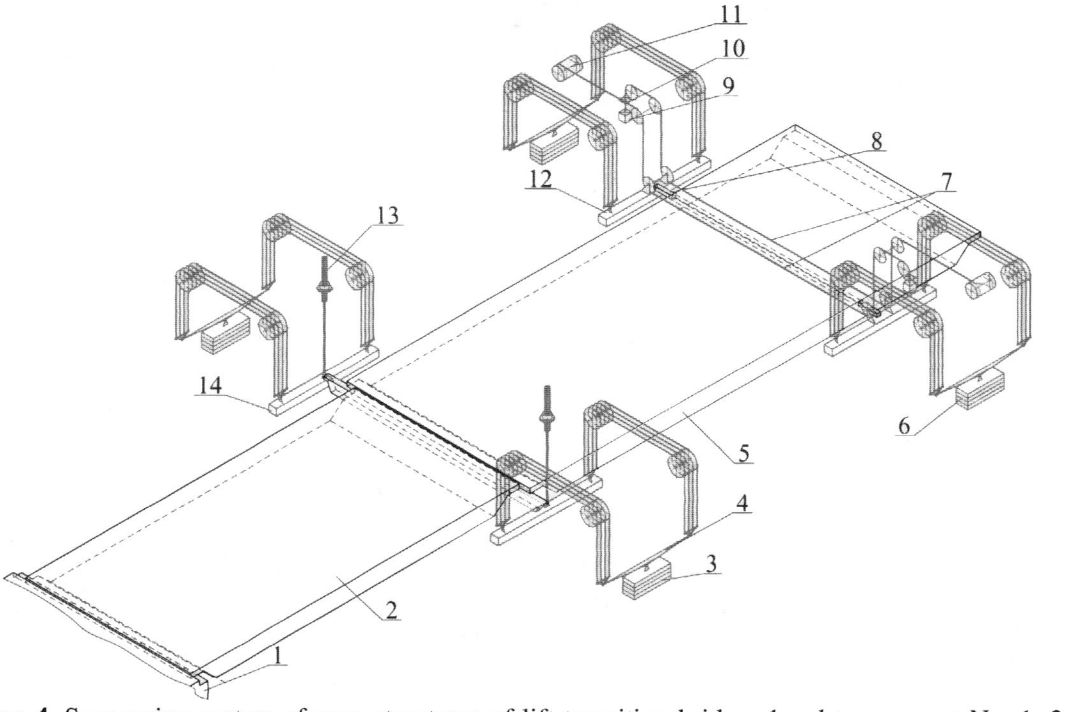

Figure 4. Suspension system of span structures of lift-transition bridge: 1 - shore support No. 1; 2 – shore superstructure; 3 - counterweight intermediate support No. 2; 4 - balancing device of the counterweight; 5 - sea span; 6 - counterweight of the sea support No. 3; 7 - lifting ropes of a sea support No. 3; 8 - lifting beam No. 3; 9 - operative counterweight; 10 - shock absorber; 11 - lifting winch of a sea support No. 3; 12 - traverse beam of the sea support No. 3; 13 - screw lifting mechanism of the intermediate support No. 2; 14 - traverse beam of the intermediate support No. 2

3. Discussion

Under conditions of long-term operation, the steel structures of the bridge underwent significant corrosion wear, as a result of which the lifting capacity of the longitudinal and transverse beams decreased by 80% and 5-15%, respectively. This led to the fact that the carrying capacity of the span was insufficient to perceive mobile loads, that is, a reduction in the weight of the railway load was required.

The project organization developed proposals that made it possible to carry out work in the shortest time and without disrupting the supply of vital goods for residents of the island region in order to ensure the safe operation of the ferry crossing before the start of its global reconstruction. The purpose of repair and restoration works is reinforcement of span elements with preservation of design lifting capacity of the structure [8-12].

4. Results

The customer accepted the option of repairing span structures with the installation of additional linings and steel bonds. As a result, the mass of the span and the amount of imbalance increased.

Difference of masses of mounted and dismantled materials of bridge span was:

- metal rolling $\Delta m_{M.PR} = 104.76 - 75.53 = 29.23$ tons;
- timber $\Delta V_{P.MAT.} = 62.25 - 60.36 = 1.89$ m^3.

Considering the density of wood $\rho \approx 0.66$ tons/m^3, we get a total increase in the mass of the sea span:

$$\Delta m_{M.PR.\Sigma} = \Delta m_{M.PR.} + \Delta V_{P.MAT.} \cdot \rho = 29.23 + 1.89 \cdot 0.66 = 30.48 \text{ tons.}$$

Accordingly, the change in weight of the sea span will be:

$$\Delta P_{M.PR.} = \Delta m_{M.PR.\Sigma} \cdot g = 30.48 \cdot 9.81 = 299.02 \text{ kN.}$$

where g - acceleration of free fall, g =9.81 m/s^2.

Difference of masses of mounted and dismantled steel structures of sea span lifting beam was:

$$\Delta m_{P.B.} = 3.13 - 1.08 = 2.05 \text{ tons;}$$

The change in the weight of the lifting beam of the marine superstructure will be:

$$\Delta P_{P.B.} = \Delta m_{P.B.} \cdot g = 2.05 \cdot 9.81 = 20.12 \text{ kN.}$$

To determine the load $\Delta P_{PR.}$ change at the sea span suspension points, we will bring changes in the weight of the sea span and the sea span lifting beam to the sea supports of the lift-transition bridge.

Let us accept the following assumptions:

- the weight of the sea span $\Delta P_{M.PR.}$ is applied in the middle of the span;
- weight of lifting beam $\Delta P_{P.B.}$ passes through axis of counterweights.

On the compiled design scheme (Figure 5) at point C, the span structure is supported on the stern of the ferry, while the span AC = 33.4 m. Point B on the scheme is the point of suspension of the span structure.

In such structures, imbalance is the weight of the span that falls on the stern of the ferry in working conditions.

The inherent imbalance of the span ensures the stability of the span during operation.

The introduction of initial imbalance prevents the possibility of spontaneous wiring under the influence of random factors, and also ensures a tight landing of the span on the support parts upon completion of aiming.

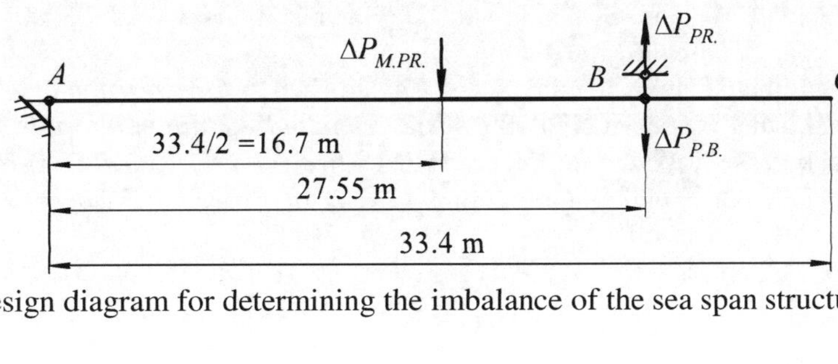

Figure 5. Design diagram for determining the imbalance of the sea span structure

When planning urgent repair and restoration work of the lift-transition bridge, the mass of the sea span was increased by 30.48 tons, a span lifting beam by 2.05 tons, which, naturally, led to the increase in imbalance on the sea support by 20.53 tons. At the same time, for each of the two suspension points of the sea span structure according to Figure 5, 10.26 tons will be additionally accounted for. Accordingly, this will necessitate an increase in the mass of the main counterweights on the sea support to maintain the design level of imbalance, which will require high material and time costs.

Obviously, it is not desirable to overstate the amount of imbalance compared to the project, since this will affect, among other things, the power characteristics of the drive of the lifting mechanism.

This circumstance led to the need to carry out a large volume of verification calculations of installed lifting mechanisms (electric winches), elements of cable-unit systems of the span structure, suspension elements of the main counterweights, etc.

It should be noted that in the process of movement of the span structure, the violation of the balance of the entire system causes an increase in the length of the supporting ropes both from the side of the span structure and from the side of the counterweight. With a small mass of span, this violation of balance is insignificant.

Verification calculations of the ropes of lifting mechanisms, as well as the ropes of counterweight devices, showed that the size of the ropes does not require changes.

5. Conclusion

The verification calculations of lifting mechanisms, span suspension and counterweight devices for the possibility of their use after the repair of the Vanino-Kholmsk ferry bridge in the Kholmsk seaport due to the increase in its mass showed that the operation of lifting mechanisms, span suspension elements and counterweight devices in the event of an increase in the mass of sea span after repair (increase) of weight of main counterweights and replacement of elements of lifting mechanisms and counterweight devices is possible.

It is advisable to increase the weight of the main counterweights on the offshore support in order to reduce the load on the lifting mechanism drive. In accordance with the breaking force of the ropes, according to our calculations, the mass of each counterweight of the sea support cannot exceed 82.35 tons. With the design weight of each of the main balances on the sea support 76 tons, in order to eliminate the problem of imbalance. It is enough to increase the weight of one counterweight on the sea support by 6.35 tons.

References

[1] Liu C, Jiang Zh and Yu H 2020 *Measurement* **151** 107169
[2] Kodur V K R and Naser M Z 2019 *J. of Constructional Steel Research* **156** 46-53
[3] Alencar G, de Jesus A, Guilherme J, da Silva S and Calçada R 2019 *Engineering Failure Analysis* **104** 154-176
[4] Andersen S-N and Tørset T 2019 *Case Studies on Transport Policy* **7(3)** 667-676
[5] Bogunovic-Jakobsen J 1997 *J. of Wind Engineering and Industrial Aerodynamics* **69–71** 795-805
[6] Milne D, Le Pen L, Watson G, Thompson D, Powrie W, Hayward M and Morley S 2018 *Transportation Geotechnics* **17(A)** 61-68
[7] Marques F, Moutinho C, Hu W, Cunha Á and Caetan E 2016 *Engineering Structures* **123** 15-29
[8] Ma H, Zhang Zh, Ding B and Tu X 2018 *Construction and Building Materials* **191** 679-691
[9] Van Puymbroeck E, Nagy W and de Backer H 2018 *Procedia Structural Integrity* **13** 920-925
[10] Dinas A, Nikolaidis Th N and Baniotopoulos C C 2017 *Procedia Environmental Sciences* **38** 578-585
[11] Plachý T, Polák M P and Ryjáček P 2017 *Procedia Engineering* **199** 3053-3058
[12] Leander J, Honfi D, Larsson-Ivanov O and Björnsson Í 2018 *Engineering Failure Analysis* **91** 306-314

IOP Conf. Series: Materials Science and Engineering 1103 (2021) 012014 doi:10.1088/1757-899X/1103/1/012014

Improved design for accelerated consolidation of subgrade when building a highway on a weak base

S N Borychev, A S Popov and D V Koloshein

Department of Construction of Engineering Structures and Mechanics, Ryazan State Agrotechnological University, 1, Kostychev Str., Ryazan, 390044, Russia

E-mail: university@rgatu.ru

Abstract. This article deals with the design of the foundation of the subgrade of a motor road on a weak foundation (peat, sapropel) in difficult climatic and engineering-geological conditions. The features of physical and mechanical properties of soft soils imply the creation of non-typical design solutions using modern technologies and building materials. As a result, the theoretical studies of the processes of accelerated consolidation of weak soils were carried out and a design of the subgrade was proposed based on the use of electro-physical and filtration properties of various aggregates, namely the cathode layer (crushed stone from blast furnace coke or conductive coal), the drainage layer (sand) and the anode layer (crushed stone from dolomite limestone or basalt). The developed structure will allow reducing the consolidation time of a weak base and providing the necessary strength of the base by increasing shear resistance, which will increase the durability and reliability of the road, as well as reduce the technological construction time.

1. Introduction

The construction of high-quality and reliable highways is an urgent task for providing the population of our country with favorable conditions of existence in modern economic relations of any industry.

The motor road is an expensive transport structure, and therefore it is necessary to shorten the technological construction time as much as possible while ensuring the regulatory requirements for it, without reducing the service life to subsequent major repairs.

This is provided for in the national project "Safe and high-quality roads" on the basis of the Decree of the President of the Russian Federation dated May 7, 2018 No. 204 "On national goals and strategic objectives of the development of the Russian Federation for the period up to 2024", in particular the section of the project design passport "Road network".

The vast territory of the country is characterized by various climatic, engineering-geological and economic conditions that create significant difficulties in the design of construction, reconstruction and repair of highways. Since most of the projected sections have difficult conditions, it becomes necessary to create non-typical projects with the latest construction technologies and improvement of the main elements of the highway.

Many research organizations and scientists both in our country and abroad have dealt with problems of designing roads in difficult conditions, including automobile ones. Starting from the middle of the 20th century, characterized by the development of new territories of Western Siberia, the discovery and development of oil and gas fields and development of the oil and gas industry led to development of the regional transport infrastructure, which was difficult in terms of engineering and geological conditions.

There is a tendency in the central regions of our country to reduce the use of valuable agricultural land as a right of way for the construction and reconstruction of highways that has led to the refusal to bypass of projected complex sections.

At the same time, the volume of earthworks associated with technologies for removing soft soil and an increase in the length of the route increased, that led to some rise in the cost and low rate of road construction, an increase in labor intensity and the duration of work. Therefore, from this period, a new direction in the design of highways in difficult conditions has been developing, based on the accumulated vast practical experience and the development of modern research in the field of soil science, engineering geology, as well as the theoretical basis of soil mechanics, rheology and physical and chemical mechanics.

Great contribution to the development of theoretical sections of soil mechanics on a weak foundation was made by Soviet scientists N.N. Maslov, N.A. Tsytovich, Yu.K. Zaretsky, N.M. Gersevanov, as well as foreigners K. Tertsagi, L. Lemb, A. Bishop, R. Peck and others [1].

From the point of view of the fundamental and practical use of the theory for specific purposes, the development of research in the field of constructing foundations of irrigation and drainage structures was of great importance, as well as industrial and civil buildings. It was carried out by B.I. Dalmatov, M.Yu. Abelev, K.P. Lundin, I.I. Vikhdyaev [2]. As for road construction, a new direction of research with the development of new devices and structures for transport structures was carried out by the following scientists: V.D. Kazarnovsky, I.E. Evgenyev, D.M. Dobrov [3].

In turn, the development of new structures and methods for calculating foundations in difficult engineering conditions was carried out in foreign countries, namely the USA (L. Casagrande, w.Web, etc.), Canada (K. Anderson, I. Mac Fariane, R. Redforth), Switzerland (A. Moos, F.J. Aeckiin, etc.), France (F. Bourges, G. Piiot, etc.), Germany (A.D. Cicker, etc.), the Netherlands, Japan and others [3].

After analyzing the existing problems of improving and building new transport facilities, it becomes necessary to create atypical projects with implementation of the latest construction technologies and improve the structures of the main elements of the highway, especially those passing through difficult natural conditions.

2. Problem statement

One of the most important elements of the road is the subgrade. Its construction constitutes the bulk of construction costs, which affects the cost of construction and the payback of the project.

At the same time, the subgrade is to conform to building standards, since the further operation of the transport structure depends on its high quality. Particularly problematic is the construction of a subgrade in swampy areas, which relate to difficult engineering and geological conditions, since the base of the structure is a weak soil, consisting mainly of peat, sapropel, marl, etc. and having special physical and mechanical properties.

Analysis and study of the features of soft soils will make it possible to create rational engineering solutions for structures of the subgrade and improve technologies for its construction.

In this case, it is necessary to take into account the influence of these properties on the stress-strain state of a weak base, the quantitative value of which depends on the following factors:

- physical and mechanical characteristics of base soils;
- the magnitudes and values of the transferred traffic load;
- engineering geological conditions of the structure of the soil layer (the level of groundwater, the outline of the roof of the underlying rocks, conditions affecting the compression of the water-soil mass);
- construction technologies and compliance with operational measures for the existence of a transport structure.

The analysis and study of the features of soft soils and their joint stress-strain interaction with the main elements of the road will make it possible to create rational engineering solutions for the structures of the roadbed and improve the technologies for its construction.

3. Materials and methods

Many researches of both native and foreign origins are devoted to constructive solutions which allow providing the necessary technological characteristics of the subgrade on a weak basis. These works deal with theoretical studies in the field of soil mechanics, in particular the system of phase interaction in the soil mass, developed by Terzaghi, which allows the use of accelerated consolidation by creating artificial channels (drains), which are designed to remove pore water under the influence of excessive pore pressure. The physical and mechanical process is based on shortening the filtration path and changing the compaction rate of weak water-saturated soil under the embankment, since it depends on the duration of water squeezing out of its pores under the influence of a constant load, thereby reducing the embankment settling time. This increases the bearing capacity and stability of the foundation, as the resistance to soil shear increases [4, 5].

When erecting a subgrade on a weak foundation, vertical and horizontal drainage is used, this depends on the strength of the weak soil under the base of the embankment. Theoretically and practically, a relationship has been established between the types of drainage, which in most cases can be considered as the spatial possibility of combining the equation for vertical and horizontal flows in two-dimensional consolidation.

Simplified engineering methods, namely the principle of vertical drainage, were most applied.

The studies in this area [4, 5, 6] have shown that for the effective operation of vertical drains, the following conditions must be met:

- the section of the drain must have high water permeability;
- the load acting on the surface of the soil being compressed must be sufficient to overcome the initial filtration gradient and carry out intensive consolidation;
- the presence of free outlet of gravity water from drains to the lower part of the embankment or to a special drainage layer.

Considering the above mentioned aspects, it is necessary to justify the influence of the hydraulic gradient, under the action of which water is squeezed out of the compressible layer when the following condition is met:

$$u \,/\, L > L_0, \tag{1}$$

where u is the pore water head; L is the filtration path; L_0 is the initial filtering gradient.

Compliance with this condition is ensured by the value of the initial filtration gradient in loose soil, which is expressed by the following dependence:

$$L_0 = C_f \cdot \lg \cdot \mathrm{p}, \tag{2}$$

where C_f is the average value of the filtration coefficient of loose soil; p is embankment pressure on the base.

Based on dependencies (1) and (2), and considering that the pressure in the pore water corresponds to the pressure of the embankment base, an expression was obtained to determine the necessary filtration path or the distance between drains to ensure complete soil consolidation:

$$\mathrm{L} > 0{,}423\,(100 - \mathrm{U_0})\,\mathrm{p} \cdot \ln \cdot \mathrm{p}/\mathrm{C_f}, \tag{3}$$

where U_0 is soil consolidation in%.

Using the obtained expression, it is possible to estimate the effect of drainage devices on acceleration of the filtration consolidation of a loose low-permeability layer, if the thickness of compacted layer H is known, the selected distance between drains is L. Then, using the ratio of the duration of the consolidation period of the formation with one drain filler t_1 to the corresponding period under the action with another filler t_2, one can choose a rational distance between drains:

$$t_1/t_2 = H/L, \tag{4}$$

As the experience of designing shows, it is difficult to fulfill this condition, since the values of L_0 are large and even the use of a temporary surcharge does not allow reaching the calculated value, thereby reducing or stopping the filtration sediment process.

4. Advanced design for accelerated consolidation

The use of filter aggregates with the required physical and mechanical properties for standard drain diameters is of great practical interest in the development of design solutions. It is possible to reduce the consolidation time and accelerate the water filtration process due to the electro-physical effect of the interaction of various materials [7, 8].

It is planned to improve the constructive solution for the accelerated consolidation of a loose base by sequent arrangement of the filtration material with different electro-physical properties in vertical drains (Fig. 1): cathode layer 5 (crushed stone from blast furnace coke or conductive coal), drainage layer 7 (sand) and anode layer 6 (crushed stone from dolomite limestone or basalt).

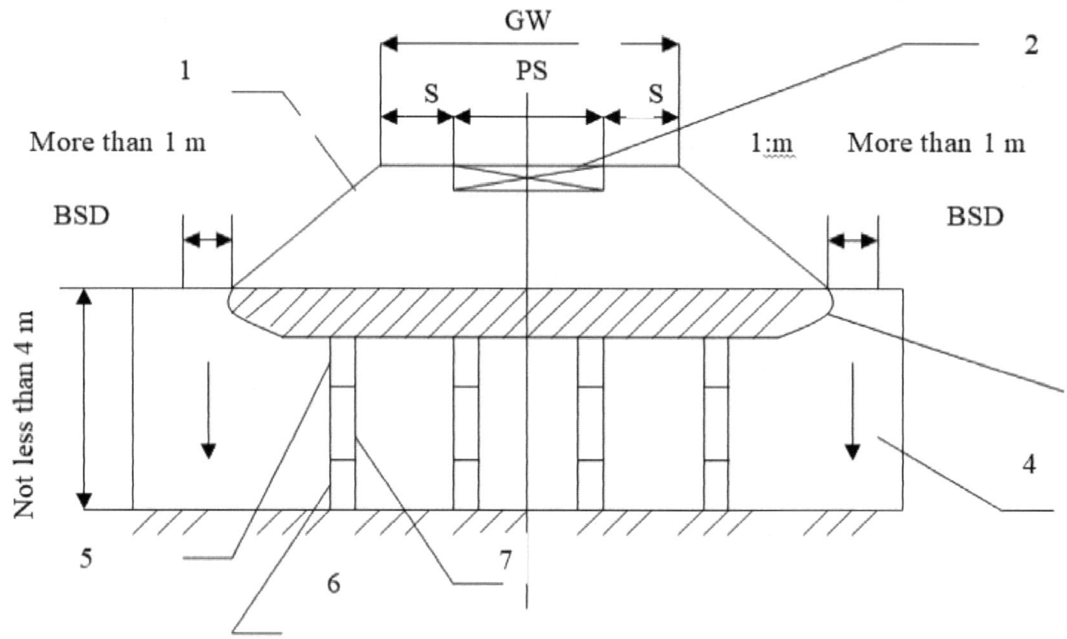

Figure 1. The design of accelerated consolidation of a loose base. BSD is the boundary of the surface diversion. PS is the pavement surfacing. S is the shoulder. GW is the grading width. 1 - embankment, 2 - pavement surfacing, 3 - bottoming, 4 - loose base, 5 - cathode layer, 6 - anode layer, 7 - drainage layer.

The principle of operation of the improved design is as follows. The placed layers of the filtration material in the drainage channels will allow the creation of electrically interconnected anode and cathode layers, which will create favorable conditions under the influence of an electrostatic field formed due to the electrical properties of materials for the transformation of cohesive water into gravitational water and its rapid removal to the lower base layer from the subgrade [9, 10]. The physical meaning is as follows: since a water molecule is a dipole and has electrical bonds, therefore, under the influence of an electrostatic field between the anode and the cathode, it will be constantly removed not only from the peat cover, but from the body of the embankment, which will accelerate the precipitation process and drainage of the subgrade.

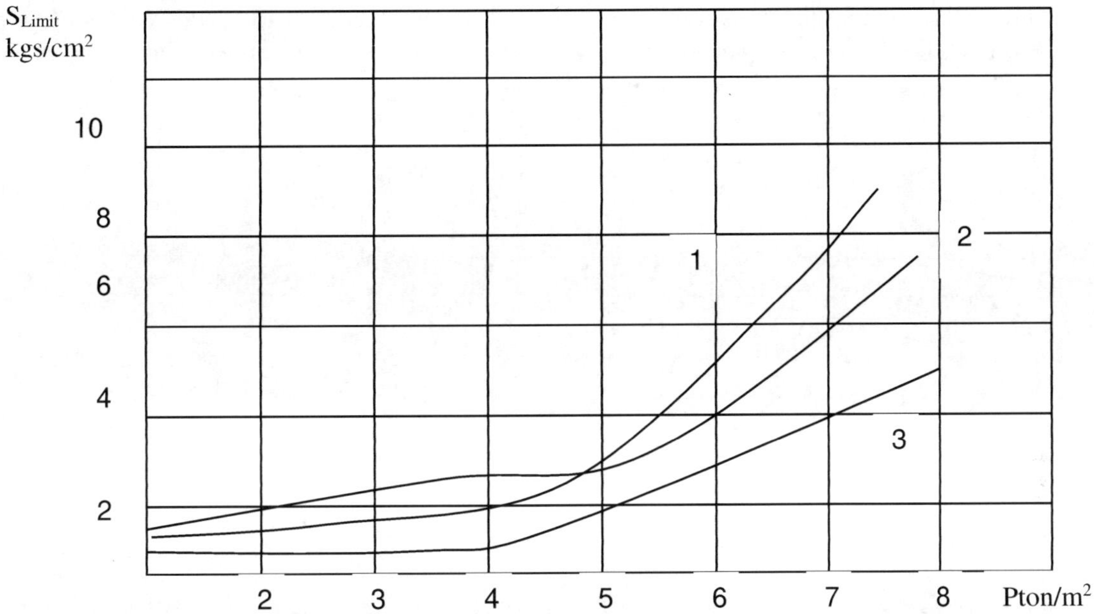

Figure 2. Influence of vertical drains with different filling on the strength of the base: 1 - combined anode-cathodic filtering filling; 2 - gravel filling; 3 - sand filling; Slimit - general resistance.

5. Conclusion

The use of a combined three-layer filling of vertical drains will not only reduce the consolidation time of a loose base, but also provide the necessary strength of the base by increasing the shear resistance, since the vertical drains are filled in the upper and lower parts with a stronger material, turning ordinary drainage slots into pile drains.

References

[1] Tsytovich N A and Ter-Martirosyan Z G 1981 *Foundations of Applied Geomechanics in Construction* (Moscow: Higher School)

[2] Dolmatov B I 2012 *Soil mechanics, foundations and foundations*: textbook, 3rd ed. (St. Petersburg: Lan)

[3] Pupatenko K V 1999 *Features of the design of a new and reconstruction of an existing subgrade on peat bases (in the conditions of the Sakhalin railway)*: candidate dissertation (St. Petersburg)

[4] Mikhalsky V V, The device for draining the subgrade of the structure: Patent No. 182595 USSR (23 June 1986)

[5] Kuhn T 1998 Density matrix theory of coherent ultrafast dynamics *Theory of Transport Properties of Semiconductor Nanostructures (Electronic Materials* vol 4*)* ed E Schöll (London: Chapman and Hall) chapter 6 pp 173–214

[6] Dorokhin S, Ivannikov V, Yakovlev K, Borychev S, Terentyev A et al 2020 Reducing the severity of a traffic accident *IOP Conf. Ser.*: *Earth Env.* **433(1)** 012014

[7] Davydova I, Mazhayskiy Y, Davydov E and Guseva T 2017 Impact of drainage on heavy metal pollution of soils and land usage regulation *Engineering for Rural Development* **18** 155-161
[8] Byshov N, Simdyankin A and Uspensky I 2017 Method of Traffic Safety Enhancement with Use of RFID Technologies and its Implementation *Transportation Research Procedia* **20** 107-111
[9] Kuraksin A, Shemyakin A and Borychev S 2017 Meso-DTA Traffic Model Technology for Evaluating Effectiveness and Quality of the Organization of Traffic in Large Cities *Transportation Research Procedia* **20** 378-383

IOP Conf. Series: Materials Science and Engineering 1103 (2021) 012015 doi:10.1088/1757-899X/1103/1/012015

Evaluation of the trailbuilder hydraulic drive thermal preparation

V V Konev

Institute of Transport, Tyumen Industrial University, 72, Melnikaite str., Tyumen, 625039, Russia

E-mail: konevvv@tyuiu.ru

Abstract. One of the main problems in operation of road-building machines in conditions of low negative ambient temperatures is the hydraulic drive thermal preparation. The hydraulic equipment is located throughout the machine and need a local thermal warming otherwise its cold start results in its intense wear. We propose the warming method that comprises circulation of a prewarmed hydraulic fluid from the hydraulic tank through the piston and rod chambers of hydraulic cylinder without rod motion. This is ensured by installation of a bypass pipe and a valve connecting the hydraulic cylinder piston and rod chambers. To evaluate operability of the hydraulic cylinder warming system we determined the fluid circulation conditions depending on the fluid flow rate and its temperature. To check these conditions, we made calculations for a trailbuilder hydraulic drive with a ripper.

1. Introduction

The research work is focused on specification requirements for the construction and installation works performed by equipment during the oil and gas pipelines construction. In the north of Tyumen region these works are performed in winter. Therefore, there are problems with the prestarting thermal preparation of working equipment [1, 2], and first of all, of its hydraulic drive. The hydraulic system is located throughout the machine, but the hydraulic cylinders (the hydraulic drive actuating elements) are located on the periphery of the system. In this regard, the cylinders have need of a local warming. The ripping trailbuilders move along the route at the end of the process flow column including such building and installation equipment as excavators, pipe laying machines, welding and taping equipment. They loosen the frozen soil and fill it into the pipeline trench. In the intershift time the equipment is stored in the process flow, in the center of which there is constantly a low-voltage power source of 30-40 V, i.e. the welding equipment and base generator which can be used to power an electric warmer.

There are practically no other possibilities to warm the machines hydraulic cylinders on a pipeline construction route. Therefore, it is necessary to consider a more universal and independent warming method. As a result of investigation of patent and technical documents, it was found out that low-voltage heating devices are not widely adopted. Additionally, the ripping trailbuilders operate autonomously, they are not used together with the welding or any other equipment having power sources to supply the warming-up systems of hydraulic drive.

Existing thermal preparation tools have certain downsides: low wattage, high specific heat flow; they are difficult-to-make. The cold start provokes an intense wear of parts and their capacity loss. To solve the above problem, it is necessary to warm up hydraulic cylinders with a fluid preheated in the hydraulic tank, where the main volume of operating fluid is located. In the context of self-sufficient operation of machines, it is recommended to use the heat of exhaust gases of the machine internal combustion engine [3, 4].

2. Research Methodology

The General Principles of the proposed heat treatment system are based on replacement of a cold working fluid inside the hydraulic cylinder with the warmed up fluid from its hydraulic tank [5]. To do that, you need a bypass pipe and a valve. The diagram of pressure distribution in the pipeline sections of the upgraded hydraulic cylinder is presented in Figure 1. To check serviceability of the proposed hydraulic cylinder warming method it is necessary to consider two variants of the fluid movement in the system: 1. Through the bypass pipe into the rod chamber of hydraulic cylinder. 2. The extended rod creates pressure in the piston chamber.

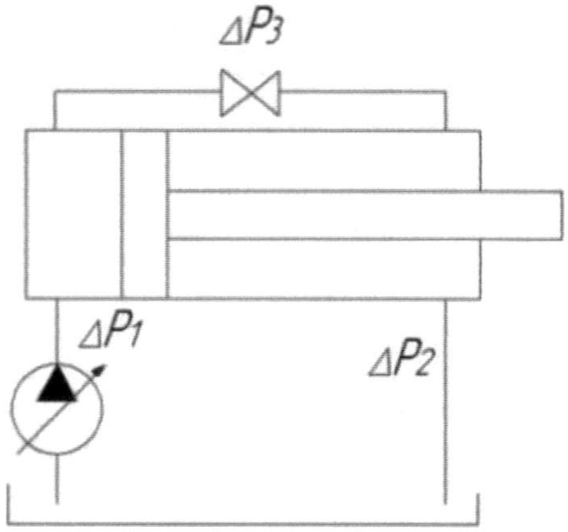

Figure 1. The diagram of pressure distribution in the pipeline sections of the upgraded hydraulic cylinder.

In the first case, the fluid overcomes resistance in the tube ΔP_3 and the drain resistance ΔP_2 (from the rod chamber to tank). The total resistance is:

- $P' = \Delta P_3 + \Delta P_2$, (kgf/cm^2).

In the second case, the drain resistance on the active area of the piston counteracts the cylinder fluid in the rod end:

- $\Delta R_2 = (F_n - F_w)/F_n$, kgf

where F_n is the piston area, (cm^2); F_w is the rod area, (cm^2).

The rod reaction normalized to area is R/F_n. The piston friction against the cylinder walls is neglected, since its force in comparison with the hydraulic pump force is significantly small. The total resistance in the second case:

- $P'' = \Delta P_2 \cdot (F_n - F_w)/F_n + R/F_n$.

The operating fluid will go through the bypass pipe when the following conditions are satisfied, (kgf/cm^2):

$$\Delta P_2 \cdot (F_n - F_w)/F_n + R/F_n > \Delta P_3 + \Delta P_2. \quad (1)$$

To prove this inequality, it is necessary to find reactions inside hydraulic cylinders and wastage of the hydraulic system at the fluid temperature first before heating (-40 °C) and then after heating (+5°C). Let us define reactions inside hydraulic cylinders (A) and (B) used for lifting and lowering of the ripping equipment and the working element of blade (Figure 2). First estimate reactions in the hydraulic cylinders of a ripping trailbuilder depending on weight of its ripper attachment. We solve it for the sum of all force moments of the equation relatively to the point A.

- $\sum M_A = -G_p \cdot l_2 + R_1 \cdot l_1 = 0$, N·m;

- $R_1 = \frac{G_p \cdot l_2}{l_1}$, N; $\frac{15249.34 \cdot 1.0115249.34 \cdot 1.01}{0.77} = 20002.4$, N.

Since the ripping load is distributed between two hydraulic cylinders, the force of one hydraulic cylinder will respectively be two times less and will be 10001.2 N = 1124 kgf.

Let us determine the reaction inside hydraulic cylinders (B) used for lifting and lowering of a blade (Figure 2). Now solve for the sum of the moments force equation relatively to the point B:

- $\sum M_B = G_o \cdot l_3 - R_2 \cdot l_4 = 0$, N·m;
- $R_2 = \frac{(G_o \cdot l_3)}{l_4} = \frac{(21966.9 \cdot 2.82)}{1.88} = 32950.4$, N.

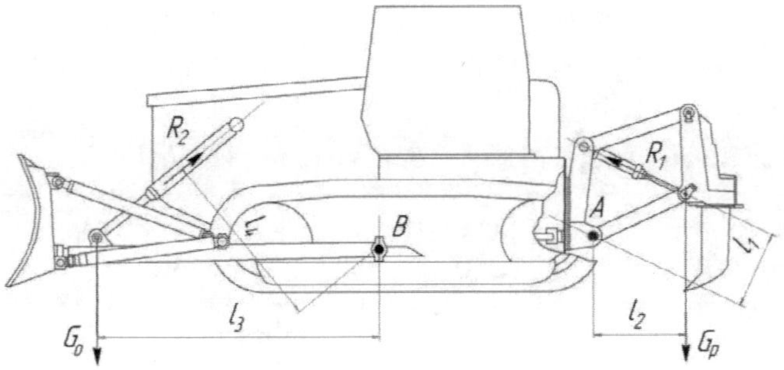

Figure 2. Design model to determine reactions in the hydraulic cylinders (A) and (B).

Since the ripping load is distributed between two hydraulic cylinders, then the force of one hydraulic cylinder will respectively be two times less and will be 16475.2 N = 1852 kgf.

3. Hydraulic losses estimation

The ripping trailbuilders DZ-117A are equipped with the steel seamless pipes and the high-pressure hoses, through which the working fluid is supplied to hydraulic units. We find the Re number from the equation [6]:

$$Re = 21.2 \cdot 10^3 \cdot Q/dv, \tag{2}$$

where, Q is a liquid flow rate, lpm; d is the internal pipe size, mm; v is a kinematic viscosity of the multigrade low-temperature hydraulic oil fluid at -40 °C, (1000 cSt).

Now solve for the Reynolds number in the pipeline with d = 30 mm and Q = 330 lpm at -40 °C (equation 2):

- $Re = 21.2 \cdot 10^3 \cdot 330/30 \cdot 1000 = 233.2.$

The dimensionless friction loss coefficient is calculated as (equation 3) [7]:

$$\lambda = 75/Re. \tag{3}$$

The friction loss factor in the hydraulic system pipeline at -40 °C and Q = 330 lpm makes (equation 3):

- $\lambda = 75/233.2 = 0.3216.$

Let us find the Reynolds number for high pressure hose provided d = 25 mm and Q = 330 lpm at -40 °C (equation 2):

- $Re = 21.2 \cdot 10^3 \cdot 330/25 \cdot 1000 = 279.84.$

The friction loss factor in the hydraulic system pipeline at -40 °C and Q = 330 lpm (equation 3) is:

- $\lambda = 75/279.84 = 0.2680.$

The calculation results for Re number and factors for friction losses in pipelines are presented in Table 1. For the fluids and represented dimensions, the friction pressure losses in a round pipe can be calculated as follows, (kgf/cm^2):

$$\Delta P = 2 \cdot 10^3 \cdot \lambda \cdot l \cdot \frac{Q^2}{d^5}. \quad (4)$$

Local pressure loss of coefficient is calculated by formula, (kgf/cm^2):

$$\Delta P = 2.1 \cdot \xi \cdot Q^2/d^4, \quad (5)$$

where ξ is the local loss coefficient.

For convenience we accept the following legend (table 2). Pressure losses on the hydraulic line route from the cylinders (A) and (B) to the hydraulic distributor are calculated for four conditions: Q = 165 lpm, t = -40 °C; Q=165 lpm, t= +5 °C; Q=330 lpm, t=-40 °C; Q=330 lpm, t= +5 °C.

Table 1. Calculation results for number Re and friction losses coefficient in the pipeline λ.

d, mm	-40 °C, Q=330 lpm, ν=1000 cSt		-40 °C, Q=165 lpm, ν=1000 cSt		+5 °C, Q=330 lpm, ν=50 cSt		+5 °C, Q=165 lpm, ν=50 cSt	
	Re	λ	Re	λ	Re	λ	Re	λ
30	232.2	0.3216	116.6	0.6432	4664	0.0161	2332	0.0322
25	279.84	0.2680	139.92	0.5360	5596.8	0.0134	2798.4	0.0268

Table 2. Losses in friction and local resistances [7].

Legend	Losses in friction and local resistances	Value
1	friction in a pipeline	λ (table 1)
2	friction in a hose-pipe	λ (table 1)
3	in the pipeline smooth knees of 90°	ξ =0.12
4	in the pipeline smooth knees less than 90°	ξ =0.04
5	in the pipeline smooth knees of 180°	ξ =0.12
6	in nozzles	ξ =0.1
7	in the pipeline smooth knees	ξ =0.12
8	inputs or outputs of the hydraulic cylinder	ξ =0.8
9	in the valve without fluid flow rotation	ξ =2

Pressure losses along the hydraulic line route from the cylinder (A) to the hydraulic distributor when Q = 165 lpm, t = -40 °C will make (equation 4):

- $\Delta P = 2 \cdot 10^3 \cdot 0.6432 \cdot 1.2 \cdot \frac{165^2}{30^5} = 1.7295$, ($kgf/cm^2$);

The local resistance pressure losses are (equation 5):

- $\Delta P = 2 \cdot 2.1 \cdot 0.12 \cdot \frac{165^2}{30^4} = 0.0169$, ($kgf/cm^2$);

The other hydraulic drive operating conditions are calculated in the same way. The calculation results are presented in the table 3. Calculation results for pressure losses from the hydraulic cylinder (B) to distributor are presented in Table 4.

Table 3. Calculation results of hydraulic line pressure losses from the hydraulic cylinder (A) to distributor.

Legend [a]	Q=165 lpm; t=-40 °C	Q=165 lpm; t=+5 °C	Q=330 lpm; t=-40 °C	Q=330 lpm; t=+5 °C
1	1.7295	0.0866	3.4590	0.1732
2	2.6897	0.1345	5.3794	0.269
3	0.0169	0.0169	0.0678	0.0678
4	0.0057	0.0057	0.0226	0.0226
6	0.1756	0.1756	0.7025	0.7025
8	0.0565	0.0565	0.2259	0.2259
ΣΔP	5.3345	0.4758	9.8572	1.461

[a] legend (table 2).

Pressure losses in the hydraulic cylinder bypass line (A) when Q = 165 lpm; t= -40 °C can be calculated as follows (equation 4):

- $\Delta P = 2 \cdot 10^3 \cdot 0.6432 \cdot 0.6 \cdot 165^2/30^5 = 0.8648$, (kgf/cm^2).

Table 4. Calculation results of the line pressure losses from the hydraulic cylinder (B) to distributor.

Legend [a]	Q=165 lpm; t=-40 °C	Q=165 lpm; t=+5 °C	Q=330 lpm; t=-40 °C	Q=330 lpm; t=+5 °C
1	6.1253	0.3067	12.2506	0.6133
2	11.6554	0.5828	23.3108	1.1655
3	0.0351	0.0351	0.1405	0.1405
4	0.0057	0.0057	0.0226	0.0226
5	0.0169	0.0169	0.0678	0.0678
6	0.1756	0.1756	0.7025	0.7025
7	0.0169	0.0169	0.0678	0.0678
8	0.1129	0.1129	0.4517	0.4517
ΣΔP	18.1438	1.2526	37.0143	3.2317

[a] legend (table 2).

Pressure losses in the electromagnetic valve of a bypass pipe when Q = 165 lpm make (equation 5):

- $\Delta P = 2.1 \cdot 2 \cdot 165^2/30^4 = 0.1412$, (kgf/cm^2).

Pressure losses in the electromagnetic valve of a bypass pipe when Q = 330 lpm are (equation 5):

- $\Delta P = 2.1 \cdot 2 \cdot \frac{330^2}{30^4} = 0.5647$, (kgf/cm^2).

The calculation results of pressure losses in a bypass hydraulic line made on the basis of the equation (4) for the other operating conditions of hydraulic drive are presented in Table 5.

Table 5. Calculation results of pressure losses in the bypass line of hydraulic cylinders (A) and (B).

Hydraulic cylinder	Q=165 lpm; t=-40°C	Q=165 lpm; t=+5°C	Q=330 lpm; t=-40°C	Q=330 lpm; t=+5°C
A	0.8648	0.0433	1.7295	0.0866
B	1.0089	0.0505	2.0177	0.101

Pressure losses in the nozzles of hydraulic cylinders (A), (B) when Q = 165 lpm make (equation 5):

- $\Delta P = 2 \cdot 2.1 \cdot 0.1 \cdot \frac{165^2}{25^4} = 0.0293$, (kgf/cm^2).

Pressure losses in the nozzles of hydraulic cylinders (A), (B) when Q = 330 lpm make (equation 5):

- $\Delta P = 2 \cdot 2.1 \cdot 0.1 \cdot 330^2/25^4 = 0.1171$, (kgf/cm^2).

Now let us check correctness of inequation for hydraulic cylinder (A) at Q = 165 lpm; t = -40 °C (Equation 1):

- $5.335 \cdot 0.6 + 7.52 > 0.477 + 5.335$, kgf/cm^2;
- $10.72 > 5.812$, kgf/cm^2.

The equation is solved, the working fluid from the cylinder (A) moves through the bypass pipe, at the given liquid flow rate and the ambient temperature. Then let us check correctness of inequation for hydraulic cylinder (A) at Q = 330 lpm; t = -40 °C (Equation 1):

- $9.8572 \cdot 0.6 + 7.52 > 0.4774 + 9.8572$, kgf/cm^2;
- $13.4343 > 10.3346$, kgf/cm^2.

The equation is solved, the working fluid from the cylinder (A) moves through the bypass pipe, at the given liquid flow rate and ambient temperature. Let us check correctness of inequation for hydraulic cylinder (B) at Q = 165 lpm; t = -40 °C (Equation 1):

- $18.1438 \cdot 0.58 + 12.03 > 0.4774 + 18.1438$, kgf/cm^2;
- 22.5518.62 , kgf/cm^2.

The equation is solved, the working fluid from the cylinder (B) moves through the bypass pipe, at the given liquid flow rate and ambient temperature. Let us check correctness of inequation for hydraulic cylinder (B) at Q = 330 lpm; t = -40 °C (Equation 1):

- $37.0143 \cdot 0.58 + 12.03 > 0.4774 + 37.0143$, kgf/cm^2;
- $33.5 > 37.4917$, kgf/cm^2.

The equation (equation 1) is not solved, the working fluid from the cylinder (B) does not move through the bypass pipe, at the given liquid flow rate and ambient temperature. That means that the piston will slide when the hydraulic cylinder is warmed through the bypass pipe.

4. Conclusion

We propose a method of warming of hydraulic cylinder which comprises circulating of the preheated hydraulic fluid from the tank through the piston and rod chambers without rod motion. As a result of calculations of working fluid pressure losses in the hydraulic systems before and after modernization, we determined that the fluid circulation conditions depend on its flow rate and temperature. Using the example of a trailbuilder hydraulic drive with rippers, we defined that the proposed warming system works. At the same time, the fluid of the hydraulic cylinders lifting and lowering the blade will not move into the bypass pipe at flow rate of 330 lpm and the liquid temperature of - 40 °C. The system works when the fluid flow rate decreases to 165 lpm. Thus, when you warm the hydraulic cylinders, it is necessary to take into account conditions of their warming-up.

References

[1] Vasilchenko V A 2006 Hydraulic drive mining machines low temperatures operation features *Mining Industry* **2**. Retrieved from: www.mining-media.ru/ru/article/newtech/1151-osobennosti

[2] Karnaukhov N N 1994 *Adaptation of construction machines to the conditions of the Russian North and Siberia* (Moscow: Nedra)

[3] Vashurkin I O 2005 *The warming of construction machines in a rigorous climate* (St. Petersburg: Nauka)

[4] Konev V, Merdanov Sh, Karnaukhov M and Borodin D 2014 Thermal preparation of the trailbuilder fluid drive, in: *Energy Production and Management in the 21st Century - The Quest for Sustainable Energy* (Southampton WIT Press) vol 1 pp 697-706
[5] Konev V V, Karnaukhov N N, Borodin D M and Polovnikov E V 2014 Mathematical modeling of thermal processes for local hydraulic drive warming *Modern problems of science and education* **5**. Retrieved from: Available at: http://www.science-education.ru/119-15076
[6] Altshul A D, Kalitzun V I and Mayranovsky F G 2013 *Hydraulic Calculation Examples* (Moscow: AljansS)
[7] Idelchik I E 1975 *Handbook on Hydraulic Resistances* (Moscow: Mechanical Engineering)

Technology of making asphalt concrete using sand slag

N S Kovalev[1], E N Otarova[2], V V Gladnev[1], E A Sadygov[1], P V Demidov[1] and I V Yaurova[1]

[1] Voronezh State Agrarian University, 1, Michurina str., Voronezh, 394087, Russia
[2] Zhukovsky–Gagarin Air Force Academy, 54a, Old Bolsheviks str., Voronezh, 394064, Russia

E-mail: main@srd.vsau.ru

Abstract. In the Central Black Earth Region, there are settlements that do not have hard-surfaced access roads, and existing access roads require major repairs. In order to reduce the cost of construction, it is proposed to use asphalt concrete mixtures from slag materials for the construction of road pavements. Slag materials have hydraulic binding properties; the processes of structure formation in them take a long time. The study of the properties of asphalt concrete was carried out depending on the time from the moment of manufacture, after prolonged water saturation and the number of freeze-thaw cycles. The obtained results indicate an increased durability of asphalt concrete road pavements made of slag materials.

1. Introduction

In many areas, rural settlements do not have paved access roads, and existing roads require major repairs. To improve the infrastructure of settlements, large budgetary funds are needed, which are not available in municipalities. To reduce the cost of construction, it is necessary to maximize the use of local building materials.

In areas with a developed metallurgical industry, local materials are slag from ferrous metallurgy. The use of asphalt concrete from slag materials in road construction will not only increase the durability of asphalt concrete pavements, but also solve socio-environmental problems by using by-products of metallurgical plants. In road construction, asphalt mixtures based on steel-smelting slags are widely used for paving devices [1-7].

Asphalt mixtures based on blast furnace slags have not yet been widely used. A feature of asphalt concrete coatings from blast-furnace slag materials is that their formation occurs during operation due to the chemical and hydraulic activity of the slag and the transformation of the structure from coagulation into coagulation-condensation [8].

2. Materials and Methods

To study the physical and mechanical properties, granulated blast furnace slags of the Novolipetsk Metallurgical Plant were used. Tests of asphalt concrete were carried out according to GOST 9128-2013. Given the hydraulic properties of slag materials, asphalt concrete properties were studied not only at the age of 2 days—as required by the standard—but also periodically from the time of manufacture of the samples after prolonged water saturation, and after repeated exposure to alternating temperatures. To study the processes of structure formation, scanning electron microscopy was used.

3. Results

The results of the study of the structural and mechanical properties of asphalt over time are presented in Table 1. Analysis of the presented results allows talking about ongoing processes of structure formation over time. The increase in compressive strength at a temperature of 20 °C can be explained from the standpoint of the diffuse theory of adhesion. The diffuse theory of adhesion easily explains this dependence by the slow diffusion of bitumen molecules and its components into the pores of slag material [9].

Table 1. Change in the physical and mechanical properties of asphalt concrete from granulated slag over time

Asphalt Composition	Age [days]	Average density, [g/cm^3]	Water saturation [vol.%]	Extension [vol.%]	Tensile strength at compression [MPa] at temperature [°C]				Water resistance coeff.
					20	50	0	water saturated	
Granulated blast furnace slag (100%), BND 60/90 bitumen (8%)	2	1.98	13.52	0.02	2.5	1.1	-	2.5	1.00
	7	2.00	12.36	0.08	2.9	1.2	-	2.9	1.00
	28	2.01	12.60	0.09	3.2	1.1	-	3.2	1.00
	60	2.00	11.99	0.08	3.5	1.1	-	3.5	1.00
	180	1.99	11.34	0.00	3.7	1.1	-	3.7	1.00

The effect of freezing and thawing on the physicomechanical properties of asphalt concrete is presented in Table 2.

Table 2. Effect of freezing-thawing on the physical and mechanical properties of asphalt concrete

Asphalt Composition	No. of freezing cycles	Tensile strength at compression [MPa] at temperature [°C]				Average density, [g/cm^3]	Water saturation [vol.%]	Extension [vol.%]	Coefficients	
		20	50	0	after water saturation				water resistance	frost resistance
Granulated blast furnace slag (100%), BND 90/130 bitumen (9%)	50	2.8	1.1	7.0	2.7	1.99	13.11	0.00	0.96	1.40
	ref.	2.0	1.2	7.0	2.0	1.98	13.10	0.00	1.00	-
	75	2.4	1.1	6.5	2.3	1.99	12.11	0.00	0.96	1.09
	ref.	2.2	1.1	7.0	2.2	1.98	13.20	0.00	1.00	-
	100	2.2	1.1	6.3	2.2	1.99	11.98	0.00	1.00	0.79
	ref.	2.8	1.1	7.2	2.7	1.98	13.30	0.00	0.97	-

Materials based on hydraulic binders have the ability to restore the original structure due to the "self-healing" of defects that occur during alternate freezing-thawing [8, 10]. In addition, the bitumen film on slag grains has a complex supramolecular (secondary) structure similar to polymer films, which indicates an increased durability of asphalt concrete.

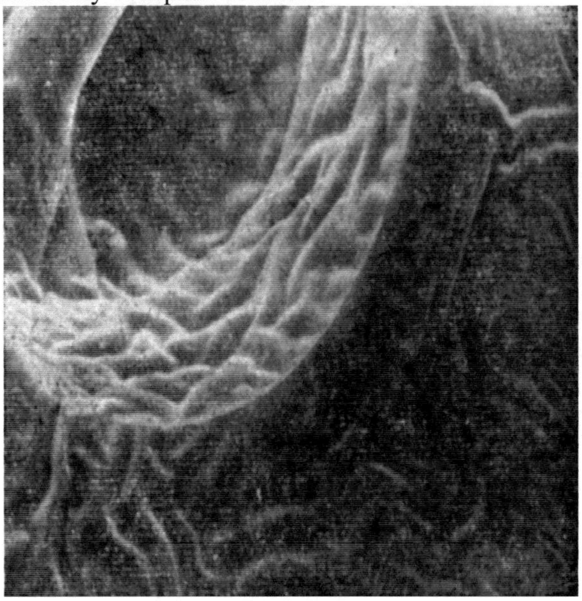

Figure 1. Supramolecular belt fibrillar structure of a bitumen film on the surface of slag materials

The stability of the structure of asphalt concrete from slag materials during prolonged water saturation (table 3) is explained by a peculiar mechanism of structure formation processes during the interaction of slag material with water. The interaction of slag materials with water begins from the moment they are produced.

Table 3. Effect of prolonged water saturation on the physical and mechanical properties of asphalt concrete

Asphalt Compo-sition	Duration of water saturation, days	Tensile strength at compression [MPa] at temperature [°C]			Ave-rage density, [g/cm³]	Water satura-tion [vol.%]	Exten-sion [vol.%]	Water resis-tance coeff.
		20	50	0				
Granulated blast furnace slag (100%), BND 90/130 bitumen (9%)	10	1.9	1.1	6.5	2.15	8.20	0.45	1.00
	ref.	1.6	1.1	7.0	2.19	5.40	0.36	0.98
	30	1.5	1.1	6.7	2.16	8.30	0.00	1.02
	ref	1.8	1.1	7.4	2.12	9.20	0.00	1.11
	60	2.0	1.1	7.2	2.17	8.10	0.00	1.01
	ref	2.0	1.1	7.7	2.18	9.50	0.00	1.20
	90	2.2	1.1	7.4	2.20	7.88	0.00	0.90
	ref	2.7	1.1	10.6	2.18	9.00	0.00	1.08
	180	2.1	1.1	8.1	2.20	7.02	0.00	1.10
	ref	3.5	1.1	10.9	2.21	9.50	0.00	1.00

During the heating of the slag material and its combination with bitumen, the processes of hydrolysis and hydration are enhanced. By the time of water saturation, the structure of asphalt concrete from slag materials is coagulating.

During the water saturation of asphalt concrete from granulated slags, slow hardening and stabilization of the structure occur, primarily due to an increase in the strength of adhesive bonds between bitumen and slag material, which is due to the modification of the surface of the slag material during hydration (Figure 2).

When using hydraulic mineral materials, water falling under a bitumen film is spent on slag hydration, and the system of pore channels draws excess moisture inside the slag grain. As a result, the structure of asphalt concrete from slag materials is transformed from coagulation into coagulation-condensation [9, 11].

3 μm

Figure 2. Surface modification of slag material after 90 days of water saturation (bitumen is extracted)

4. Conclusion

- The processes of structure formation immediately from production to prolonged water saturation and a number of freeze-thaw cycles were investigated.
- The use of asphalt concrete from slag materials in road construction will not only reduce the cost of construction and increase the durability of asphalt concrete pavements, but also solve socio-environmental problems by using by-products of metallurgical plants.

References

[1] Shiha M, El-Badawy S and Gabr A 2020 Modeling and performance evaluation of asphalt mixtures and aggregate bases containing steel slag *Construction and Building Materials* **248**

[2] Zumrawi M and Khalill F 2015 Experimental study of steel slag used as aggregate in asphalt mixture *Int. J. Civ. Environ. Struct. Constr. Arch. Eng.* **9(6)**

[3] Wen H, Wu S and Bhusal S 2016 Performance evaluation of asphalt Mixes containing steel slag aggregate as a measure to resist studded tire wear *J. of Materials in Civil Engineering* **28(5)**

[4] Pasetto M and Baldo N 2010 Experimental evaluation of high performance base course and road base asphalt concrete with electric arc furnace steel slags *J. of Hazardous Materials* **181(1-3)** 938-948

[5] Liu W, Li H, Zhu H and Xu P 2019 Properties of a steel slag-permeable asphalt mixture and the reaction of the steel slag-asphalt interface *Materials* **12(21)**

[6] Behiry A. E A E-M 2013 Evaluation of steel slag and crushed limestone mixtures as subbase material in flexible pavement *Ain Shams Engineering J.* **4(1)** 43-53

[7] Maghool F, Arulrajah A, Horpibulsuk S and Du Y-J 2017 Laboratory evaluation of ladle furnace slag in unbound pavement-base/subbase applications *J. of Materials in Civil Engineering* **29(2)**

[8] Kovalev N S 2012 *Scientific and practical principles of frost resistance and crack resistance of asphalt concrete coatings from slag materials* (Voronezh: Voronezh State Agrarian University) p 270

[9] Kovalev N S and Bykova Y A 2008 Investigation of the physico-chemical interaction of slag materials with bitumen *Bulletin of the Volgograd State University of Architecture and Civil Engineering. Series: Construction and Architecture* **11(30)** 81-87

[10] Shestoperov S V 1966 *Durability of concrete transport structures* (Moscow: Transport)

[11] Bratchun V I, Gritsuk Yu V, Bespalov V L, Pakter M K and Gubar A V 2014 Asphalt slag concrete on anionic bitumen emulsions *Science and technology in the road industry* **4** 22-26

Modeling the stress-strain state of crane metal structures

A G Lenivtsev and I V Dudanov

Samara State Technical University, Academy of Architecture and Civil Engineering, 194, Molodogvardeyskaya St., Samara, 443001, Russia

E-mail: lenivtsev.ag@samgtu.ru

Abstract. The intensification of construction production imposes very strict requirements on the safety of construction machines. When carrying out lifting and hoisting operations using overhead cranes, production safety is primarily determined by the technical condition of the undercarriage and crane runways, as well as metal constructions. The authors deal with the issues related to modeling and research of the stress-strain state of metal structures of an overhead crane. A model of the crane metal structure is proposed, which includes the main and end box-section crane girders. To study possible geometric movements of the crane, a model of an overhead crane was designed in KOMPAS-3D software environment. The model contains main and end box-section girders. The model allows determining the maximum possible misalignment of the crane and runout of one of the sides. To determine the bending, tensile and compressive stresses in the bridge truss and the deformations of the crane metal structures, depending on the working loads, a 3D model of the bridge crane was created in the ADW Winmachine software environment. With its help, the bending stiffness of the crane bridge was established. These values are necessary to compile a mathematical model of an overhead crane as a control object in the synthesis of an automatic control system.

1. Introduction

The intensification of construction production imposes very strict requirements on the safety of construction machines. When carrying out lifting and hoisting operations using overhead cranes, production safety is primarily determined by the technical condition of the undercarriage and crane runways. Malfunctions in the operation of the undercarriage or changes in the geometry of the crane metal structures due to imprecise control significantly reduce the service life of the rail track from the design life of 8 years (on average) to overhaul - down to 2-3 years. Travel wheels have lower performance properties and their service life is further reduced. In addition to high economic costs, such violations in crane operation are associated with the occurrence of emergency situations - crane derailment [1].

The effective elimination of these problems is associated with the automation of an overhead crane - the development of automatic control systems [2]. The synthesis of automatic control systems is preceded by the stage of modeling the dynamics of an overhead crane as a control object, the parameters of the computational model of which must fully correspond to the actual operating crane [3, 4].

2. Materials and methods

At the *Samara Stroyfarfor LLC* enterprise (Samara), a double-girder overhead crane is operated at the raw material unloading process area. It has the following main technical characteristics:

IOP Conf. Series: Materials Science and Engineering 1103 (2021) 012017 doi:10.1088/1757-899X/1103/1/012017

1. Crane dimensions: crane span - 34500 mm, crane base – 5000 mm, wheel diameter – 500 mm, distance between bumpers (extreme points of the crane) - 6710 mm;
2. Characteristics of the drive of the travel mechanism: Electric motor - MTF-311-6, 11kW, 945 rpm - 2 pcs. Reducer - 1Ц2У-250-12.5-12/21 reduction ratio - 12.5 - 2 pcs.
3. Crane mass (design) - 31.5t, trolley mass - 3.5t, cab mass - 1t, grab mass - 1.9t, lifting capacity - 5t.
4. Maximum load of the crane wheel on the rail (calculated) - 185 kN.

To study possible geometric movements of the crane, a model of an overhead crane, used at Samara Stroyfarfor LLC, was designed in KOMPAS-3D software environment (Fig. 1). The model is built on a scale of M1:1 and contains two main and two end box-section girders. Traveling wheels are installed on the end girders (wheel flanges are shown by rectangles). The model allows determining the maximum possible misalignment of the crane (when the flanges of the diagonally located wheels touch the rails) - in the angular dimension of 16', while the runout of one side is 165mm.

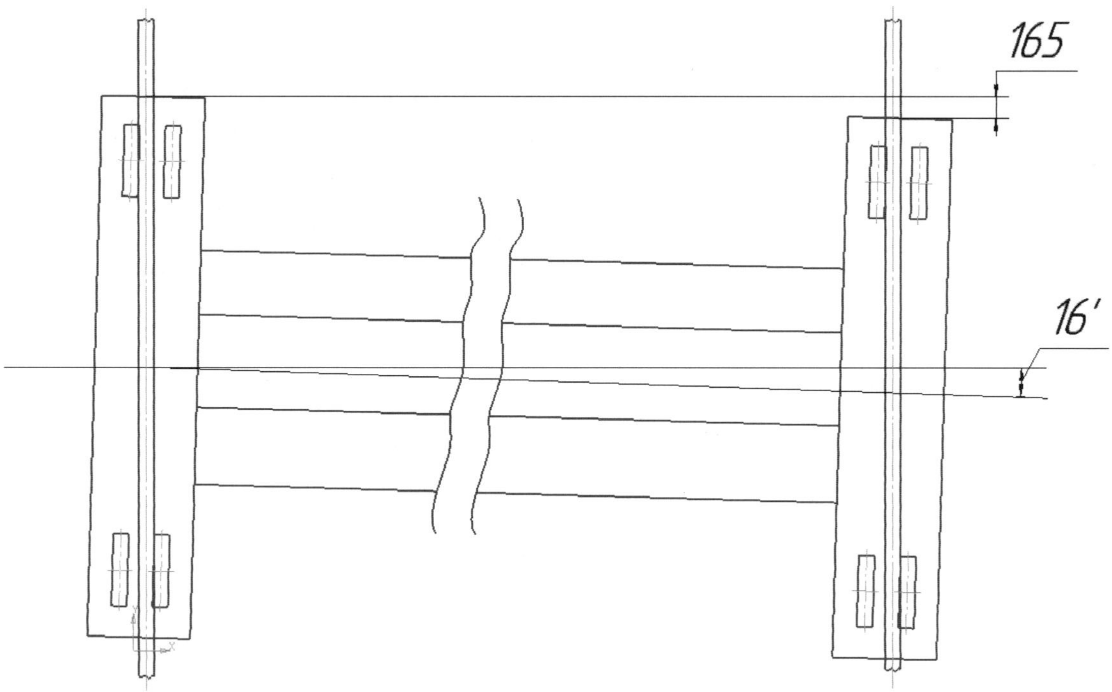

Figure 1. Overhead crane model

Figure 2 shows a model of a running wheel for determining the working clearances between the flanges of traveling wheels with a diameter of 500 mm and the side surfaces of the rail head of the KP-70 brand. Under normal operating conditions of an overhead crane, if the crane runway is in good condition, there should be no contact between the flange and the side surface of the KR-70 rail. The model is also designed to determine the contact area between the flange and the side surface of the rail.

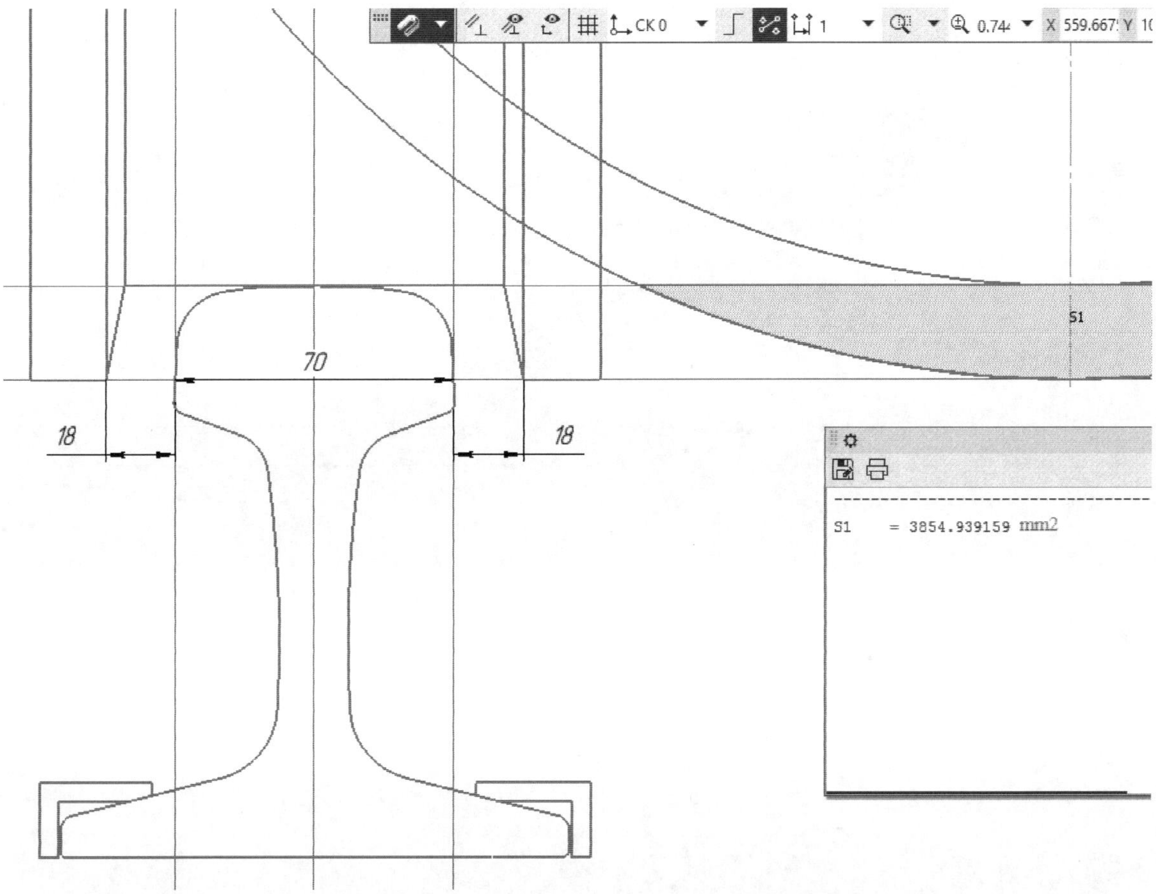

Figure 2. Traveling wheel of an overhead crane

3. Computational experiment

To determine the bending, tensile and compressive stresses in the bridge truss and the deformations of the crane metal structures, depending on the working loads, a 3D model of the bridge crane was created in the ADW Winmachine software environment [5, 6].

In addition to the dead weight of the crane metal structures, the following loads were used in calculations:

1. The load from the trolley weight, grab bucket and the mass of the material being lifted (60-100 kN) is applied to the rails of the cargo trolley on the upper part of the main girders;
2. Load from the weight of the crane operator's cab (10kN), applied to the main girder;
3. Bending moment (9kN·m), creating the maximum possible runout of one side of the crane 165mm from the uneven operation of the crane bridge drives.

The calculations show that the maximum misalignment of the crane (165 mm) occurs under the action of an additional force of 265 N applied to one of the end girders.

When the mass of the crane grab is varied, a load is created in the range from 60 kN (empty bucket) to 100 kN (full bucket), which causes deformation of the main girders by 37 mm and 60 mm, respectively (Fig. 4).

IOP Conf. Series: Materials Science and Engineering 1103 (2021) 012017 doi:10.1088/1757-899X/1103/1/012017

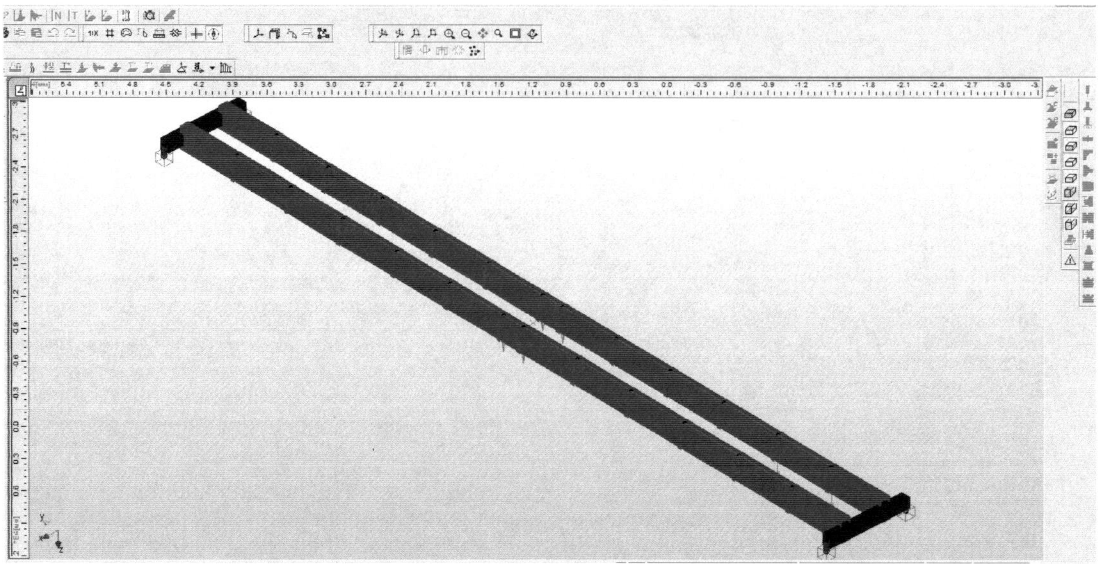

Figure 3. ADW Winmachine overhead crane 3D model.

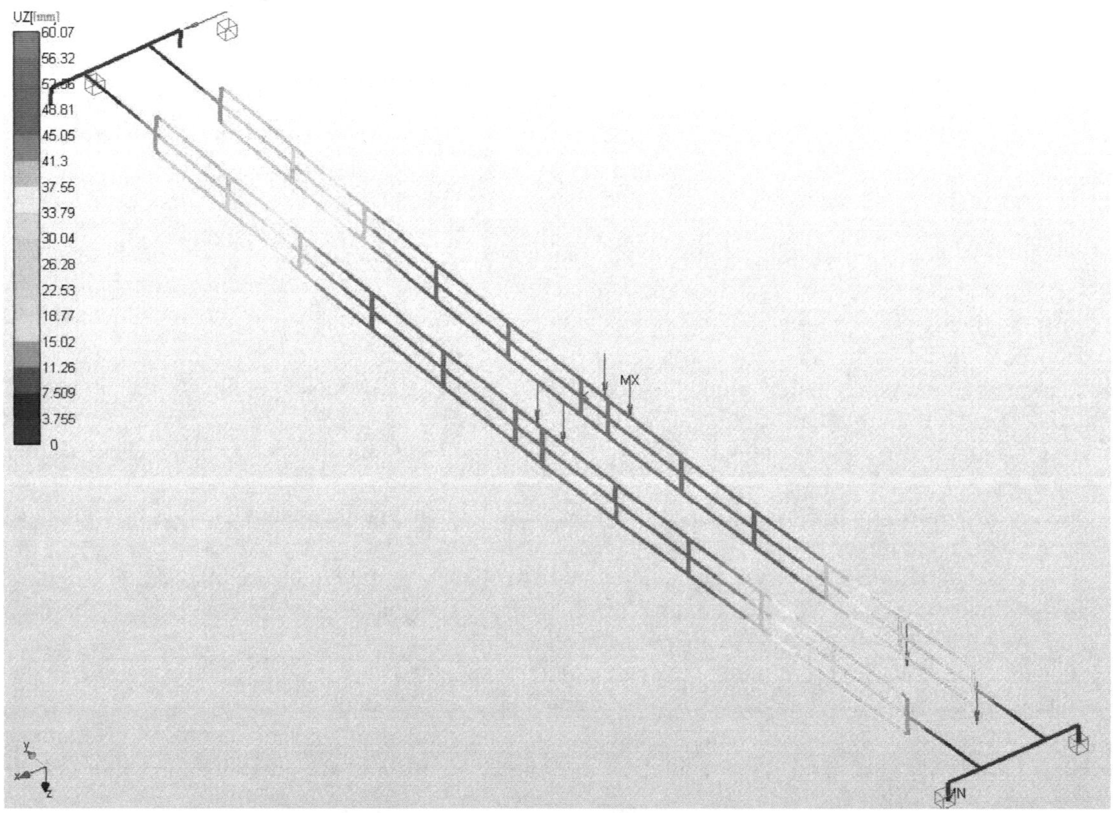

Figure 4. Crane bridge travel map along the Z-axis with 100 kN load.

The results obtained make it possible to determine the deflection stiffness of the crane bridge $C=\Delta N/\Delta Z$, where ΔN - change in the applied load; ΔZ - travel of the crane bridge in the vertical plane:

C=1.76·10^6 N/m. Table 1 shows the deflection of the bridge from the applied load in the operating range.

Table 1. The deflection of the crane bridge depending on the applied load.

Load (kN)	Bridge deflection (mm)
60	37.29
80	49.57
100	60.07

4. Discussion

The main reason for the increased wear of the wheels of cranes with a separate movement mechanism is the non-parallelism of the axes of the crane rails [3, 4]. Unfortunately, even a perfect rail installation will not eliminate the bend of the crane bridge while traveling. This situation is caused by a mismatch in the characteristics of the drive motors and the predominantly off-center position of the transported load. Moving the load away from the center of the crane increases the load on the corresponding motor and it reduces its rotation speed. Rotation of the left and right motors at different angular speeds leads to runout of one side of the crane relative to the other. When the crane moves with a bend, the flanges of the traveling wheels constantly rub against the rails. This leads to intense wear of the flange and rail side surfaces.

Using a 3D model of an overhead crane, it was experimentally established that the maximum possible misalignment of the crane is in an angular dimension of 16 ', while the runout of one side is 165 mm. When the mass of the crane grab is varied, a load is created in the range from 60 kN (empty bucket) to 100 kN (full bucket), which causes deformation of the main girders by 37 mm and 60 mm, respectively. The results obtained make it possible to determine the deflection stiffness of the crane bridge C=1.76·10^6 N/m.

5. Conclusion

The computational experiments carried out and the results obtained make it possible to assess the nature and level of the bridge deflection from the applied load in the operating range, to determine the deflection stiffness of the crane bridge. These values are necessary to compile a mathematical model of an overhead crane as a control object [3, 4] in the synthesis of an automatic control system [2, 7, 8].

References

[1] Dukelsky A I 1973 *Crane Handbook* (Leningrad: Mechanical engineering)
[2] Telichko L Ya and Dorofeev A A 2012 Electric Drive Control System for Smooth Movement of Overhead Cranes *Voronezh State Technical University Bull.* **3(6)** 113
[3] Galitskov S Ya and Ionov A A 2010 Manipulator of the Pallet Storage Area as a Control Object *Samara State Technical University Bull.* **2(27)** 18
[4] Blinchikov O I and Galitskov S Ya 2008 Gantry Manipulator with Suspended Traverse as a Control Object *Samara State Technical University Bull.* **2(22)** 11
[5] Shelofast V V 2005 *Basics of Machines Engineering* (Moscow: APM)
[6] Dudanov I V, Lenivtsev A G and Fadeev A S 2019 Dynamic Load Simulation of the Excavator Hydraulic Drive *IEEE Xplore Digital Library* 1-4
[7] Galitskov S Ya, Galitskov K S and Samohvalov O V 2020 Computer Modeling of the Dynamics of Energy Consumption during Expanded Clay Burning *IEEE Xplore Digital Library* 401-406
[8] Galitskov S Ya, Galitskov K S and Dudanov I V 2018 Synthesis of High-Precision Hydraulic Control Systems of an Excavator *MATEC Web Conf.* **196** 6

Mathematical modeling of overhead crane under conditions of stress-strain state of metal structure as a control object

A G Lenivtsev, I V Dudanov and A S Fadeev

Samara State Technical University, Academy of Architecture and Civil Engineering, 194, Molodogvardeyskaya St., Samara, 443001, Russia

E-mail: lenivtsev.ag@samgtu.ru

Abstract. The article considers an example of the application of the existing mathematical model of an overhead crane to equipment with irreversible structural defects. The course of the solution of the problem is considered: obtaining primary data, finding the existing mathematical model of the crane, adapting it to the problem in question, building a computational model, modeling and analyzing the results obtained. The peculiarity of a crane as a control object is the non-typical state of the metal structure. The overhead crane is in this state after prolonged operation of the equipment with incorrect loading and control modes. As a result of such operating modes, significant wear of the wheels and the rail track take place. Worn elements must be replaced. The frequency of wheel replacement is very intensive, and is about twice a year. The process of replacing wheels with corner boxes is a very expensive procedure – it is necessary to buy equipment, stop the warehouse for a few days and attract staff. To eliminate such costs, it is proposed to conduct a simulation of the crane in order to identify methods of controlling it. The result of the first stage of modeling is the answer to the question: is it possible to operate the crane that is deformed using the correctly selected control mode.

1. Introduction

The intensification of construction production imposes very strict requirements on the safety of construction machines. When carrying out lifting and hoisting operations using overhead cranes, production safety is primarily determined by the technical condition of the crane mounting and crane runways. Malfunctions of the undercarriage or changes in the geometry of the crane metal structures, happening due to imprecise control procedures, significantly reduce the service life of the rail track. Travel wheels have lower performance properties and their service life is further reduced. In addition to high economic costs, such violations in crane operation are associated with the occurrence of emergency situations - crane derailment.

2. Problem statement

Crane construction. The authors adopt a two-girder overhead crane (Fig. 1), working at the technological stage of unloading raw materials at a plant producing building materials, for consideration. It has the following main technical characteristics [1]:

- crane physical dimensions: crane span - 34500 mm, crane base – 5000 mm, wheel diameter – 500 mm, distance between bumpers (extreme points) of the crane - 6710 mm;

- drive characteristic of the travel mechanism: two electric motors MTF-311-6 (power - 11 kW, rotational speed - 945 rpm), two gearboxes 1Ц2У-250-12.5-12 / 21 (reduction ratio 12.5);
- crane mass (design) - 21 t, trolley mass - 1.5 t, cab mass - 1 t, grab mass - 1.9 t, crane lifting capacity - 10 t;
- maximum load of the crane wheel on the rail (calculated) - 185 kN.

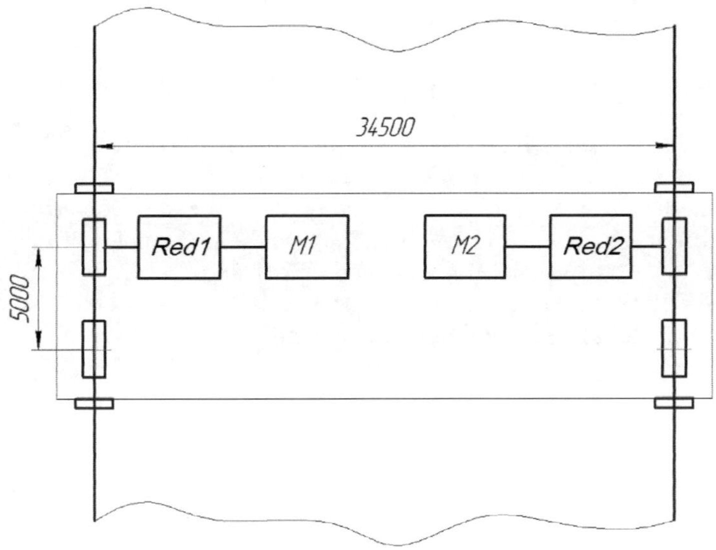

Figure 1. Design scheme of an overhead crane.

Analysis of the condition of the overhead crane. Inspection of the crane condition showed: there is significant wear on the flanges of one of the driving wheels, wear on the flanges of the second driving wheel is within normal limits; wear of the flanges of driven wheels is even, exceeding the norm; wear of rails - along one of the end parts of the rail head.

After analyzing the survey results, it is not difficult to come to the following conclusions. First, there is a violation of the crane geometry, leading to the uneven distribution of the load from the crane weight on three wheels, the fourth wheel is less loaded. Secondly, intensive wear of the flanges on one side of the running wheels is caused by the uniformity of the travel of the loaded grab. Thirdly, long-term, significant misalignment of the crane led to the wedging of the bridge and to the irreversible stretching of the elements of the metal structure.

Problem statement. For the overhead crane described above, it is required to carry out mathematical modeling of the undercarriage, taking into account the stress-strain state of the metal structure of the crane frame. Moreover, it is required to identify the most rational methods for controlling the movement of the crane, which can reduce wear on wheels and rails, on the obtained model.

When setting the problem, the existing devices and methods for controlling such cranes were analyzed. The analysis of these solutions showed the impossibility of their application on an overhead crane with a large span, or their low energy or functional efficiency.

3. Results and discussion

The ***mathematical model*** of an overhead crane of a pallet storage area was adopted as a prototype [2]. This model is similar to the considered crane. The undercarriage has the same design, but the differences are in the bridge design and the hooking of the moving load.

In the construction of the bridge, elastic-dissipative elements were taken into account, which were calculated by the method of finite-difference modeling in the previously considered structure. For modeling, the load will be considered concentrated and located in the middle of the bridge.

The mathematical model of the crane given in [2] and [3] is supplemented by the preliminary deformation of the crane bridge, which is expressed by the displacement of the position of one girder relative to the other. Further, these positions are indicated by coordinates x_1 and x_2. In addition, for the adequacy of the model with a deformed bridge, the form of nonlinearities has been changed. In the model of an overhead crane [2], nonlinearities with symmetry relative to zero (in the range from 0 to $+\infty$) are applied, in the model corresponding to the crane under consideration, total nonlinearities in the range from $-\infty$ to $+\infty$ are taken into account.

Based on the mathematical description and block diagram (Fig. 2), given in [2, 4], as well as taking into account the previously described changes, a computational model was built, shown in Fig. 3.

Figure 2 shows the block diagram of an overhead crane as a control object. Moreover, at that the control object is multidimensional [5]. Here, ω_1 and ω_2 are the frequencies of the voltages supplied to asynchronous motors with transmission mechanisms DPM1 and DPM2, F_{T1} and F_{T2} are the traction forces for DPM1 and DPM2 supplied to the mechanical part - MCh. F_{TP1} and F_{TP2} - friction forces. The rest of the output coordinates are required to set up the model.

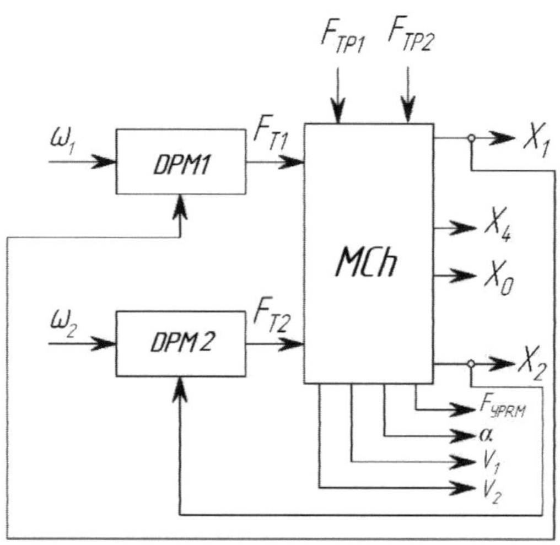

Figure 2. Block diagram of an overhead crane.

Figure 3 shows a ***computational model*** of an overhead crane as a control object. It can be seen here that DPM1 and DPM2 are represented by asynchronous motors (AD1, AD2), gearboxes with appropriate reduction ratios (ired, ired3) and wheel model (1/Rk, 1/Rk3). Chains Jdv, Derivative1, Derivative, ired1, 1/Rk1 and Jdv1, Derivative3, Derivative2, ired2, 1/Rk2 are necessary to form the static load torque of the linearized mathematical model of an asynchronous motor [6]. Scope blocks are designed to display graphs of changes in the corresponding values: F - traction forces, X - overhead crane displacements, Mdv - motor torques, Wdv - motor shaft rotation frequencies, Wz - frequency difference in voltages supplied to motors, Wk - crane wheel rotation frequencies, dX - crane bridge misalignment, Scope3 - frequency difference in voltages supplied to motors. Defining parameters: w_o is the base voltage frequency supplied to the motors, Constant is the initial frequency mismatch of the supply voltages of the two motors, Step 3 is the frequency difference of the supply voltages of the motors.

The mathematical model of an asynchronous motor was adopted on the basis of paper [6], its computational model is shown in Figure 4. The computational model of the mechanical part, based on the description [2] and the previously described changes, is shown in Figure 5. Here the Constant block defines the initial misalignment of the crane bridge.

The parameters of the variables shown in Figures 3-5 are calculated according to known laws [2, 6, 7]. The technical characteristics of the crane were taken as the initial data.

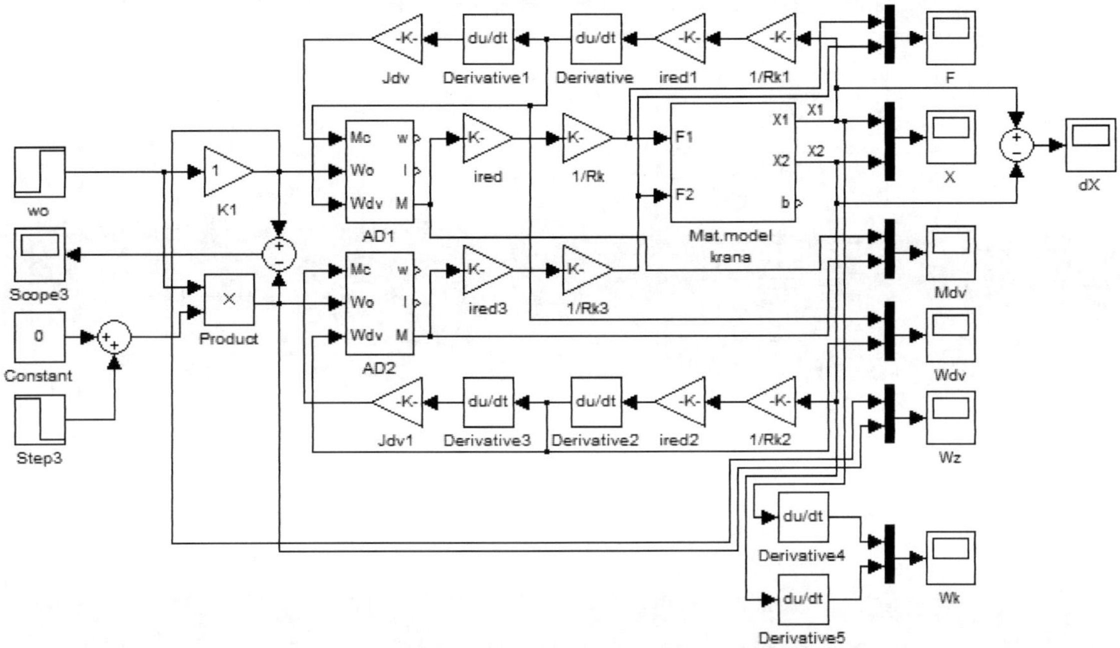

Figure 3. Computational scheme of an overhead crane.

Modeling

Initially, the experiments were carried out confirming the adequacy of the model, which was evaluated by the speed of the crane, the time of its acceleration and deceleration and the response to input parameters. The model was validated. Further, the existing methods of managing multidimensional control objects were analyzed and a series of experiments were conducted [8].

In the first series of experiments, the frequencies of the supply voltages on the motors are equal and amount to 241 rad/s. The initial displacement ΔX is set at first at 0.15 m, then at 0.5 m. At 30 s, the voltage frequency of the second motor is increased by 20%. In this case, the motors torques change and come to a steady-state value of 300 Nm, the displacement ΔX changes sign, and is 0.18 m, which signals the rotation of the crane bridge and friction of the wheel flanges against the rails.

In the second series of experiments, the frequencies of the supply voltages on the motors are equal and amount to 241 rad/s. At 30 s, the frequency of the supply voltage of the second motor is decreased by Δω (5, 10, 15, 20, 30, 35%) Initial displacement ΔX = 0 m.

The experiment determined torques for the motors M, bridge misalignment ΔX, transient time t_t and readjustment σ. The results of the experiment are summarized in Table 1, and, for convenience, graphs are drawn for M (Δω) (Fig. 6a), ΔX (Δω) (Fig. 6b), t_t (Δω) (Fig. 6c) and σ (Δω) (Fig. 6d). With a further increase in Δω, stable self-induced vibrations appear.

In the third series of experiments, we will introduce non-linearity of the "dead zone" type into one of the drives. This nonlinearity will simulate bridge wedging when the joints are irreversibly stretched. In this case, we will supply equal voltage frequencies to the motors (241 rad/s). At 30 s, the frequency of the supply voltage of the second motor will be increased by Δω = 30%. Initial displacement ΔX = - 0.15 m.

The experiment showed that there is no change in the position of the bridge (ΔX remains at the same level).

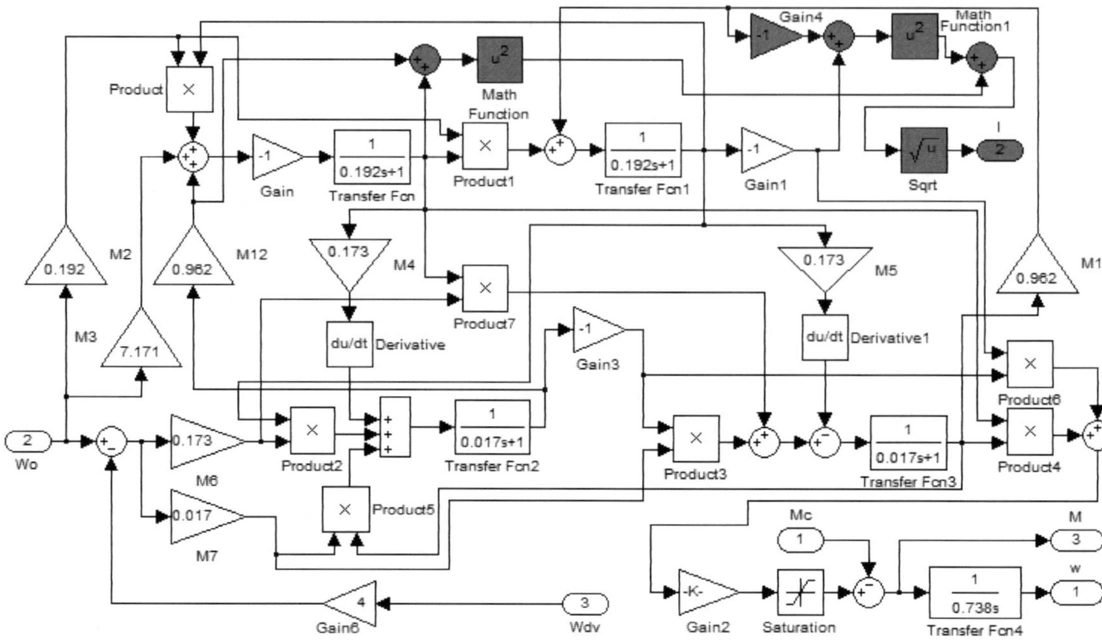

Figure 4. Computing scheme of subsystems AD1 and AD2

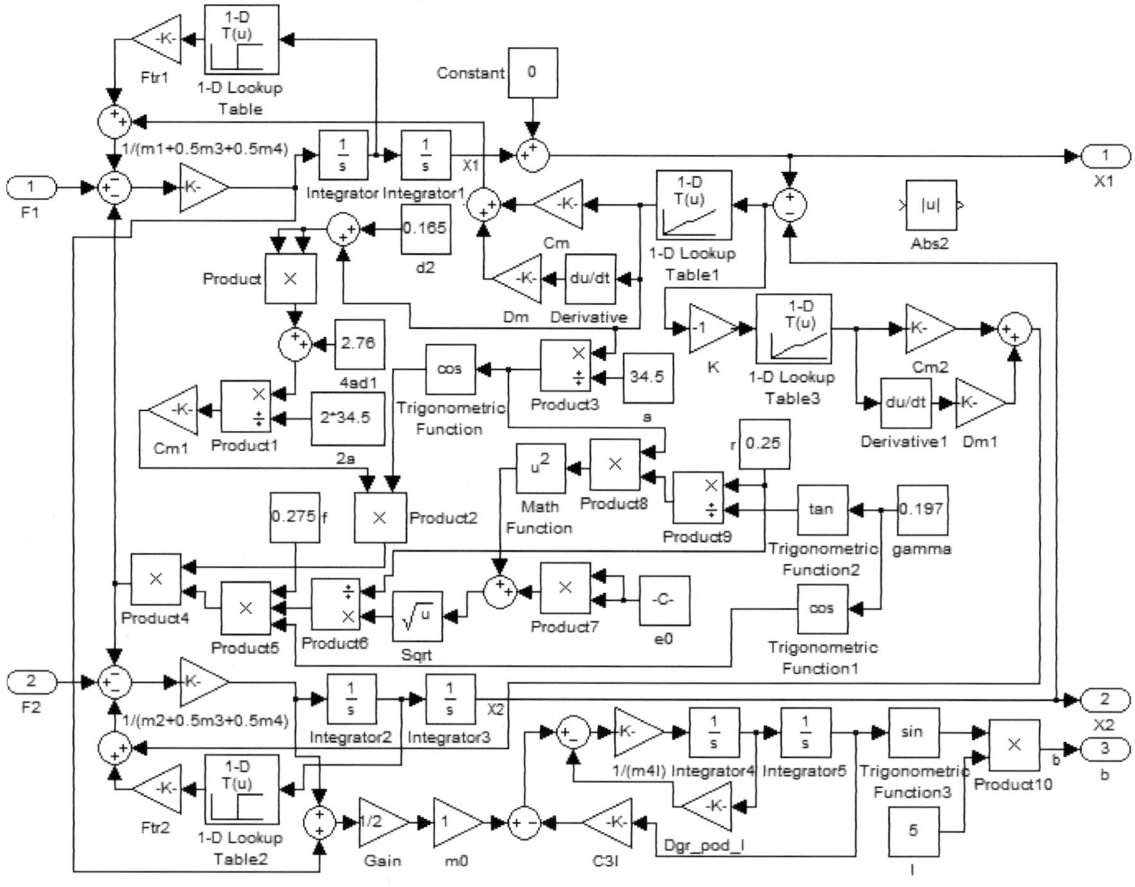

Figure 5. Computational scheme of the mathematical model of crane subsystem

Table 1. Results of the second series of experiments

$\Delta\omega$, rad/s	M, N m		ΔX, m	t_t, s	σ, %
0.05	-224	240	-0.1814	8	0.00
0.1	-437	449	-0.197	6	0.00
0.15	-623	630	-0.21	4	0.00
0.2	-770	785	-0.22	3	0.00
0.3	-937	950	-0.233	4	0.13
0.35	-957	963	-0.234	10	1.71

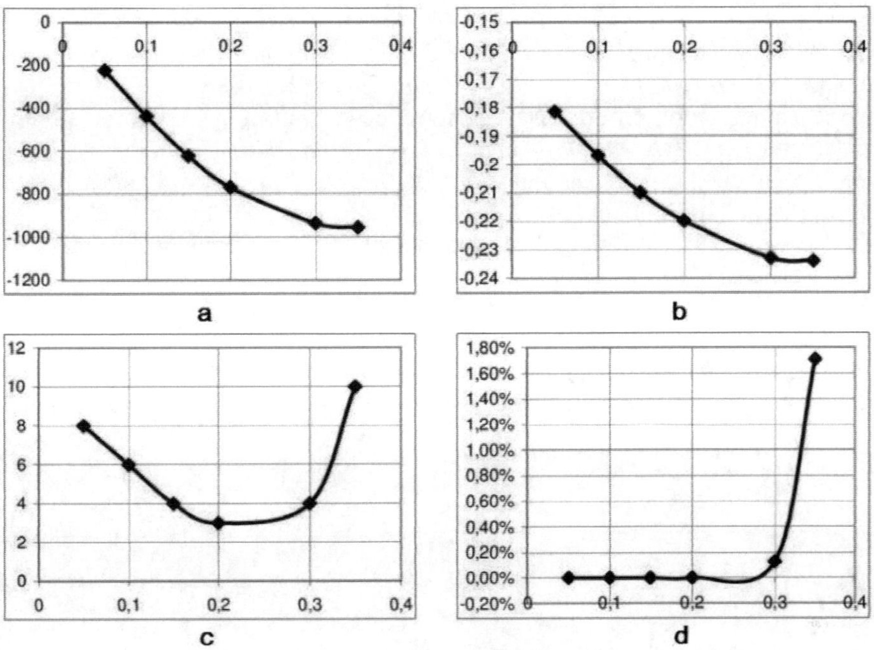

Figure 6. Results of the second series of experiments

Based on the results of the *first series of experiments,* it can be concluded that the crane is controllable under normal operating conditions and is capable of returning to its original state.

Based on the results of the *second series of experiments,* it can be concluded that:

- the crane, as a control object, achieves the best speed of operation at a speed difference in the range of 15 ... 30%;
- with an increase in the frequency mismatch of the supply voltages of the drives, the torque increases, which then leads to wear of the rails and wheel flanges;
- when the frequency mismatch of the supply voltages of the drives is more than 35%, stable self-induced vibrations occur and the crane does not straighten.

According to the results of the *third series of experiments,* it can be concluded that when modeling crane jamming, the change in the position of the bridge does not occur (ΔX = const), and therefore the effort developed by the motor with a difference in the frequencies of the supply voltages of 30% is not enough to overcome the obstacles created by the deformation of the joints.

4. Conclusion

The method of modeling an overhead crane considered in the article based on the results of observation and keeping records of the state of individual nodes of crane equipment (wheels and rail track) is of great practical importance in the operation of lifting and transporting equipment. Namely, the application of the described method gives an answer to the question of the need for the fastest repair of mechanical equipment and metal structures. This is achieved by the fact that during the modeling, it is possible to test various modes of crane control and some of them can ensure the operation of the deformed crane in a safe mode, with the minimum possible wear of crane undercarriage components. In this case, it is possible to ensure the operation of the crane up to the scheduled maintenance and repair time.

References

[1] Dukelsky A I 1973 *Crane Handbook* (Leningrad: Mechanical engineering)
[2] Galitskov S Ya and Ionov A A 2010 Manipulator of the Pallet Storage Area as a Control Object *Samara State Technical University Bull.* **2(27)** 18
[3] Blinchikov O I and Galitskov S Ya 2008 Gantry manipulator with suspended traverse as a control object *Samara State Technical University Bull.* **2(22)** 11
[4] Telichko L Ya and Dorofeev A A 2012 Electric drive control system for smooth movement of overhead cranes *Voronezh State Technical University Bull.* **3(6)** 113
[5] Rapoport E Ya 2003 *Structural modeling of objects and control systems with distributed parameters* (Moscow: High School)
[6] Galitskov S Ya, Galitskov K S and Maslyanitsyn A P 2002 *Dynamics of the Induction motor* (Samara: SSUCCE)
[7] Krupovich V I 1982 *Reference design of automated electric drive and control systems of technological processes* (Moscow: Energoizdat)
[8] Alpatov Yu N and Boliakno V A 2016 Synthesis of multidimensional control systems to solve two-dimensional and three-dimensional systems *System. Methods. Technologies* **3(31)** 111-115

IOP Conf. Series: Materials Science and Engineering 1103 (2021) 012019 doi:10.1088/1757-899X/1103/1/012019

Mathematical modeling of thermal processes in building enclosures

A P Maslyanitsyn, K S Galitskov and A S Fadeev

Samara State Technical University, 244, Molodogvardeyskaya Str., Samara, 443001, Russia

E-mail: maes@samgtu.ru

Abstract. In the development of automatic control systems (ACS) for heat supply of buildings, mathematical models of their control objects play a significant role, one of the most important elements of which being the enclosing structures. The existing models of thermal processes indoors either do not take into account the dynamic effect of the enclosing structures, or are complex analytical expressions, the use of which in the engineering practice of the synthesis of ACS for the indoor thermal regime is difficult. The purpose of the study was to create a linear mathematical model of the thermal interaction of the external and internal air environment through the enclosing structures. To achieve this goal, a functional diagram of a heated room as a control object was developed, which describes the mechanism of thermal interaction of its elements. On its basis, design schemes of thermal processes for external and internal enclosing structures were developed. They were used to create systems of equations linking temperatures on the surfaces of layers of external enclosing structures and air temperatures outside and inside the premises. At the same time, the assumption was introduced that the dynamic temperature distribution over the thickness of the enclosing structure is linear. To find the temperatures at the boundaries of the layers, the average temperature value over the thickness of the structure was used. As a result of the study, equations for the dynamics of heat transfer processes were obtained, and a generalized structure of a model of multilayer external enclosing structures of a room as an element of a control object was developed. The resulting model can be used to develop a generalized mathematical model of the thermal regime of both individual rooms and the building as a whole.

1. Introduction

The process of optimization of heat energy consumption, provided that comfortable microclimate conditions are ensured, is one of the most important tasks in the operation of residential and public buildings. Its solution is closely related to the use of automation tools that control and regulate heating, air conditioning and ventilation systems [1, 2]. The most important role in ensuring comfortable temperature regime in rooms is played by heating systems.

There is a wide variety of heating systems in buildings. They differ in the types of heat supply sources, methods of delivery and distribution of the heating medium between consumers, etc. One of the common options for organizing heat supply to buildings is a centralized system in which the source of heat energy is a boiler plant (thermal station), and the heating medium is water.

The development of any control system requires knowledge of the mathematical model of its control object. In our case, we consider the processes of heat exchange between the external environment, building structural elements, the heating system and the air of the building as such control object. The constituent elements of the control object have spatial distribution, therefore their dynamics is described by differential equations in partial derivatives [3].

Currently, there is a number of mathematical models of heated rooms and enclosing structures, in particular [3, 4]. As a rule, they are aimed at solving computational problems, i.e. determination of thermal regime parameters for various combinations of external and internal conditions in which the premises of the buildings are located. The high complexity of these models complicates their adaptation and application in the engineering practice of the synthesis of heat supply control systems. Therefore, the purpose of this study is to develop a linear mathematical model of thermal processes in enclosing structures that is convenient for use in problems of synthesis involving systems for automatic control of thermal conditions of buildings.

2. Mathematical Modeling of Enclosing Structures as a Control Object Element

2.1. Definition of the control object

The internal energy of the room air is determined by the dynamic balance of heat fluxes arising from its interaction with the heating device, enclosing structures and the outdoor environment. Therefore, we take the total of the heating device HD, external (ExS) and internal (InS) enclosing structures and room air (RA) as the control object of the room air temperature control system (Figure 1).

The output coordinate of the control object is the room air temperature T_{ra}, the control action is the heat flux rate of the heat carrier of the heating device G_1, and the main disturbance is the temperature of the outside air T_{ext}.

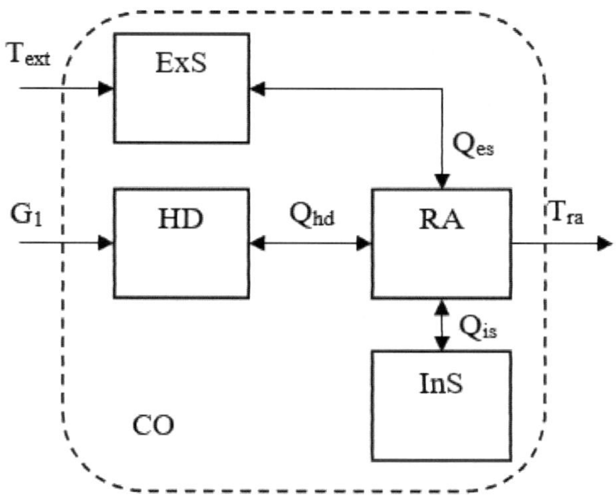

Figure 1. Functional diagram of the control object

2.2. Assumptions made when developing a mathematical model

Indoor thermal processes are characterized by a complex dynamic interaction of spatially distributed elements. Moreover, some of their thermophysical parameters are nonlinearly dependent on temperature, which significantly complicates the mathematical model. Therefore, when developing a linear mathematical model focused on the use of synthesis of automatic control systems in analytical methods, it is necessary to introduce the following assumptions:

1. All spatially distributed parameters of the control object are replaced by their average values. This allows us to proceed to the mathematical model of the control object with lumped parameters.
2. We assume that the densities and heat capacities of air and water are constant and independent of temperature.
3. The modeled room is located in the middle of the building and has only one outside wall.
4. The thermal regimes of the modeled room and those adjacent to it coincide.

5. The temperature distribution over the thickness of structures is linear, both in statics and dynamics.
6. There are no infiltration, insolation or household heat fluxes in the room.

2.3. Mathematical model of external and internal enclosing structures

External enclosing structures can include both transparent elements (windows, stained-glass windows) and opaque elements (walls). Walls play the main role in thermal protection of premises from the external environment and, as a rule, are multilayer. We will assume that external enclosing structures consist of four layers: textured facade layer 1, insulation 2, cellular concrete 3 and cement mortar 4.

In order to compile a system of equations describing the dynamics of external enclosing structures, a design scheme has been developed (Figure 2), showing the mechanism of heat exchange between the inside air environment and the outside air through a multi-layer enclosing structure.

The upper part of the design scheme shows heat fluxes in different layers of the enclosing structure. The heat flux of each layer includes two components: static (transmission) and dynamic, which appears only when the temperature head of the layer changes. In stationary mode, $Q_{0c} = Q_{1c} = Q_{2c} = Q_{3c} = Q_{4c} = Q_{st}$, i.e. the same transmission heat flux passes through all layers.

In the design scheme, the dashed line describes the stationary temperature distribution in different layers of the building enclosure. The horizontal solid lines in the diagram correspond to the approximation of the spatial temperature distribution within the layer by its average value.

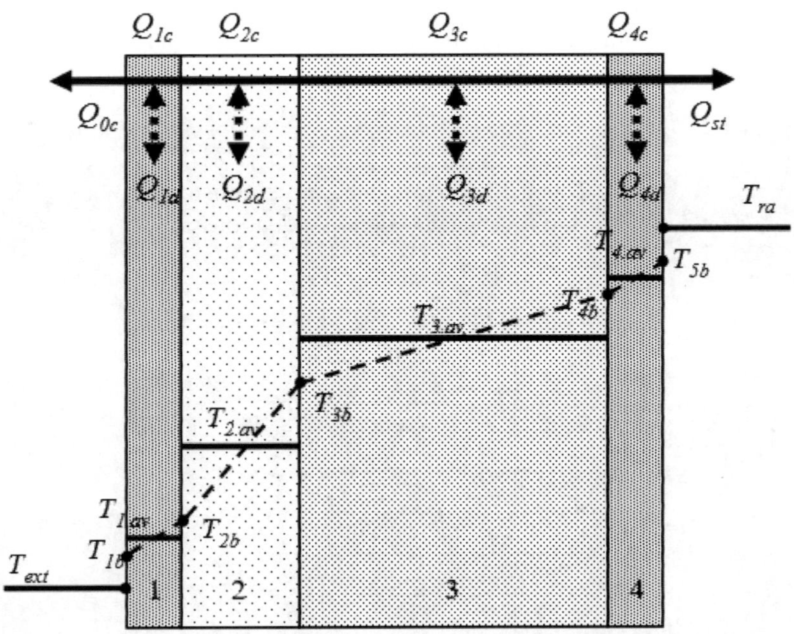

Figure 2. Calculation scheme of thermal processes in a four-layer enclosing structure

Outside air with a temperature T_{ext} affects the outer surface of the wall, which has a temperature T_{1b}, by convection. If $T_{1b} \neq T_{ext}$, then under the influence of the temperature head $(T_{ext} - T_{1b})$ there occurs a heat flux Q_{0c} [5]

$$Q_{0c} = (T_{ext} - T_{1b}) \cdot \alpha_1 \cdot F_c \tag{1}$$

Where F_c is the wall area, α_1 is the coefficient of heat transfer by convection at the outer surface of the wall, which is usually determined by the Frank formula [5]

$$\alpha_1 = 7{,}34 \cdot e^{0.656 \cdot \upsilon} + 3{,}78 \cdot e^{-1.91 \cdot \upsilon}, \tag{2}$$

here υ is the air velocity. In order to preserve the linearity of the equation (1), we will assume that $\upsilon=const$.

Heat flux Q_{0c} causes a change in the internal energy of layer 1, which can be described by the expression for the average temperature

$$T_{1.av} = \frac{1}{C_1 \cdot m_1} \int Q_{1d} dt, \tag{3}$$

Where C_1 and m_1 – the specific heat of the material and the mass of layer 1 of the enclosing structure, Q_{1d} - dynamic heat flux

$$Q_{1d} = Q_{0c} - Q_{1c}. \tag{4}$$

Transmission heat flux Q_{1c} can be found by equation [5]

$$Q_{1c} = \left(T_{1.av} - T_{2b}\right) \cdot \frac{2}{R_1} \cdot \frac{1}{F_c}, \tag{5}$$

where T_{2b} is the temperature on the inner surface between layers 1 and 2, R_1 is the thermal resistance of layer 1

$$R_1 = \frac{\delta_1}{\lambda_1}, \tag{6}$$

here δ_1 and λ_1 – thickness and thermal conductivity of layer 1 material.

If we assume that both in statics and dynamics, the temperature distribution over the layer thickness is linear, then the average value of the temperature of layer 1 $T_{1.av}$ can be defined as the arithmetic mean of temperature values T_{1b} and T_{2b} on its borders

$$T_{1.av} = \frac{T_{1b} + T_{2b}}{2}. \tag{7}$$

Thermal processes in layers 2 and 3 are similar to those considered above, therefore, based on expressions (3) - (7) one can have the following system of equations

$$\begin{cases} Q_{1c} = Q_{2c} + Q_{2d}, \\ Q_{2c} = \left(T_{2.av} - T_{3b}\right) \cdot \dfrac{2}{R_2} \cdot \dfrac{1}{F_c}, \\ T_{2.av} = \dfrac{1}{C_2 \cdot m_2} \displaystyle\int Q_{2d} dt, \\ T_{2.av} = \dfrac{T_{2b} + T_{3b}}{2}, \\ Q_{2c} = Q_{3c} + Q_{3d}, \\ Q_{3c} = \left(T_{3.av} - T_{4b}\right) \cdot \dfrac{2}{R_3} \cdot \dfrac{1}{F_c}, \\ T_{3.av} = \dfrac{1}{C_3 \cdot m_3} \displaystyle\int Q_{3d} dt, \\ T_{3.av} = \dfrac{T_{3b} + T_{4b}}{2}. \end{cases} \tag{8}$$

Where Q_{2c}, Q_{3c} – transmission heat fluxes of layers 2 and 3, Q_{2d}, Q_{3d} – dynamic heat fluxes, $T_{2.av}$, $T_{3.av}$ – average values of temperatures, C_2, C_3 and m_2, m_3 – specific heat capacities of materials and masses of layers 2 and 3, T_{3b}, T_{4b} – temperatures at the boundaries of layers 2-3 and 3-4, R_2, R_3 - thermal resistance of layers 2 and 3.

Thermal processes in the last 4th layer of the building enclosure can be described by the equations

$$\begin{cases} Q_{3c} = Q_{4c} + Q_{4d}, \\ Q_{4c} = \left(T_{4.av} - T_{5b}\right) \cdot \dfrac{2}{R_4} \cdot \dfrac{1}{F_c}, \\ T_{4.av} = \dfrac{1}{C_4 \cdot m_4} \int Q_{4d} dt, \\ T_{4.av} = \dfrac{T_{4b} + T_{5b}}{2}. \end{cases} \quad (9)$$

Where Q_{4c} – transmission heat flux of layer 4, Q_{4d} – dynamic heat flux, $T_{4.av}$ – average temperature value, C_4 and m_4 – specific heat capacity of the material and mass of layer 4, T_{5b} is the temperature on the inner surface of the enclosing structure, R_4 is the thermal resistance of layer 4.

If $T_{5b} \neq T_{ra}$, where T_{ra} is the average room air temperature, then a convective heat flux will appear between the wall surface and the room air.

$$Q_{4c} = Q_{st} = \left(T_{5b} - T_{ra}\right) \cdot \alpha_2 \cdot F_c, \quad (10)$$

here α_2 is the convection heat transfer coefficient at the inner wall surface [5]

$$\alpha_2 = 1.66 \cdot \sqrt[3]{\left(T_{ra} - T_{5b}\right)}. \quad (11)$$

In order to preserve the linearity of the equation (10), we will assume that $\left(T_{ra} - T_{5b}\right) = \Delta t_1 = const$, then

$$\alpha_2 = 1.66 \cdot \sqrt[3]{\Delta t_1}. \quad (12)$$

If we generalize the above equations and represent them in operator form, we can obtain a mathematical model of thermal processes for an n-layer enclosing structure

$$\begin{cases} Q_{0c}(p) = \left(T_{ext} - T_{1b}(p)\right) \cdot \alpha_1 \cdot F_c, \\ Q_{(i-1)c}(p) = Q_{ic}(p) + Q_{id}(p), \\ Q_{ic}(p) = \left(T_{i.av}(p) - T_{(i+1)b}(p)\right) \cdot \dfrac{2}{R_i} \cdot \dfrac{1}{F_c}, \\ T_{i.av}(p) = \dfrac{1}{C_i \cdot m_i \cdot p} Q_{id}(p) \\ T_{i.av}(p) = \dfrac{T_{ib}(p) + T_{(i+1)b}(p)}{2}, \\ Q_{nc} = Q_{st}(p) = \left(T_{(n+1)b}(p) - T_{ra}(p)\right) \cdot \alpha_2 \cdot F_c, \end{cases} \quad (13)$$

Where $i \in 1 \ldots n$ is the ordinal number of the layer, p is the Laplace operator.

The system of equations (13) allows synthesizing the generalized structure of the mathematical model of thermal processes in a multilayer enclosing structure (Figure 3). It is a sequential connection of operators S, each of which describes thermal processes at the internal and external boundaries of the enclosing structure. Operators S_1 (Figure 4) and S_n (Figure 5) describe convective heat transfer processes on the outer surfaces of the enclosing structure. Operators S_2 ... S_i ... S_{n-1} (Figure 6) are a mathematical description of conductive thermal processes at the internal boundaries between the layers of the enclosing structure.

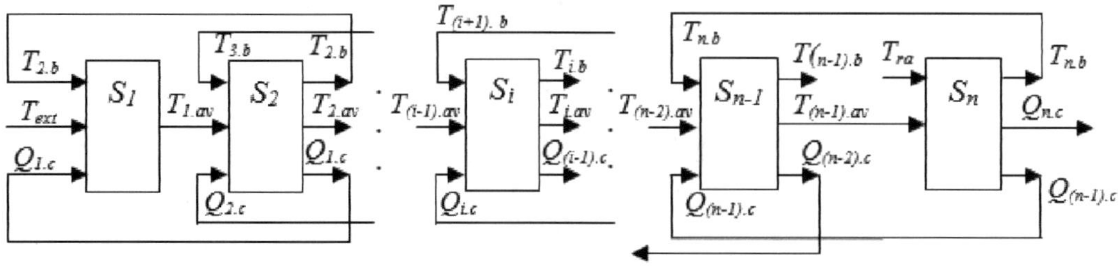

Figure 3. Structure of the mathematical model of a multilayer enclosing structure

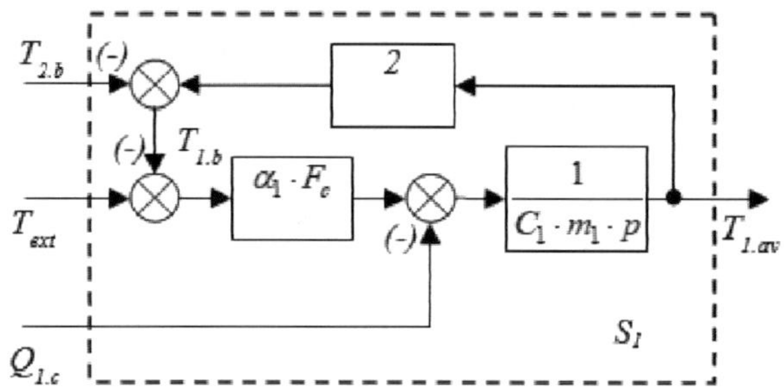

Figure 4. Structure of the mathematical model of the 1st (outer) layer of the enclosing structure

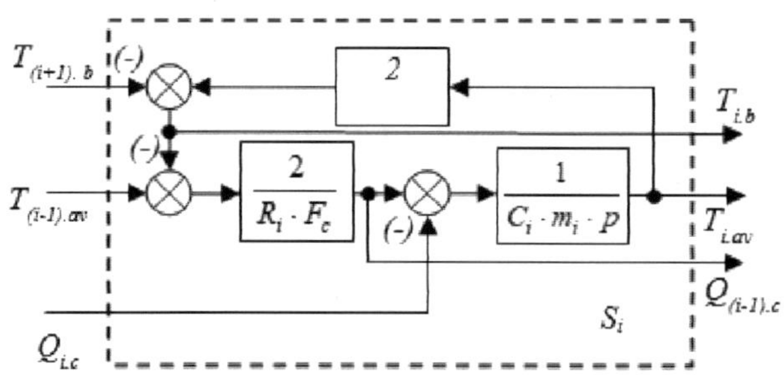

Figure 5. Structure of the mathematical model of the i-th layer of the enclosing structure

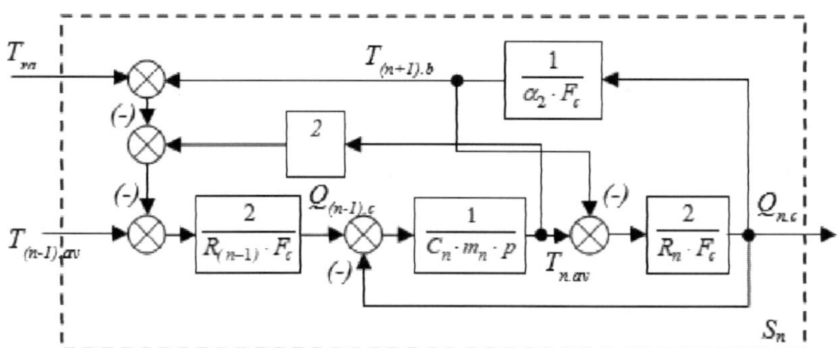

Figure 6. Structure of the mathematical model of the last (internal) layer of the enclosing structure

The transparent part of the enclosing structure (window, stained-glass window) belongs to non-heat-consuming structures, therefore the thermal processes in it can be described by the expression for transmission heat losses

$$Q_w(p)=\left(T_{ra}(p)-T_{ext}\right)\cdot\frac{1}{R_w}\cdot\frac{1}{F_w}, \tag{14}$$

where R_w and F_w – thermal resistance and area of the window structure.

Then the total heat flux of the external enclosing structure

$$Q_{es}(p)=Q_{st}(p)+Q_w(p). \tag{15}$$

To describe the thermal processes of internal enclosing structures (internal partitions, floors), a design scheme has been developed (Figure 7). According to it, the volume of each internal structure is divided into two equal zones of influence from the air inside the modeled room and the air from rooms adjacent to it. In accordance with the assumption that the thermal regimes of neighboring rooms are identical, there are no transmission heat fluxes through the internal enclosing structures.

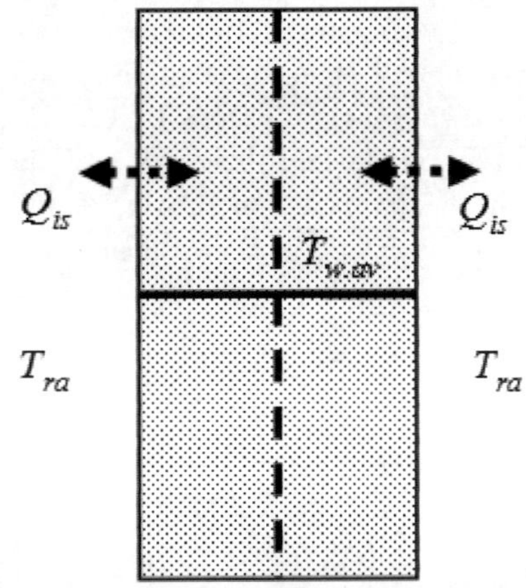

Figure 7. Calculation scheme of thermal processes in an internal enclosing structure

When the heating system is turned on (off), the room air temperature $T_в$ begins to change. As a result thermal pressure and dynamic heat flow arise between the air and the surface of the internal structures $\left(T_{ra}-T_{w.av}\right)$

$$Q_{is}=\left(T_{ra}-T_{w.av}\right)\cdot\alpha_2\cdot F_{ins}, \tag{16}$$

where F_{ins} is the area of internal enclosing structures, $T_{w.av}$ – average temperature of the internal enclosing structure

$$T_{w.av}=\frac{2}{C_w\cdot m_w}\int Q_{is}dt, \tag{17}$$

where C_w, m_w – specific heat capacity of the material and the mass of the structure.

Let us represent the equations (16) and (17) in operator form

$$\begin{cases} Q_{is}(p)=\left(T_{ra}(p)-T_{w.av}(p)\right)\cdot\alpha_2\cdot F_{ins}, \\ T_{w.av}(p)=\dfrac{2}{C_w\cdot m_w\cdot p}Q_{is}(p). \end{cases} \tag{18}$$

2.4. Mathematical model of thermal processes for room air

Room air is an accumulator of heat flows passing through external and internal enclosing structures, as well as of heat gains from internal heat sources (people, household and heating appliances). The air temperature in the heated room has a certain spatial distribution. However, supposing that natural air circulation leads to its even distribution, we assume that the room air temperature is a lumped parameter and is equal to the average value. We should also assume the absence of infiltration, insolation, and household heat flows. Then the average room temperature is

$$T_{ra} = \frac{1}{C_a \cdot m_a} \int (Q_{es} + Q_{is} + Q_{hd})dt ,$$

or in operator form

$$T_{ra}(p) = \frac{1}{C_a \cdot m_a \cdot p}(Q_{es}(p) + Q_{is}(p) + Q_{hd}(p)), \quad (19)$$

where C_a, m_a - specific heat and mass of room air.

3. Results and discussion

Let us combine the previously developed models in order to obtain a structural scheme of a generalized control object. In this case, we assume that the opaque part of the external enclosing structure has four layers and contains an opening filled with a transparent window structure. As a result, the structure of a generalized control object is obtained (shown in Figure 8). In it, the model of the opaque part of the enclosing structure is represented by blocks S_1 - S_4 , the model of the opening transparent part (window) – by the S_w block, and the model of the internal enclosing structures – by the S_{ins} block. The operation of the heater is taken into account with the heat flow Q_{hd}, while we assume that its value is constant and does not depend on the air temperature in the room.

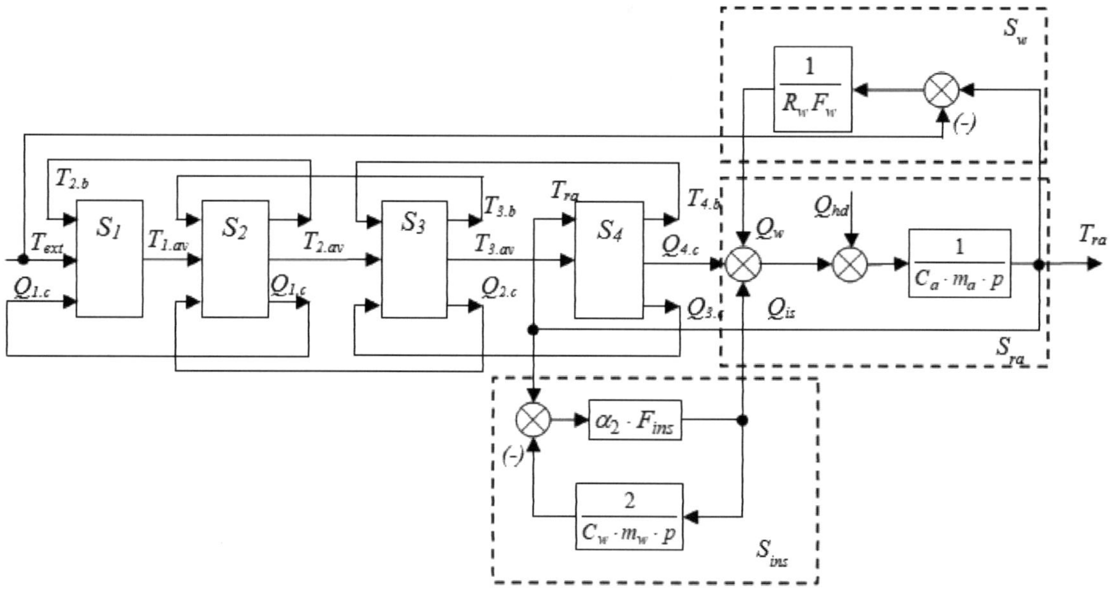

Figure 8. The structure of the mathematical model of a multilayer enclosing structure

The advantage of the resulting structure is its linearity, which allows us to obtain transfer functions connecting the variable values of the control object by the method of structural transformations. In the future, they can be used in the synthesis of automatic control systems for the thermal regimes of buildings.

4. Conclusion

Multilayer building enclosures have a complex dynamic mechanism of thermal interaction with the external and internal air environments of the premises. The developed mathematical model of these processes can be specified for any number of layers. It is presented in the form of three types of blocks, combining which you can easily synthesize the setup of the model of an n-layer structure. In this case, blocks can simulate individual physical layers of a structure or correspond to groups of intermediate layers into which physical layers can be divided to improve the accuracy of modeling. In the latter case, it is necessary to additionally solve the problem of minimizing the number of intermediate layers for the purpose of ensuring acceptable calculation accuracy.

The important feature of the obtained model is its orientation towards application in the field of synthesis of systems for automatic control of air temperature in a room, where the apparatus of transfer functions is widely used. The linearity of the created model makes it possible to obtain transfer functions connecting the variables of the generalized control object, either by solving a system of operator equations or by the method of structural transformations. In the future, these can be used in the synthesis of automatic control systems for the thermal regime of a building.

References

[1] Skanavi A N and Makhov L M 2002 *Heating* (Moscow: ASV) p 576
[2] Kalmakov A A, Kuvshinov Yu Ya, Romanova S S and Shchelkunov S A 1986 *Automation of Heat and Gas Supply and Ventilation Systems* V N Bogoslovsky, ed (Moscow: Stroyizdat) p 479
[3] Chistovich S A, Averyanov V K, Tempel Yu Ya and Bykov S I 1987 *Automated Heat Supply and Heating Systems* (Leningrad: Stroyizdat, Leningrad branch) p 248
[4] Bogoslovsky V N 1979 *Thermal Regime of the Building* (Moscow: Stroyizdat) p 248
[5] Fokin K F 2006 *Construction Heat Engineering of Building Enclosures* Yu A Tabunshchikova and V G Gagarin, Eds (Moscow: AVOK-PRESS) p 256

IOP Conf. Series: Materials Science and Engineering 1103 (2021) 012020 doi:10.1088/1757-899X/1103/1/012020

Research of robotics and automation of construction processes in Russia

G V Mikheev, E V Soloveva, E D Yanovskaya, A Y Guzhev and A R Barna

Kuban State Technological University, 2, Moskovskaya st., Krasnodar, 350072, Russia

E-mail: jpt@kubstu.ru

Abstract. An important role in the construction industry is played by the robotics of individual processes in construction. However, despite a wide range of scientific and design developments in the field of construction, the level of automation and robotization processes remains quite low. This is due to the need to systematize the performed research and development, to carry out complex research and development activities. The paper presents data on the technological prerequisites for robotization of construction processes in Russia. The factors hindering the use of industrial robots in Russia are identified. The distinctive parameters and characteristics of robots in the construction of unique buildings and structures are considered. The main tasks for robotization of technological processes at construction enterprises are formulated. The analysis of the levels injuries at construction sites is carried out. The possibility of introducing robotic technology in the construction of unique buildings with complex technical features is considered. As a result of the research carried out, the scientific problem of analysis and synthesis specialized robots for the construction of unique buildings and structures has been considered. The solution to this problem is important, since it contributes to the development of highly effective means of complex mechanization and automation of construction processes that increase the productivity and quality of installation, finishing and plastering works, reduce their labor intensity, free people from harmful and dangerous working conditions.

1. Introduction

Construction is one of the leading spheres of human activity. The variety of ongoing construction projects and construction processes is increasing every day.

Construction processes are characterized by increased complexity and danger. The success of any construction project is determined by three indicators: cost, quality, timing.

To improve the quality of indicators in construction, it is necessary to improve construction production with the help of progressive and innovative technologies. Robotization technology accelerates construction processes, improves the quality and safety of work performed, and also has prerequisites for reducing the proportion of manual labor.

2. Materials and methods

The goal of research consists in analyzing and assessing the effectiveness of using robotic technology in construction and the possibility of its implementation in Russia.

Since the mid-80s of the 20th century, the tasks of robotization and automation of construction work have become relevant to scientists and specialists from research and construction organizations, such as the Massachusetts Institute of Technology, Munich University of Technology, Moscow State University of Civil Engineering and other scientific organizations. In this study, we have analyzed scientific developments to solve the problem of automation and robotization of construction processes [1–12]. However, despite a wide range of scientific and design developments in the field of construction robotics, the level of automation and robotization of construction operations remains rather low. It is required to systematize the performed research and development, to carry out complex research and development activities.

3. Research and analysis of technological prerequisites for the robotization of construction processes in Russia

Analyzing the technological features of production in the construction industry and taking into account the extensive experience of robotization at enterprises, we have formulated the following main tasks of robotization processes at construction enterprises:

- development of methods for analyzing the technology of production building products and materials, conducting on their basis a survey of enterprises and drawing up standard comprehensive scientific substantiated plans for creating certain types of technological processes;
- determination of priority technological operations and implementation of industrial manipulators and robots on them;
- development of special manipulators and robots for the construction industry;
- carrying out work on the unification and typification of robotics equipment for construction industry enterprises;
- development technological equipment for robotization of operations in the construction industry;
- carrying out work to improve technological processes and individual operations, taking into account the requirements of automation and robotization for effective implementation;
- creation and implementation of robotic systems, sections and lines, development of standard layout elements for various production processes;
- preparation of new industries for the introduction of flexible technologies and the creation of production systems (lines, sections and workshops) on their basis;
- solution of technical and organizational aspects of operation at enterprises of the construction industry.

Analysis of the main types of construction processes shows that individual construction processes can be automated using robotics. Automation and robotization will reduce the workload on builders engaged in manual labor in the construction of unique buildings and structures. Industrial robots allow a more rational use of labor resources and provide important social and economic benefits. The increase in labor productivity is also facilitated by an increase in the quality of work and a decrease in rejects in the production of building materials and products due to the elimination of individual and subjective factors. The scale of the construction industry and its dependence on manual labor create economic potential for automation and robotization of construction processes.

It is important to note that the price of housing in Russia does not depend on the cost of labor. The introduction of robots will not solve the problem of rising house prices. However, today some cutting-edge manufacturing enterprises are beginning to consider the frequent use of technological robots.

We have identified factors constraining the use of industrial robots in Russia: a) the lack of Russian enterprises' own experience in using robots, methodological materials of the technical and economic complex of robotic technologies; b) lack of qualified specialists able to ensure the operation of robots; c) a limited number of specialists capable of designing robotic areas for the implementation of robots and their technological training during construction.

4. Distinctive parameters and characteristics of robots in construction

Technological progress in the development of robots was aimed at improving control systems. The first generation industrial robots were programmable. The second generation of robots is equipped with sensor systems and a vision function. The third generation of robots has intelligent control.

Small dimensions and weight characteristics of robotic equipment make it profitable and convenient for transportation. Robotic equipment can be operated by one qualified operator. Therefore, this technology is appropriate in situations that are potentially hazardous to the health of workers. Analyzing construction manipulators and robots on the market can be divided into several groups.

The first group consists of manipulators, robots for performing assembly work. Robotics of the first group are characterized by a wide range of carrying capacity, mobility, and increased positioning accuracy. This group includes manipulators and robots for assembling building elements, partitions, assembling equipment and scaffolding.

The second group consists of construction manipulators and robots for concrete work. This group is subdivided into three subgroups: manipulators and robots for placing and compacting concrete mixture, means of robotics for reinforcement work at the facility, handling equipment for erection and dismantling of construction formwork.

The third group of construction manipulators and robots is made up of robotics for finishing works. This group includes manipulators and robots for painting, plastering, facing works, applying insulating mastics to roofs, as well as for installing monolithic floors. This group is distinguished by a low carrying capacity of mechanisms, mobility of structures, the use of program control and the need for sensor devices.

The fourth group consists of manipulators and robots for excavation work. This group includes multifunctional handling equipment designed for the construction of excavations, backfilling and compaction of soil, laying communications. Equipment in this group is supplied with remote, software or combined control.

The fifth group includes manipulators and robots for loading and unloading operations. This group includes manipulators and robots of various carrying capacities for loading and unloading building structures and elements, containers and packages, as well as small-piece cargo. Such manipulators can be installed on vehicles and self-propelled chassis.

There are a sufficient number of models on the market of robots that perform many operations in the construction, reconstruction of buildings and structures. Industrial and construction robots are characterized by active and purposeful interaction with the external environment. This is achieved due to the structural organization of the robot, which is determined by the purpose, the nature of technological operations, the type of movements in the working area, the requirements for carrying capacity and positioning accuracy.

5. Injury rate as one of the main aspects of the use of robotics

Construction according to working conditions is a potentially hazardous industry. The temporary nature of workplaces, the need to perform significant amounts of work at height and in difficult climatic conditions lead to the emergence of hazardous and harmful factors that pose a potential threat to the life and health of workers and others, which necessitates increased attention to labor protection issues.

During the construction of buildings and structures, being on the construction site itself, workers are faced with dangerous factors: the presence of moving mechanisms and machines, an increased noise level, an increased vibration level, the possibility of getting burned when working with hot mastics or a blowtorch, the danger of electric shock, physical overload, presence of gas-hazardous and flammable substances, unstable stacks of stored products, high or low air temperature of the working area depending on the season, risk of falling on slippery floors, moving parts of production equipment, location of the workplace at a considerable height relative to the floor surface.

Let us highlight the main causes of industrial injuries during construction: a) lack of supervision over the correct and safe conduct of work; b) operation of faulty equipment; c) gross violation of safety requirements; d) violation of technological regulations, lack of construction organization projects, work production projects and other documentation or their poor quality development; e) work in difficult climatic conditions; f) violation of hygienic standards (increased content of harmful substances in the air of working areas; g) insufficient or irrational lighting, increased noise levels, vibration, unfavorable meteorological conditions, the presence of various radiation above permissible values); h) psychophysiological factor (physical and neuropsychic overload of the worker).

The reason for the terrible trend of injuries lies, first of all, in non-compliance with labor safety requirements in the construction industry. Neglect of norms is a subjective reason, depending on many factors: the physical and moral state of the employee, his responsibility, discipline, attentiveness. One of the factors to reduce subjectivity can be the introduction of robots and manipulators into the technology and organization of construction production.

6. Conclusion

The unique combination of compactness and power of robots finds its application in the most difficult conditions and inaccessible places on the construction site, while robotization provides a significant reduction in construction time. The emergence of robotic technology is associated with the desire of builders to shorten work times and increase profits.

The conducted research allows us to formulate the following conclusions: a) digital technologies allow the construction industry to function effectively and efficiently; b) robotization allows one to drastically reduce the construction time of the facility; c) automation and robotization makes it possible to build various unique structures at construction sites; d) the introduction of robotic technology will make it possible to build houses of non-standard shape with complex facades, arches and unusual curved interiors; e) the main positive factors of automation and robotization during construction are increased productivity, the ability to work in hard-to-reach and dangerous places for finding a person; f) robots save a lot of building materials.

As a result of the research carried out, the scientific problem of analysis and synthesis of specialized robots for the construction of unique buildings and structures has been considered. The solution to this problem is important, since it contributes to the development of highly effective means of complex mechanization and automation of construction processes that increase the productivity and quality of installation, finishing and plastering works, reduce their labor intensity, free people from harmful and dangerous working conditions.

References

[1] Aghimien D O, Aigbavboa C O, Oke A E and Thwala W D 2019 Mapping out research focus for robotics and automation research in construction-related studies: A bibliometric approach *J. of Engineering, Design and Technology* **18(5)** 1063-1079

[2] Solovyova E V, Sekisov A N, Gura D A, Mikheev G V and Kovtunenko M G 2020 Developing Optimization Modelling Methodology for Production Costs Generation *Helix, Singapore* **10(2)** 63-71

[3] Bulgakov A, Bock T and Otto J 2019 Robot Manipulators for Plastering Work *2019 Int. Multi-Conf. on Industrial Engineering and Modern Technologies (FarEastCon)* (Vladivostok, Russia) pp 1-5. DOI: 10.1109/FarEastCon.2019.8934132

[4] Sekisov A N 2019 Improving the efficiency of the organization of construction production based on the use of BIM-technologies *IOP Conf. Ser.: Mater. Sci. Eng.* **698** 066005

[5] Knyazeva N 2017 Information Support for Automation Systems for the Organization of Construction *Int. Review of Civil Engineering (IRECE)* **8(4)** 125-132. DOI: https://doi.org/10.15866/irece.v8i4.12344

[6] Nehaj R, Molotkov G, Rudchenko I, Grinev A and Sekisov A 2017 Algorithm of composing the schedule of construction and installation works *IOP Conf. Ser.: Earth Env. Sci.* **90** 012019.

[7] Mocerino C 2018 Digital revolution in efficient self-organization of buildings: towards intelligent robotics *2018 Energy and Sustainability for Small Developing Economies (ES2DE)* (Funchal) pp

1-6. DOI: 10.1109/ES2DE.2018.8494237.

[8] Popov R, Sekisov A, Gura D, Ivanov L and Shipilova N 2020 Genesis of organizational and technological planning in Russian building practice *Turismo: Estudos &Práticas* (UERN), Mossoró/RN, Caderno Suplementar **04**. Retrieved from: http://natal.uern.br/periodicos/index.php/RTEP/article/view/2370, http://natal.uern.br/periodicos/index.php/RTEP/article/view/2370/2145

[9] Sekisov A, Ovchinnikova S, Norenko M and Matevosyan V 2020 Prospects for the solar energy development in housing construction: an assessment of economic efficiency *IOP Conf. Ser.: Mater. Sci. Eng.* **913** 042051. DOI: 10.1088/1757-899X/913/4/042051)

[10] Shcherbakov V, Karpik A and Barsuk M 2020 Automation of Railroad Construction Technology Using Surveying Methods, in: Popovic Z, Manakov A, Breskich V (eds) *VIII Int. Sci. Siberian Transport Forum. TransSiberia 2019. Advances in Intelligent Systems and Computing* **1116** (Springer, Cham) DOI: https://doi.org/10.1007/978-3-030-37919-3_19

[11] Mikheev G V, Sekisov A N, Gura D A, Abazyan A G and Kuznetsova O A 2019 Economic and marketing adaptation of business processes in the modern Russian real estate market *Revista Inclusiones* **6(2)** 119-124

[12] Tuskaeva Z, Aslanov G and Alikova Z 2018 Automation of Calculations for Selecting Construction Equipment Purchasing Techniques, in: Murgul V, Popovic Z (eds) *Int. Sci. Conf. Energy Management of Municipal Transportation Facilities and Transport EMMFT 2017. Advances in Intelligent Systems and Computing* **692** (Springer, Cham) DOI: https://doi.org/10.1007/978-3-319-70987-1_133

IOP Conf. Series: Materials Science and Engineering 1103 (2021) 012021 doi:10.1088/1757-899X/1103/1/012021

Studying the dynamics of the automatic control system for the elevation position of a motor grader blade with an alternative position sensor

K S Galitskov, M A Nazarov and A S Lukyanov

Samara State Technical University, 244, Molodogvardeyskaya St., 443100, Samara, Russia

E-mail: maes@samgtu.ru

Abstract. The existing problems of controlling the elevation position of a motor grader blade when leveling a roadbed, caused by arising measurement errors associated with both the operation of the control systems themselves and disturbing influences, are considered. It is shown that the optimally tuned regulators in the existing control system do not provide the required quality indicators for the operation of a construction machine. It is proposed to equip the automatic control system for the elevation position of the grader blade with an alternative position sensor, the signal from which is compared with the reference action and the difference is introduced as a correction to the control system. The developed system allows us to increase the accuracy of automatic elevation positioning of the motor grader blade up to the quality required by regulatory documents.

1. Introduction

One of the most important stages in road construction is the precise grading of the roadbed by motor graders. In the process of work, the spatial position of the grader blade should correspond as closely as possible to the longitudinal and transverse profiles of the road project design. The solution to this problem is carried out with the help of automatic control systems for the elevation position, angle and tilt and cutting of the blade in accordance with the project. There are several principles for constructing automatic blade control systems, among which the most effective is the use of the global navigation satellite system (GNSS). GNSS operation is subject to a lot of interference, therefore, in order to reduce the measurement error, the method of differential corrections is used, as a result of which the positioning accuracy according to the GNSS equipment manufacturers is 15-25 mm. The results of the conducted studies of the automatic control system for the elevation position of the motor grader blade (BEP ACS) Leica iCON grade 3D showed that the errors in measuring the spatial position of a stationary object using GNSS in Real Time Kinetic RTK-mode are significantly variable in time. They lead to a quasi-harmonic change in the ΔZ value of the measured coordinates Z. It has been experimentally established that the average value of the oscillation period ΔZ is 17 min, and their amplitude reaches 20-30 mm. Such indicators are unsatisfactory from the point of view of the arrangement of the roadbed with the quality required by regulatory documents. Therefore, the development of effective methods and means for controlling the elevation position of the motor grader blade, which provide the required performance of the construction machine, is relevant.

2. Mathematical model of the blade elevation sensor

In the mathematical modeling of the BEP sensor as a dynamic link of the system under study, we will take into account that the time of transmission of data obtained with the help of GNSS to the ACS BEP is on average 0.1 s [1]. Then the output signal Z_{oc} of the feedback sensor can be represented, in operator form, as a sum of signals Z'_{oc} and ΔZ (Fig. 1)

$$Z_{oc}(p)=Z(p)\cdot e^{-\tau_u p}+\Delta Z(p)=Z'_{oc}(p)+\Delta Z(p) \tag{1}$$

where Z is the measured coordinate; τ_u is the delay time; ΔZ - quasi-harmonic interference; Z'_{oc} - feedback signal in the absence of interference. We assume that the ΔZ signal is generated by some interference generator (IG).

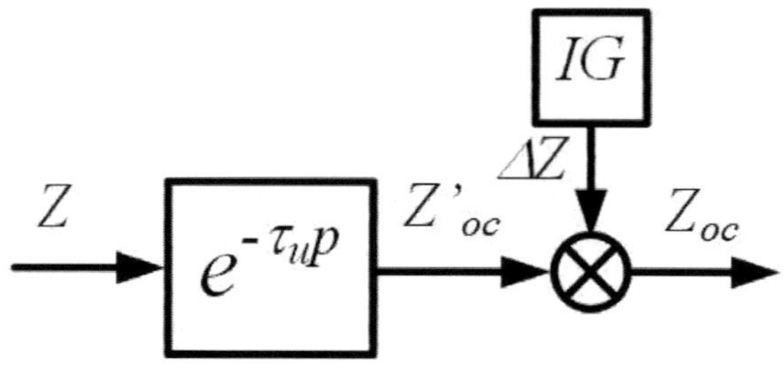

Figure 1. Block diagram of the mathematical model of the blade elevation sensor

It is proposed to analyze the operation of the BEP ACS under the action of two types of signal ΔZ coming from the interference generator (IG) (Fig.1), which simulates the inaccuracies of the GNSS operation:

- type I

$$\Delta Z(p)=A_1\cdot\frac{\omega_1}{p^2+\omega_1^2}+A_2\cdot\frac{\omega_2}{p^2+\omega_2^2}; \tag{2}$$

- type II

$$\Delta Z(p)=A_1\cdot\frac{\omega_1}{p^2+\omega_1^2}. \tag{3}$$

Here A_1, A_2 and ω_1, ω_2 are the amplitudes and frequencies of the first and second harmonics of the interference signal, respectively.

The conducted studies of the existing BEP ACS, taking into account the adopted model of the feedback channel [2], showed that the error ΔZ, summed up with the reference signal Z_r, causes a mismatch between the position Z of the blade and the required elevation position Z_r even under the conditions of a correctly selected regulator. Furthermore, the maximum deviation from the nominal value of Z_r of the position of working body reaches 25 mm, which substantially exceeds the allowable indicators set by regulations.

3. Mathematical model of ACS of the hydraulic drive of a motor grader blade taking into account the correction device

To improve the accuracy of the automatic control system for the elevation position of the blade (BEP ACS) in conditions of large errors in the operation of the GNSS, it is proposed to use the technology of dynamic correction. Its principle is to periodically use a more accurate measuring device (alternative AS_Z sensor) and calculate the difference between its reading Z_{AS} and the signal setting Z_r at certain time intervals T_i:

$$\Delta_i=Z_r-Z_{AS}(T_i). \tag{4}$$

Subsequently Δ_i is entered into the control system as a correction to the current measurements.

A functional diagram of the BEP ACS with a correction device has been developed (Fig. 2). A distinctive feature of the proposed approach to the construction of BEP ACS, different from the traditional one, is the discreteness of the calculation of the correction in combination with the use of a memory element, the function of which is the formation of a correcting signal in the time interval between connections to the reference device for measuring BEP. The approach under consideration makes it possible to carry out the technical implementation of the dynamic correction system (DCS), based on various types of sensors for measuring the spatial position of objects.

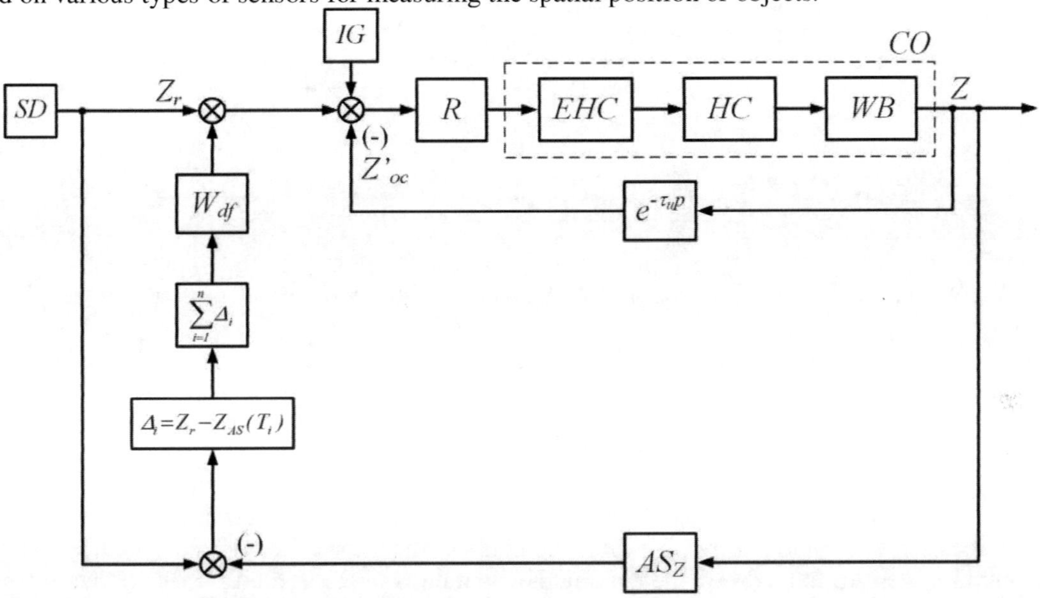

Figure 2. Functional diagram of the BEP ACS with a correction device: SD - setting device; R - regulator; EHC - electrohydraulic converter; HC - hydraulic cylinder; WB - working body (blade); IG - interference generator; AS_Z - alternative sensor; W_{df} - dynamic filter.

4. Studying the influence of the discreteness of information obtained from an alternative sensor of the position of the motor grader blade on the ACS error

To assess the influence of the discreteness of data obtained from an alternative blade position sensor on the error of the BEP ACS in the Matlab software environment, its computational model was developed (Fig. 3). Here, the hydraulic drive model developed in [3-5] is used as a control object (CO). The correction device is modeled by a combination of a comparator, an adder, and Zero-Order Hold and Memory blocks. The reference signal is fed to the direct input of the comparator, and the signal from the alternative BEP sensor AS_Z is fed to its inverse input. The memory block is modeled by elements of the Simulink library of the Matlab program – an adder, "Zero-Order Hold" and "Memory".

When simulating the operation of the BEP ACS with a correction device, it is assumed that the movement of the motor grader when leveling the roadway is constant and υ_{mg} = 1 m/s, and the sampling step Δt of the difference signal Δ_i is determined by:

$$\Delta t = \frac{\Delta L}{\upsilon_{mg}}$$

where ΔL is the distance between alternative sensors.

During the research, the following options for the location of alternative sensors were taken $\Delta L \in 10, 25, 50, 75, 100, 150$ m. Then the sampling step, in accordance with (5), is $\Delta t \in 10, 25, 50, 75, 100, 150$ s.

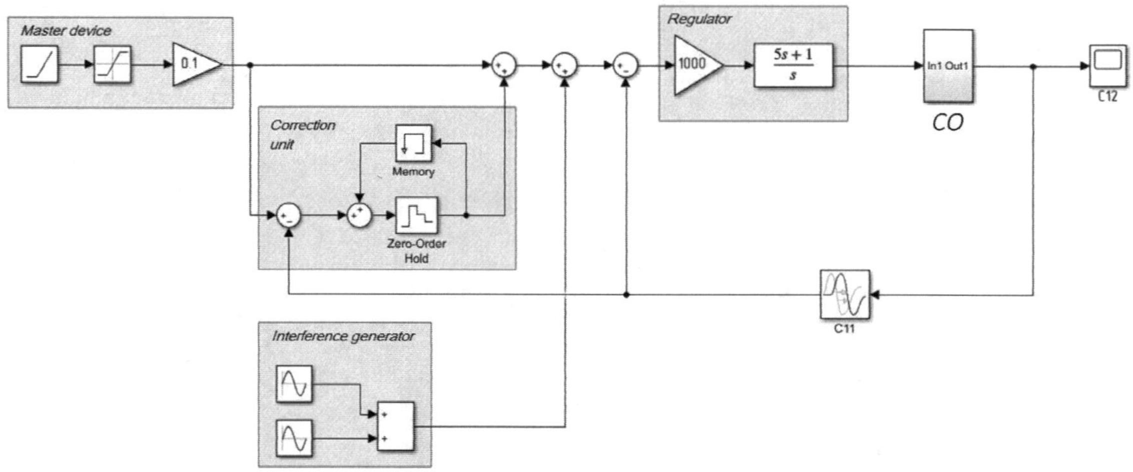

Figure 3. Computational model of BEP ACS taking into account the correction device

Taking into account the accepted conditions mentioned above, it is proposed to conditionally divide computational experiments on the study of performance indicators of BEP ACS into two stages. At first stage we take into account the interference of the inaccurate operation of GNSS of the first type (2), at second stage - the type II interference (3).

To analyze the experimental results obtained at the first stage, let us select the most characteristic curves at Δt = 10, 50, 100 s, respectively (Fig. 4). Comparison of those presented in Fig. 4 characteristics shows a natural decrease in the magnitude of the error with a decrease in the magnitude of the sampling step Δt. In this case, the maximum deviation is 0.022 m (at Δt = 100 s), the minimum is 0.002 m (at Δt = 10 s).

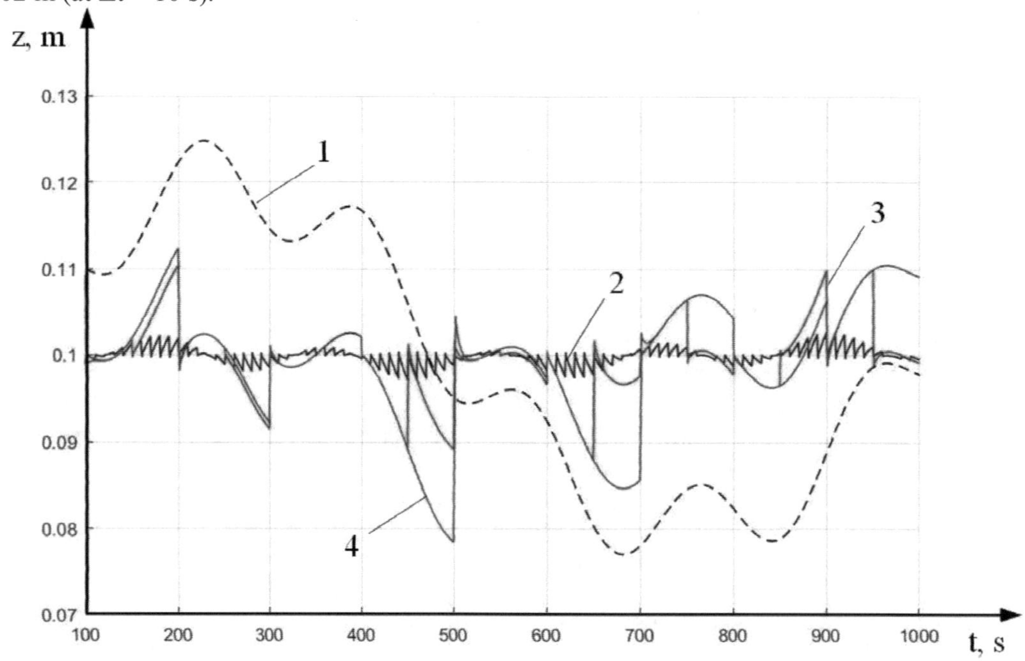

Figure 4. The graph of the change in the position of the working body under the action of type I GNSS interference: 1 - in a system without a correction device; 2, 3, 4 - with a correction device at Δt = 10, 50, 100 s, respectively

IOP Conf. Series: Materials Science and Engineering 1103 (2021) 012021 doi:10.1088/1757-899X/1103/1/012021

Analysis of the results of computational experiments involving the second type of interference showed that, as with the type I interference, a decrease in the error value is observed with a decrease in the sampling step. The maximum deviation is observed at Δt = 150 s and is 0.018 m, the minimum error (0.0005 m) corresponds to Δt = 10 s.

Conclusion

Summing up what has been said, the developed method for the dynamic correction of the BEP ACS of the motor grader and the analysis of the results of the experimental studies of the accuracy of the blade position on the created computational model of the ACS, in which the GNSS receiver is used. We showed satisfactory results and the expediency of its practical application to improve the accuracy of the elevation position of the motor grader blade.

References

[1] Kozlov, A V 2010. *Some Methods and Algorithms for Determining the Position Using Inertial and Satellite Navigation Systems* (Moscow: Moscow State University named after M.V. Lomonosov)
[2] Galitskov S Ya, Lukyanov A S and Fadeev A S 2019 Analysis of the Causes of Errors in the Elevation Positioning of a Motor Grader Blade by an Automatic Control System Based on the Receiver of the Global Navigation Satellite System *Mechanization and Automation of Construction,* S Ya Galitskov (ed) (Samara: SamSTU) pp 119-127
[3] Kononykhin B D 1990 *Laser Control Systems for Road Construction Machines* (Moscow: Mechanical engineering)
[4] Shcherbakov V S, Belyaev N V and Belyaev V V 2009 *Automation of Preliminary Technical Design of a Motor Grader* (Omsk: Publishing house SibADI)
[5] Shcherbakov V S, Lazuta I V and Denisova E F 2012 *Automation of the Design of the Main Parameters of the Control Device for the Working Body of the Bulldozer Unit* (Omsk: Publishing house SibADI)

IOP Conf. Series: Materials Science and Engineering 1103 (2021) 012022 doi:10.1088/1757-899X/1103/1/012022

Increase in corrosion resistance of carbon steel surface due to plasma surface modification with powder mixtures of NiCrBSi alloy and chrome carbide

Van Chie Nguyen[1], **A E Balanovskiy**[1], **A N Baranov**[1], **E A Guseva**[1], **V V Kondrat'ev**[1], **M G Shtayger**[2] **and A I Karlina**[1]

[1] Irkutsk National Research Technical University, 83, Lermontova Street, Irkutsk, 664074, Russia
[2] MC Mechel Steel, 1, Krasnoarmeysky Street, Moscow, 125167, Russia

E-mail: karlinat@mail.ru

Abstract. The studies assess the possibility of increasing the corrosion resistance, wear resistance of the surface of steel St3 using the method of plasma surface modification. The addition of OK 84.78 electrode coating powder in mixtures containing NiCrBSi leads to the increase of surface hardness. It is shown that the higher the content of OK 84.78 coating powder in the modifying mixture, the higher the probability of occurrence of defects in the fused surface layer (cracks, pores, heterogeneity). Studies of corrosion processes in a 3% NaCl solution show that coatings containing alloyed elements Cr, Ni have a higher passivation in a corrosive environment. NiCrBSi + 20% OK 84.78 coating is considered to be high corrosion-resistant.

1. Introduction

In recent years, the type of hardening of corrosion-resistant coatings based on Ni alloy has been well studied [1, 2]. In order to achieve the expected coating properties, there has been a trend towards adding other additional components like WC, TiC, SiC, which are successfully used in various surface treatments to obtain multicomponent coatings. [3-6]. Chromium carbides (CrxCy) advantages are chosen due to their popularity, low cost and high hardness. Chromium carbide is a potentially suitable ceramic phase for use in cermet due to the excellent oxidation resistance of its three polymorphs - cubic ($Cr_{23}C_6$), hexagonal (Cr_7C_3) and rhombic (Cr_3C_2).

Considering that the radius of chromium and iron is almost the same, therefore it diffuses more easily into steel. The combination of Ni-based alloy and CrxCy additive is found in the using of laser cladding [5], vacuum brazing [7], plasma spraying [8], HVOF [9]. NiCrBSi is a self-fluxing nickel-based alloy that is widely used for surface modification using highly concentrated sources of heating [10-12]. The presence of alloyed elements Ni, Cr leads to passivation in corrosive environments due to the formation of their oxides.

Several studies have noted that the coatings of the NiCrBSi system have high corrosion resistance [13, 14]. In their work [15], Shabana and his colleagues obtained two types of NiCrBSi, Cr_3C_2-NiCr coatings by thermal spraying (HVOF). Corrosion tests carried out in a 3.5% NaCl solution showed that the corrosion resistance of NiCrBSi was slightly higher than coating Cr_3C_2-NiCr. The authors believe that Cr_3C_2-NiCr can be used for applications where corrosion and wear are of paramount importance. In an environment with a 5% solution of $H_2(SO)_4$, Cr_3C_2-NiCr, the coating showed better corrosion resistance behavior than NiCrBSi [16]. Corrosion study [17] in 3.5% NaCl solution showed that using the HVAF-Sprayed method, the resulting NiCr coating has a more negative potential than the Cr_3C_2-NiCr coating. The pores and carbides inherent in Cr_3C_2-NiCr coatings significantly reduce corrosion resistance, the former by providing preferred paths for ion diffusion, and the latter by the formation of cathode sites in galvanic vapors (between NiCr and Cr_3C_2).

Previously, coatings were obtained in the heating of modification mixtures containing NiCrBSi by an electric arc in a protective gas [18]. It is shown that high-quality coatings are obtained with a thin powder pre-layer and only in the mode of reflow of the main substrate, but their hardness is not very high. The method of surface modification of metals using plasma heating has long been used in many industries [19, 20]. In this method, successful, coatings are obtained by chromium plating, carburizing and chromium-nickel plating, but the production of plasma coatings based on a mixture of NiCrBSi and chrome carbide has not yet been sufficiently studied. It is known that the OK 84.78 welding electrode is widely used for surfacing of hardening layers operating under conditions of intense abrasive wear. Its matrix is made of austenitic cast iron saturated with coarse chromium carbides. The weld metal also has high corrosion resistance in contact with corrosive media and scale resistance.

The purpose of this work is to evaluate the corrosion resistance of a plasma coating made of powder mixtures of alloy NiCrBSi and an OK 78.84 electrode coating.

2. Materials and methods of study

2.1. Materials and methods

The chemical composition of steel St3 consists of C (0.14-0.22%), Si (0.05-0.17%), Mn (0.4-0.65%), Ni, Cu, Cr (≤ 0, 3%), As (≤ 0.08%), S, P (≤ 0.05%). For preparing the paste, the following powders were used: alloy PR-N80X13S2R with granules of 40 - 100 microns; electrode OK 84.78; office glue (silicate). The chemical composition of the alloy PR-N80X13S2R, OK 84.78 and their content in the initial compositions are shown in Table 1.

Table 1. Percentage of individual elements in different mixtures.

	1	2		3		4		5		6
el.	**NiCrBSi 100%**	**NiCrBSi 90%**	**OK 78.84 10%**	**NiCrBSi 80%**	**OK 78.84 20%**	**NiCrBSi 70%**	**OK 78.84 30%**	**NiCrBSi 60%**	**OK 78.84 40%**	**OK 78.84 100%**
Ni	80	72		64		56		48		40
Cr	14	20.87		27.74		34.61		41.48		48.35
Fe	1.8	1.62		1.44		1.26		1.08		0.9
B	1.5	1.35		1.2		1.05		0.9		0.75
Si	2.4	2.36		2.32		2.28		2.24		2.2
C	0.3	1.4		2.5		3.6		4.7		5.8
Mn	0	0.38		0.76		1.14		1.52		1.9

The preparation of the surface coating consisted of applying a paste from the mixtures according to Table 1 using glue, drying at 50-60 °C for 1 hour. After drying, the coating thickness of all samples was about 0.25 mm. Plasma treatment modes: current 140A, consumption of argon 10 l/min; the gap between the tungsten electrode and the surface of substrate is 5 mm; the moving speed of the samples is 2.7 mm/s.

The samples processed after plasma modification were cut using a "Polilab P100A" machine and then pressed using a "Polilab S50A" press. The surface of the plasma samples was polished on a "Polilab P12M" machine to shine, after that, they continued to poison in a 1% solution of nitric acid with ethanol. Observing their microstructure using an optical microscope MET-2. The microhardness was measured on a device of hardness testers of the HMM - G series.

2.2. Corrosion control method
To control the corrosion behavior of the test samples, a 3% NaCl solution was used. The steel was preliminarily etched in hydrochloric acid to a pure metal, then washed in distilled water and degreased with ethyl alcohol 99%. The coated samples were polished with 1000 grit Eagle waterproof abrasive paper, then washed with distilled water and finally ethyl alcohol. Before measurements, the area of one cm^2 was fixed on the surface. The remaining surface was covered with insulating tape and waterproof glue so that the shielded surface were not exposed to the electrolyte. The prepared samples were placed in a 500 ml beaker so that the selected area was completely immersed in the working solution. Used potentiostat-galvanostat "PI-50-Pro" for corrosion control. Operating mode: Starting potential -1400 mV; Maximum potential 100 mV; Minimum potential -1500 mV; final potential 100 mV; current range 2000 mA; potential range 50 V; sweep speed 26.932 mV/s. Work with the PS_Pack2 program according to the "Linear sweep" experiment type.

3. Results and its discussion

3.1. Study of the microstructure and microhardness of the obtained coatings
After plasma treatment of the samples, the cross-sectional views of the surface coating are shown in Table 2.

It can be seen that high-quality coatings are No. 1, 2, 3, 6, and No. 4, 5, which have several defects (cracks, pores, heterogeneity). Coating No. 1 has a great depth, the others have a depth of about 850-900 microns. The large difference in depth between coating No. 1 and the rest is explained by the fact that the NiCrBSi alloy is more ductile than the OK 84.78 powder alloy, which has more hard phases (carbides). This prevents heat transfer to the inner sublayers, because of which the substrate dissolves less and the bath has a small volume.

At the same time, the coating with a high chromium content (48.35%) in the original composition No. 6, having the radius of chromium and iron is almost the same, therefore, the composition of OK 84.78 is more saturated in the steel substrate. Other coatings containing the compositions OK 84.78 and NiCrBSi bring inhomogeneous matrices of coating than single alloys (NiCrBSi or OK 84.78), even the presence of NiCrBSi still adds various elements, and therefore these compositions complicate the dilution of the substrate and the diffusion of alloyed elements. As a result, it is obvious that thin sublayers appear on the surface in coatings No. 1, 2, and cracks in others. The microhardness values of the coatings are presented by their distribution in Figure 1.

Table 2. Characteristics of the cross-section of coatings.

№	Initial composition	Type of coatings	Note
1	NiCrBSi		Uniform weld pool Homogeneous layer Depth 1300 microns
2	NiCrBSi + 10% OK 84.78	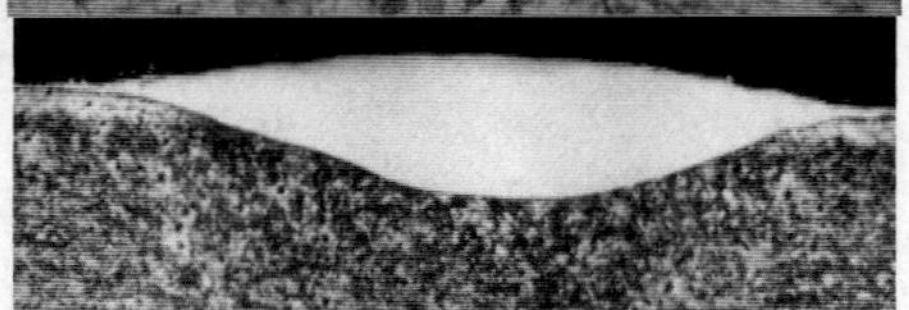	Uniform weld pool Homogeneous layer Depth 900 microns
3	NiCrBSi + 20% OK 84.78		Uneven weld pool Homogeneous layer Depth 850 microns
4	NiCrBSi + 30% OK 84.78	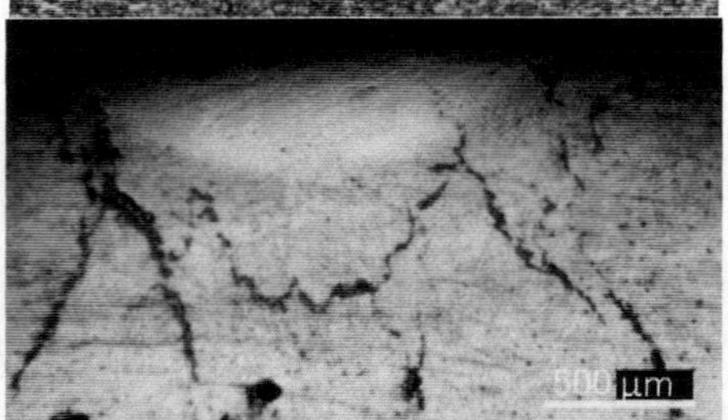	Pores, Cracks Inhomogeneous layer
5	NiCrBSi + 40% OK 84.78	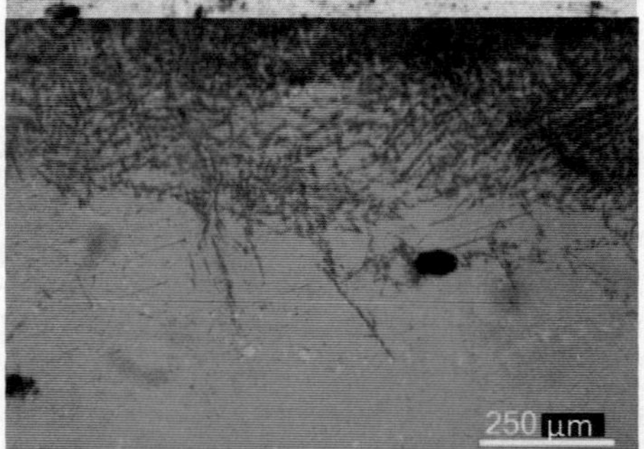	Pores Inhomogeneous layer
6	OK 84.78	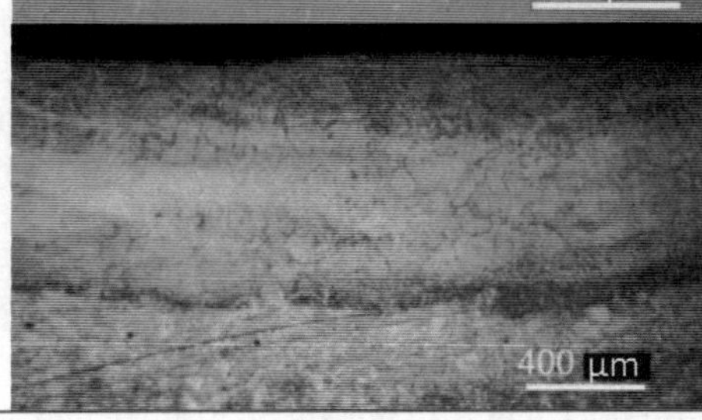	Homogeneous layer Depth 910 microns

IOP Conf. Series: Materials Science and Engineering 1103 (2021) 012022 doi:10.1088/1757-899X/1103/1/012022

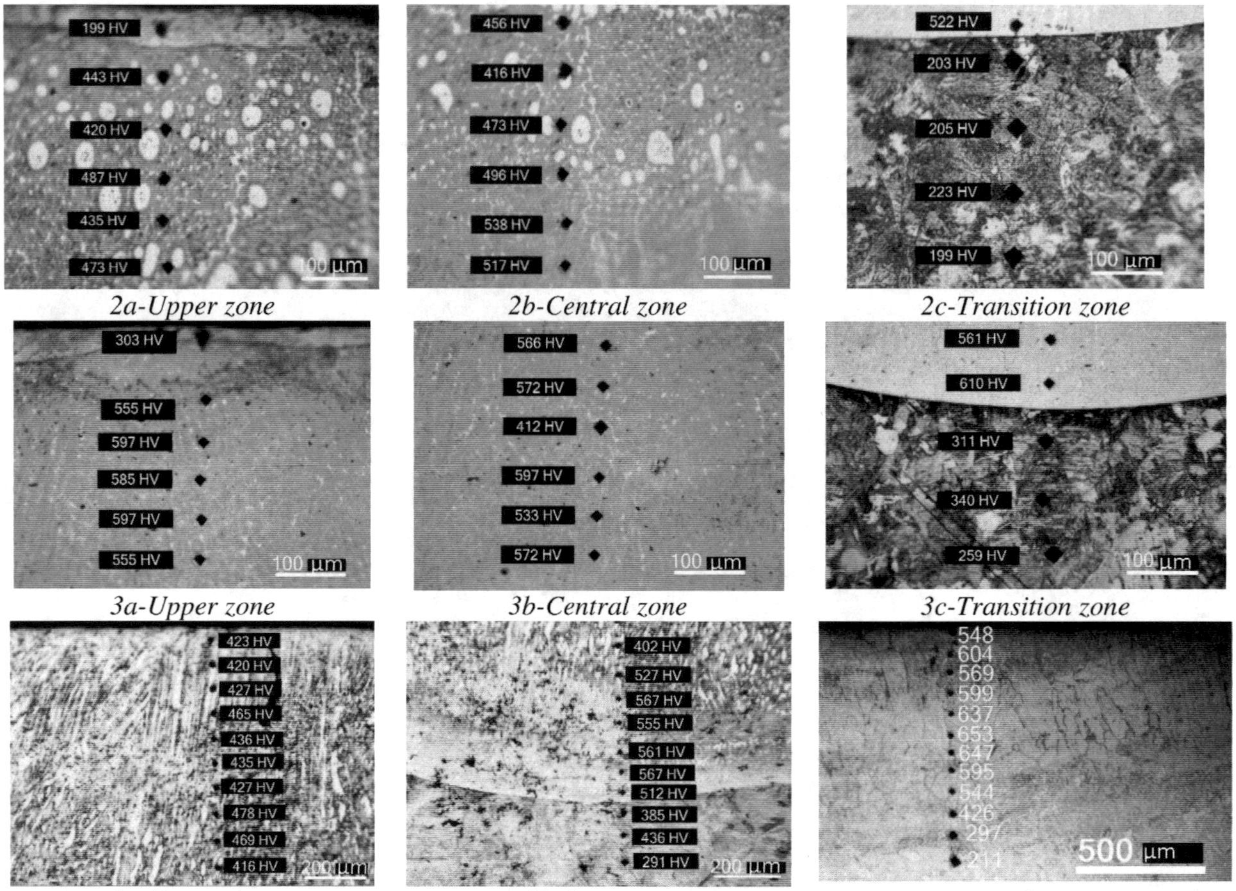

2a-Upper zone *2b-Central zone* *2c-Transition zone*

3a-Upper zone *3b-Central zone* *3c-Transition zone*

1a-Upper zone *1b-Low zone* *6*

Figure 1. Depth distribution of the microhardness of coatings: 1- NiCrBSi; 2- NiCrBSi+10% OK 84.78; 3- NiCrBSi+20% OK 84.78; 6- OK 84.78.

Coatings No. 2, 3 have thin sublayers (thickness about 50 microns). In these sublayers, the structure differs from the rest of the coverage areas. If the first points of hardness measurement on coatings No. 2, 3 (199 HV and 303 HV) are not counted, then the dispersion of the microhardness values of the main coating zone is presented by the diagram in Figure 2. It can be seen that in coating No. 2 the range of the confidence interval of measurement errors is narrow, i.e. the microhardness values are concentrated, and the average measurement values coincide with the median and close to the average value of the microhardness of coating No. 1.

This can be explained by the fact that the addition of 10% OK 84.78 does not increase the hardness, but with an increase in the chromium and carbon content and a corresponding decrease in the nickel content, which leads to a decrease in the formation of the γ-Ni phase, the hardness of which is lower than that of other phases in the coating. At the same time, the concentration of hardness values of coating No. 3 is lower, but most of its values (75%) are higher than the maximum for coating No. 2. All hardness values of coating No. 3 are distributed with a similar concentration as for coating No. 1, but more than 50% of the value is higher than the maximum for coating No. 1. This means that an increase in the weight of OK 84.78 by another 10% can noticeably increase the hardness of the surface coating, i.e. the role of chromium and carbon in the formation of harder phases with an increase in their content was confirmed. When using only the coating composition of OK 84.78, the resulting coating has high microhardness values in the range of 300-650 HV. Low values (297 and 426 HV) at the last points near the coating / substrate boundary may be caused by insignificant diffusional carbon. Because of these points, the concentration of microhardness values becomes the lowest.

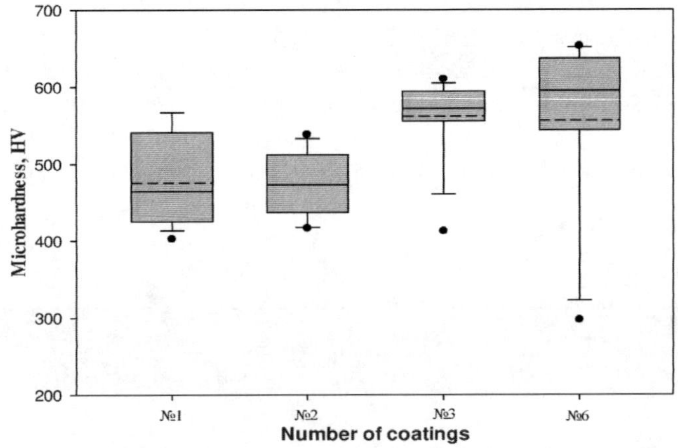

Figure 2. Dispersion of microhardness values of the cross-section of coatings: No. 1 - NiCrBSi; No. 2 - NiCrBSi + 10% OK 84.78; No. 3 - NiCrBSi + 20% OK 84.78; No. 6- OK 84.78.

3.2. *Study of corrosion behaviour in 3% NaCl solution*

For each alloy and St3 steel, the corrosion control was repeated 6 times. The dependences of the current (mA) on the potential (mV) relative to the silver chloride reference electrode are shown in Figure 4. The experimental error for St3 steel is from 1.0 to 10.3%, for NiCrBSi, OK 84.78, NiCrBSi + 10% OK 84.78, NiCrBSi + 20% OK 84.78 coatings from 0.32 to 9.03%, and for NiCrBSi coatings + 30% OK 84.78, NiCrBSi + 40% OK 84.78 from 0.4 to 15.5%. The dependence lines for all coatings have a section in the middle, parallel to the horizontal axis, and then the corrosion current sharply increases for St3 steel at a potential of -500 mV, for the rest, the potential increases to -300, -200 mV. The shape of these lines proved that all coatings have a higher corrosion resistance and the highest resistance for the alloy NiCrBSi + 20% OK 84.78.

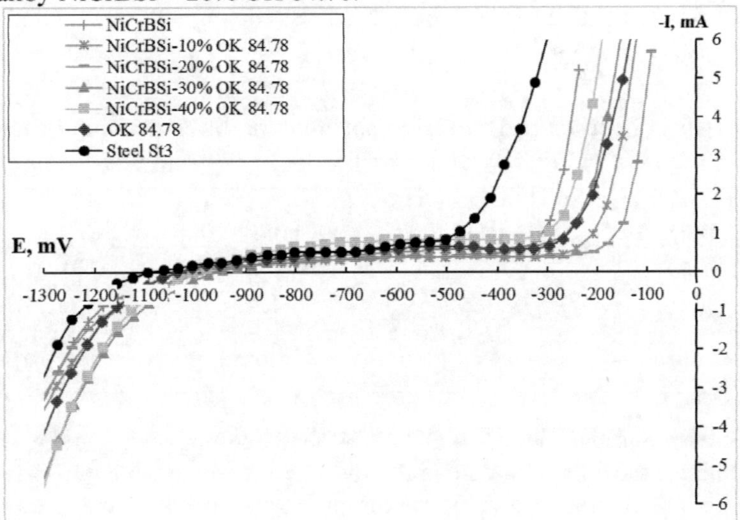

Figure 3. Dependencies of the current (mA) on potential (mV) relative to the silver chloride reference electrode for NiCrBSi, NiCrBSi-10% OK 84.78, NiCrBSi-20% OK 84.78, NiCrBSi-30% OK 84.78, NiCrBSi-40% OK 84.78, 100% OK 84.78 and steel St3.

IOP Conf. Series: Materials Science and Engineering 1103 (2021) 012022 doi:10.1088/1757-899X/1103/1/012022

Typical polarization curves are shown in Figure 4. This graph gives examples of the line of voltage versus logarithm of current density for St3 steel and NiCrBSi + 10% OK 84.78 coating. It can be seen that the coating based on a mixture of NiCrBSi and 10% OK 84.78 is more passive against the corrosive environment than steels. The coating passivation line will extend from -1000 to -300 mV, and for steel, it slopes earlier from the horizontal. A typical graph represents the role of alloying elements, especially chromium. Its presence leads to an increase in corrosion resistance due to the formation of oxides, and so will greatly reduce the current density. However, for a significant comparison between them, the construction of an approximating curve with the value of the reliability of the approximation gives the current and the corrosion potential. The result is shown in Table 3.

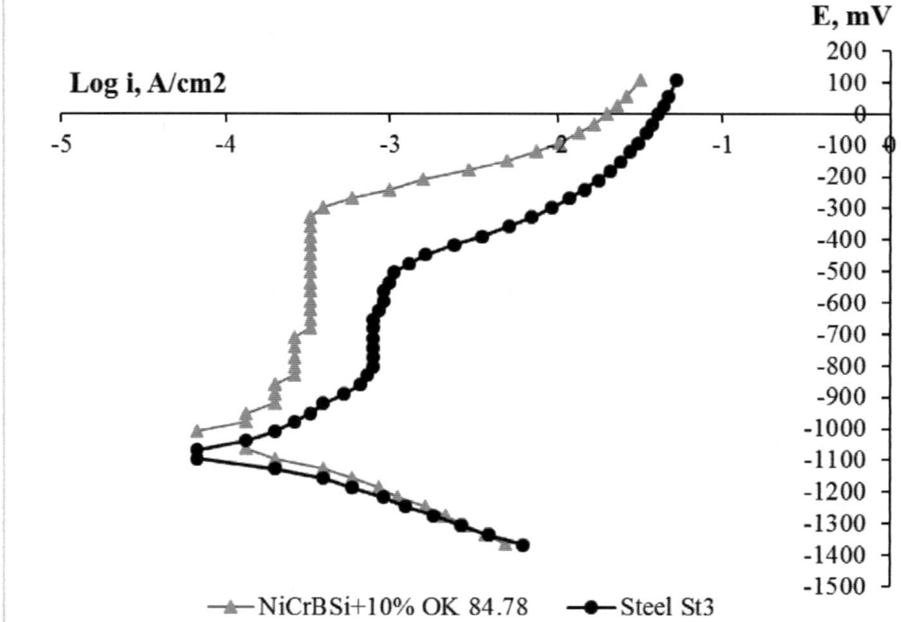

Figure 4. Voltage dependencies on logarithm of current density for anodic and cathodic polarization curves for a St3 steel sample and a NiCrBSi coating + 10% OK 84.78 in a 3% NaCl solution.

Table 3. Average value of current and corrosion potential for different compositions and steel St3.

	1	2		3		4		5		6	7
Val.	**NiCrBSi**	**NiCrBSi**	**OK 78.84**	**NiCrBSi**	**OK 78.84**	**NiCrBSi**	**OK 78.84**	**NiCrBSi**	**OK 78.84**	**OK 78.84**	**St3**
	100%	90%	10%	80%	20%	70%	30%	60%	40%	100%	
I, A/m^2	1.082	1.167		1.180		1.420		1.776		1.258	1.304
E, mV	-985	-957		-952		-931		-950		-1014	-1088

First, with an increasing in the content of OC 84.78, i.e., a corresponding increase in the content of chromium in the coatings, the corrosion potential moves in a more positive, but from 30% to 100% in a more negative value. For coatings No. 2, 3, the current and corrosion potential does not differ much, but diverges for coating No. 4, 5. A strong change can be caused by defects, including pores, cracks and heterogeneity, because of which it leads to pitting corrosion. Coating No.1 has the lowest current, but the corrosion potential is more negative than multicomponent coatings.

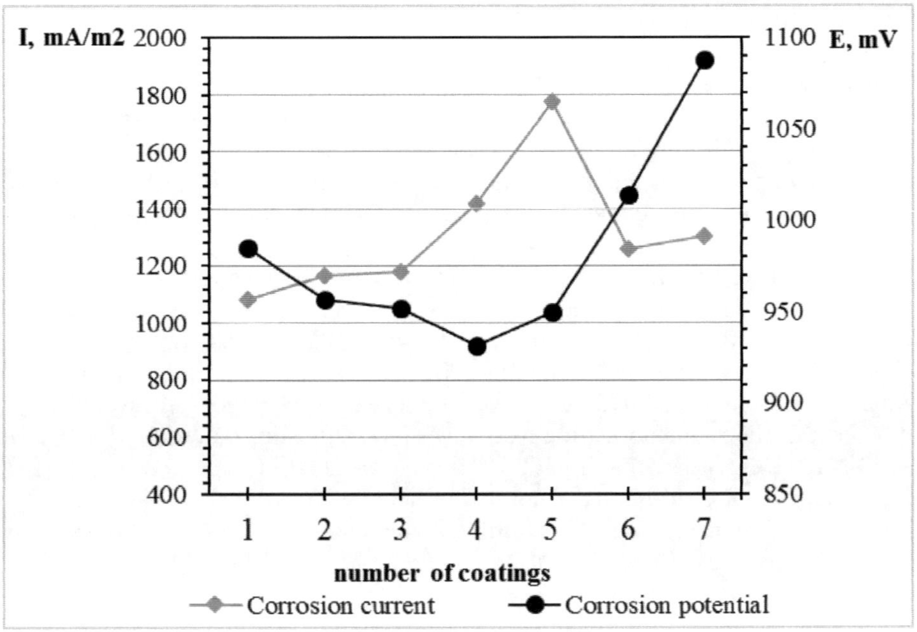

Figure 5. Distribution of average values of corrosion current and potential by composition: 1 – NiCrBSi; 2 – NiCrBSi/OK 84.78 (90%-10%); 3 – NiCrBSi/OK 84.78 (90%-10%); 4 – NiCrBSi/OK 84.78 (80%-20%); 5 – NiCrBSi/OK 84.78 (70%-30%); 6 – NiCrBSi/OK 84.78 (60%-40%); 7- St3.

4. Conclusion

The formation of coatings from a series of mixtures NiCrBSi and OK 85.78 in the plasma-heating mode was established: power 140 A, sample speed 2.7 mm / s, gas flow rate 10 l / min). The depth of the NiCrBSi coating is 1300 μm, the others have a thickness of about 800-900 μm.

The study of the microstructure has shown that for the same coating thickness, single alloys are more saturated inside the steel. With the increase in the content of OK 85.78, i.e. chromium, carbon, defects (pores, cracks, heterogeneity) appear in the coatings, especially with compositions of 30, 40%.

Microhardness measurements showed that there is a possibility of increasing the hardness of the NiCrBSi coating with the addition of OK 84.78. The spread of microhardness values with a composition of 10% OK 84.78, the values of hardness are more concentrated near the average value, and with its 20% in the composition, it is much higher than that of the coating NiCrBSi and (NiCrBSi + 10% OK 84.78).

When studying corrosion in a 3% NaCl solution, the addition of OK 84.78 to the main mixture leads to a decrease in the corrosion current, except for coatings that have defects. With compositions having 10, 20% OK 84.78, the difference in the corrosion current between them and the NiCrBSi coating is not large. According to the value of the microhardness of the cross-section and the corrosion current strength, the coating of the mixed composition (NiCrBSi + 20% OK 84.78) is more corrosion-resistant.

References

[1] Jing Yu and Bo Song 2018 Friction and wear behavior of a Ni-based alloy coating fabricated using a multistep induction cladding technique *Results in Physics* **11** 105-111. DOI: https://doi.org/10.1016/j.rinp.2018.08.049

[2] Wen Sun, Ayan Bhowmik, Adrian Wei Yee Tan, Fei Xue, Iulian Marinescu, Feng Li and Erjia Liu 2019 Strategy of incorporating Ni-based braze alloy in cold sprayed Inconel 718coating *Surface & Coatings Technology* **358** 1006-1012. DOI:

https://doi.org/10.1016/j.surfcoat.2018.12.050

[3] Saeedi R, Shoja-Razavi R, Bakhshi S-R, Erfanmanesh M and Ahmadi-Bani A 2020 Optimization and characterization of laser cladding of NiCr and NiCr–TiC composite coatings on AISI 420 stainless steel *Ceramics International* DOI: https://doi.org/10.1016/j.ceramint.2020.09.284.

[4] Bhosale D G, Rathod W S and Rukhande S W 2020 Effect of counter faces on sliding wear behavior of WC-Cr3C2-Ni composite coating deposited by high velocity oxy fuel *Materials Today: ProceedingsIn press* 1-6 DOI: https://doi.org/10.1016/j.matpr.2020.08.466

[5] Peng Wang, Shuang Li, Xin Geng, Jipeng Zhang and Chuncheng Wei 2020 Microstructure, surface stress and surface temperature response of ZrB2–SiC based coatings *J. of Alloys and Compounds* **843** 156084 DOI: https://doi.org/10.1016/j.jallcom.2020.156084

[6] Mao-Sheng Yang, Xiu-Bo Liua, Ji-Wei Fan, Xiang-Ming He, Shi-Hong Shi, Ge-Yan Fu, Ming-Di Wang and Shu-Fa Chen 2012 Microstructure and wear behaviors of laser clad $NiCr/Cr_3C_2$–WS_2 high temperature self-lubricating wear-resistant composite coating *Applied Surface Science* **258** 3757-3762. DOI: https://doi.org/10.1016/j.apsusc.2011.12.021

[7] De Wang, Yujiang Xie, Yanhong Yang, Mingsheng Wang, Wenqin Wang, Xiangping Cheng and Deping Lu 2016 Influence of Cr addition on microstructure of vacuum brazed NiCr – Cr_3C_2 composite coatings *Materials Characterization* **115** 46-54. DOI: https://doi.org/10.1016/j.matchar.2016.03.016

[8] Liua Q, Wang Y, Bai Y, Li Z D, Tan G L, Bao M Y, Li X J, Zhan H, Sun Y W, Chong N J, Wang R J and Ma Y S 2020 Formation mechanism of gas phase in supersonic atmospheric plasma sprayed NiCr-Cr_3C_2 cermet coatings *Surface & Coatings Technology* **397** 126052. DOI: https://doi.org/10.1016/j.surfcoat.2020.126052

[9] Shankar R, Balasubramanian K R and Sivapirakasam S P 2020 Hot corrosion behavior of nanostructured and conventional HVOF Cr_3C_2 – NiCrBSi coatings on super alloy *Materials Today: Proceedings In press, corrected proof.* 1-8. DOI: https://doi.org/10.1016/j.matpr.2020.04.005

[10] Meng J, Shi X, Zhang S, Wang M, Xue F, Liu B, Cui W and Bian L 2019 Friction and wear properties of TiN-TiB2-Ni based composite coatings by argon arc cladding technology *Surf. Coat. Technol.* **374** 437-447

[11] Makarov A V, Soboleva N N, Malygina I Yu and Kharanzhevskiy E V 2019 Improving the properties of a rapidly crystallized NiCrBSi laser clad coating by high-temperature processing *J. of Crystal Growth* **525** 125200. DOI: https://doi.org/10.1016/j.jcrysgro.2019.125200

[12] Rachida Rachidi, Bachir El Kihel and Fabienne Delaunois 2019 Microstructure and mechanical characterization of NiCrBSi alloy and NiCrBSi-WC composite coatings produced by flame spraying *Materials Science & Engineering B* **241** 13-21. DOI: https://doi.org/10.1016/j.mseb.2019.02.002

[13] Zhaia L L, Bana C Y and Zhangb J W 2019 Microstructure, microhardness and corrosion resistance of NiCrBSi coatings under electromagnetic field auxiliary laser cladding *Surface & Coatings Technology* **358** 531-538. DOI: https://doi.org/10.1016/j.surfcoat.2018.11.034

[14] Serres N, Hlawka F, Costil S, Langlade C and Machi F 2011 Corrosion properties of in situ laser remelted NiCrBSi coatings comparison with hard chromium coatings *J. of Materials Processing Technology* **211** 133-140. DOI: https://doi.org/10.1016/j.jmatprotec.2010.09.005

[15] Shabana, Sarcar M M M, Suman K N S and Kamaluddin S 2015 Tribological and Corrosion behavior of HVOF Sprayed WC-Co, NiCrBSi and Cr3C2-NiCr Coatings and analysis using Design of Experiments *Materials Today: Proceedings* **2(4-5)** 2654-2665

[16] Sidhu T S, Prakash S and Agrawal R D 2007 A Comparative Study of Hot Corrosion Resistance of HVOF Sprayed NiCrBSi and Stellite-6 Coated Ni-Base Super alloy at 900°C *Materials Science & Engineering A* 210-218

[17] Sadeghimeresht E, Markocsan N and Nyle´n P 2017 Microstructure Effect of Intermediate Coat Layer on Corrosion Behavior of HVAF-Sprayed Bi-Layer Coatings *J. Therm. Spray Tech.* **26**

243-253
[18] Nguyen Van Trieu 2020 Increasing the hardness of the surface layer of low-carbon steel due to plasma treatment of the modifying coating *ISTU Bulletin* **24(1)** 52-63
[19] Balanovskii A E 2019 New mechanism of interaction between a welding-arc discharge of reverse-polarity direct current and an aluminum surface *High Temperature* **57(6)** 784–797
[20] Balanovskii A E and Vu V H 2017 Plasma surface carburizing with graphite paste *Letters on materials* **7(2)** 175-179

IOP Conf. Series: Materials Science and Engineering 1103 (2021) 012023 doi:10.1088/1757-899X/1103/1/012023

Design of prefabricated reinforced concrete structures: comparative analysis of prefabricated reinforced concrete floor slab

A V Evstratova[1], A K Krishtalevich[1], V V Belov[2] and S E Nikitin[1]

[1] AFRY, Jaakonkatu 3, Vantaa, 01620, Finland
2 AO Atomproject, Savuchkina 82, lit. A, Saint-Petersburg, 197183, Russia

E-mail: evstratova.a.v@yandex.ru; alexey.krishtalevich@gmail.com; vvbelov@atomproekt.com; nse85@mail.ru

Abstract. The article describes application of precast, cast-in-situ and composite structures on construction site. Importance of Analysis of Russian (SP) and European design standards (EuroCodes) is underlined. Historical review of Russian codes for composite reinforced concrete structures is given. Methods of composite structures calculating according to current Russian Codes are given. Ultimate and service limits states are listed for describing structures. Advantages and disadvantages of cast-in-situ, prefabricated and complex technologies are discussed. Comparison of prefabricated, monolithic and complex technologies is given on example of intermediate floor slab. Examples of sections of prefabricated reinforced concrete composite structures are summarized. Composite structures detailing is given on example of real project of Parking building of Okhta-Mall shopping and entertainment center, Saint-Petersburg. Constructive solutions are presented that increase the bearing capacity and resistance to progressive collapse of structures. Manufacturing photos of prefabricated ready-made reinforced concrete slabs as a part of future composite slab are presented. Examples of detail drawings for composite concrete slabs, executed in accordance to Eurocode, are given. Suggestions for improving the existing Code of Practice are given. Reference to similar scientific papers is presented related to composite reinforced concrete elements. Article is developed with financial and technical information support of AFRY Oy Finland.

1. Introduction

Currently, it is important to improve the quality of construction installation works in combination with an increase of technical safety of design solutions, a reduction of construction period and minimization of operations on the building site. One of the promising areas is construction of prefabricated reinforced concrete structures. The article provides an overview of the domestic and foreign design regulatory framework and gives recommendations for its improvement, makes a comparative analysis of precast reinforced concrete structures, offers modern design solutions and protection measures against progressive destruction. Relevance of the current topic is confirmed by wide amount of researches in field on composite concrete structures [1–8].

2. Current state of the regulatory framework

The remain in force documents of the Russian Federation of the design of precast reinforced concrete structures are the Design Codes "Precast reinforced concrete structures. “Design rules” [9] and

"Concrete and reinforced concrete structures. Guidelines" [10]. Previously, the guidelines of the calculation of the precast reinforced concrete structures were presented by Design Code [11] and Guide [12], which also contains examples of structural analysis.

The main European standards for the design of precast reinforced concrete structures are the Eurocode "Design of concrete structures. General rules and rules for buildings" [13] and the Eurocode for precast reinforced concrete structures. For example, EN 13474: 2005 + A2 applies to the design of precast slab consist of precast prestressed slabs and cast-in-situ concrete on the top of the structure. According to Eurocode's Appendix F, the strength and stability of composite slabs is determined by the adhesion of the precast elements to the cast-in-situ concrete. If shear forces are transmitted through the contact area of the precast and cast parts, the design of the floor is carried out similarly to the design of cast-in-situ slabs.

According to the Design Code [10], the calculation is presented for two stages of construction work:

— the calculation of the prefabricated element (before cast-in-situ part reach the certain strength) is carried out according to SP 63.13330, taking into account the loads of the cast part and assembling loads;
— the calculation of a precast reinforced concrete structure (after cast-in-situ part reach the certain strength) is carried out according to the recommendations of SP 337.1325800 for working loads, taking into account stresses and deformations, which arose before the cast-in-situ concrete achieve exact strength.

The guidelines of the calculation for the Design Code [10]:

— calculation of strength for normal and inclined sections (taking into account the different concrete strength of precast element and cast-in-situ part);
— calculation of the contact joint strength between the concrete of precast element and cast-in-situ part of the structure (calculation of the contact joints is based on the limiting state along the joint's surface, limited by inclined cracks);
— calculation of the formation and opening of cracks;
— calculation of the deformations.

At the same time, both in Russia and in the European Union countries, the most serious attention is paid to the problem of protecting public and residential buildings from progressive collapse [14, 15]. Science groups are working with developing of current approaches [16–24].

3. Comparative analysis of prefabricated, precast reinforced concrete and cast-in-situ structures

The choice of building technology is one of the most important stages affecting on the assembling technology of structures, quality of elements, installation speed and many other characteristics of future building. Consider the advantages and disadvantages of various technologies for construction of prefabricated, cast-in-situ and prefabricated reinforced concrete frames.

Construction experience shows that the producing of structures at the factory ensures their quality, high work mechanization, production speed and operational processes control. The advantages also include high speed and installation of structures in different weather conditions. The disadvantages of prefabricated structures are lower resistance to progressive collapse, necessity of elements transportation from the factory, lack of flexibility in case of changes in the design, arrangement of technological holes, installation of embedded parts and anchors.

Cast-in-situ frame provides freer planning solutions by using structures of complex shapes, the possibility of making changes during construction and reconstruction, higher fault-tolerance and survivability.

The disadvantages of cast-in-situ structures are their large weight, significant increase in the labor intensity of work on construction site, lack of favorable weather conditions for large amount of concrete work or necessity temporary shelters, including winter conditions. Quality control of cast-in-situ structures at the site is significantly complicated and requires more skilled workers.

IOP Conf. Series: Materials Science and Engineering 1103 (2021) 012023 doi:10.1088/1757-899X/1103/1/012023

Recently, in search of optimal solutions, prefabricated reinforced concrete technology has been increasingly used, combining the best qualities of prefabricated and cast-in-situ structures and providing a solution to the following tasks:

— implementation of the main bearing elements at the factory (comfortable working conditions, high productivity and quality of products);
— protection of the structure from progressive collapse;
— reduction of construction time by formwork decrease when prefabricated elements are used as permanent formwork (Figure 1);
— reducing weight of the structure;
— layout flexibility;
— perception of significant life loads;
— flexible execution of openings and installation of embedded parts and anchors;
— the aesthetic role of prefabricated elements in interior design.

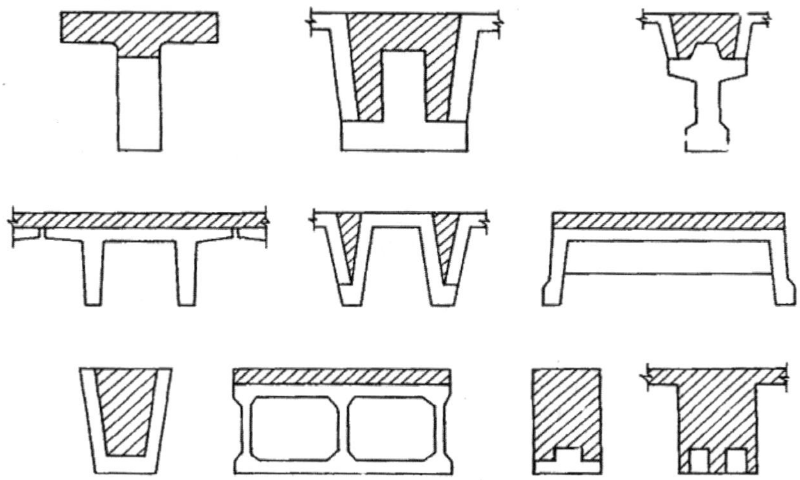

Figure 1. Examples of sections of prefabricated reinforced concrete structures (cast-in-situ part of the concrete is shaded).

4. Structural solutions for prefabricated reinforced concrete structures

Consider precast reinforced concrete floor (Figure 2), which used in the Okhta Mall Parking building in the Saint-Petersburg. Project was designed by AFRY. The floor structure is 120 mm prefabricated slab and 100 mm cast-in-situ upper part (Figure 3). Precast slab has a rough surface and starter bars into cast-in-situ part of floor to ensure their joint work. The slab rests on the cantilevers of cast-in-place poststressed beams, located with 8.4 m step. Upper reinforcement is located in cast-in-situ part of floor. It passes between adjacent slabs above the beam by that ensures joint operation of the slab as a continuous beam and protection against progressive collapse.

From an aesthetic point of view, lower part of the floor is smooth and corresponds to the A1 surface finish class in accordance with GOST 13015.2012, which does not require additional processing on the construction site. Precast slab also works as a permanent formwork that does not require installation of mounting racks. It is greatly simplifies process of cast-in-situ work. On top of the finished structure provides a finish reinforced concrete screed 70 mm. Coating allow obtaining required class of finishing of upper slab surface, to hide seams and irregularities of the casted floor and to protect the upper reinforcement over the beam.

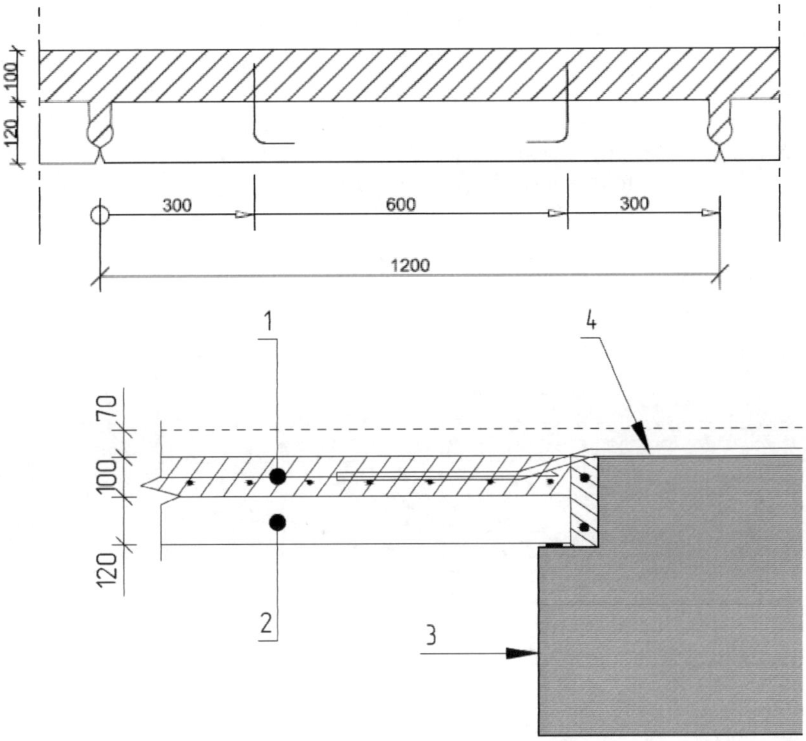

Figure 2. Scheme of prefabricated reinforced concrete floor slab of the Okhta Mall Parking: 1 – cast-in-situ part of the slab, 2 - precast slab (reinforcement is not shown conventionally), 3 - poststressed beam, 4 - reinforcement for prevent of progressive slab collapse.

Figure 3. Precast slab of the Okhta Mall Parking floor structure.

Structure of precast reinforced concrete floor by using prefabricated solid slabs was also used in construction of the Mega shopping malls in St. Petersburg, Moscow, Novosibirsk, etc.

The important requirement to connection of prefabricated reinforced concrete structures is design solutions that prevent progressive collapse. In the case of floor failure in the span area, support reinforcement is provided above the beams in the middle floor supports (Figure 2) and starter bars from wall panels or beams along to the end supports (Figure 4).

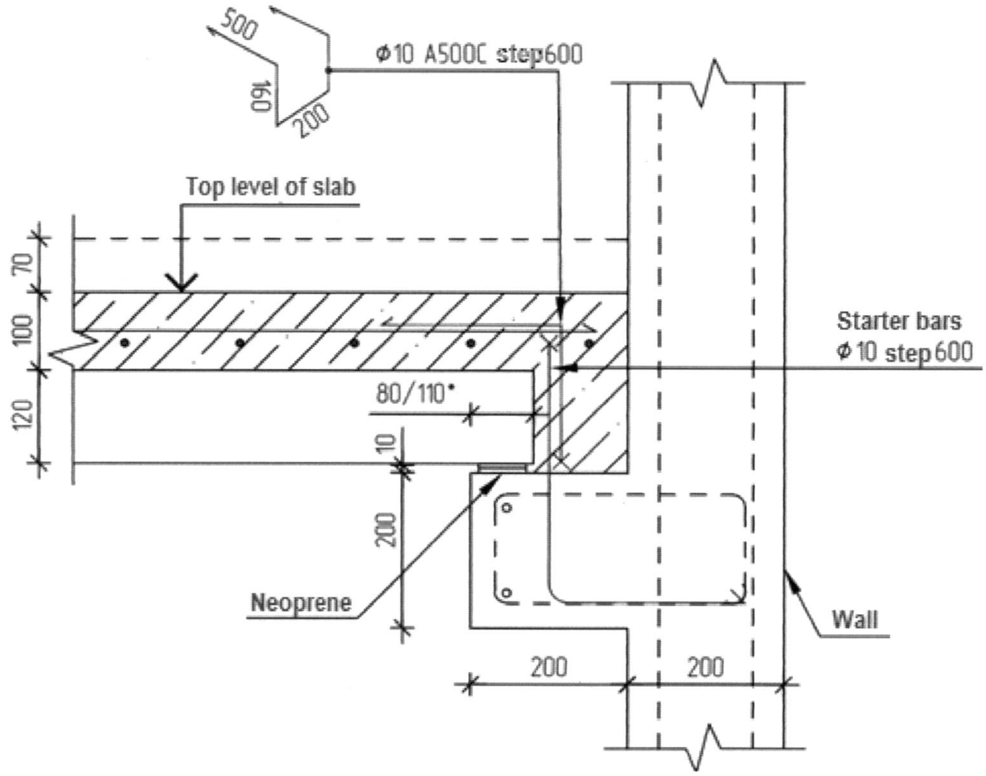

Figure 4. Scheme of end reinforcement of floor.

Collaboration of beams and floor slabs is ensured by outlets from the beam into the cast-in-situ part of the floor slab (Figure 5). The formation of the T-section further increases the bearing capacity of the beam.

If opening in the floor is required, cast-in-situ part allows installation bordering reinforcement. It provides support of clipped prefabricated slabs over the adjacent ones (Figure 6).

A nuance in the design of these precast slabs is also the use of outlets for upper monolithic part as lifting loops (Figure 7).

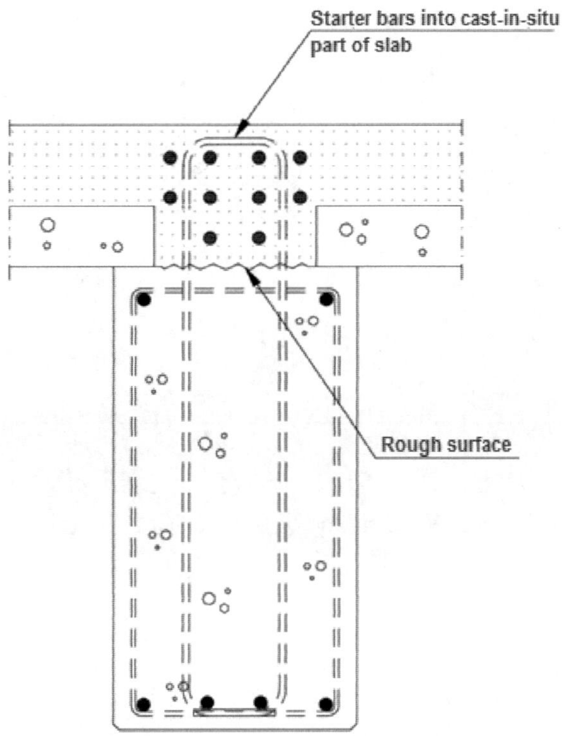

Figure 5. Scheme of reinforcement of slab and beam connection.

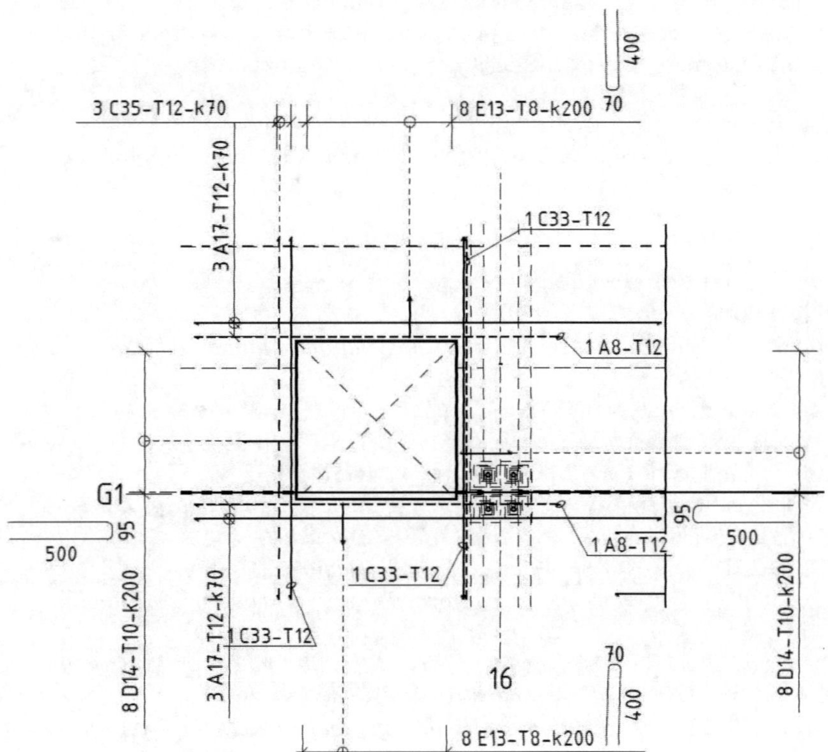

Figure 6. Scheme of reinforcement of floor openings. An example of the drawing according to Eurocode.

IOP Conf. Series: Materials Science and Engineering 1103 (2021) 012023 doi:10.1088/1757-899X/1103/1/012023

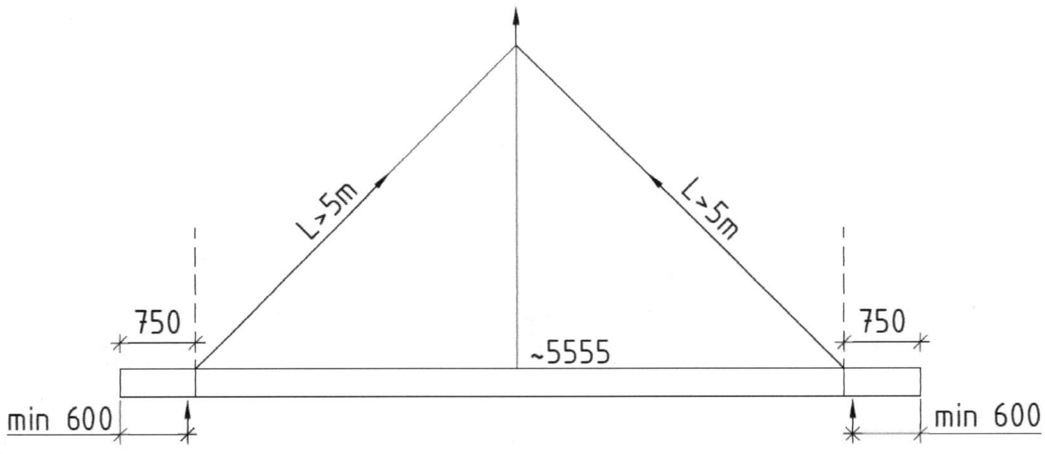

Figure 7. Slinging scheme of the precast part of the slab.

5. Conclusion

Based on the experience of design, operation and requirements of European standards, we can talk about necessity similar requirements in the mandatory sections of domestic Design Codes. Frequent publications [7-9] and development of recommendations [10-12] on the subject of progressive collapse confirm this.

Each of the reinforced concrete technologies has its own limits of applicability and, if properly implemented, it can bring significant economic, planning, aesthetic and other benefits. The prefabricated reinforced concrete frame is a promising direction in construction and attracts more and more attention from customers, architects, designers and developers.

Acknowledgments

We would like to express our gratitude to AFRY Oy Finland for financial and technical information support.

References

[1] Yakubov E N 2011 *StroyProfil* **8** 8-11
[2] Gurov E P 2010 *StroyProfil* **4** 5-10
[3] Nedviga E S and Vinogradova N A 2016 *Construction of unique buildings and structures* **4** 87-102
[4] Abramyan S G and Gnatyuk D V 2017 *Naukovodenie* **9(1)**
[5] Koyankin A A and Mitasov V M 2015 *Vestnik MGSU* **9** 28-35
[6] Koyankin A A and Mitasov V M 2015 *Vestnik MGSU* **10** 32-39
[7] Rudenko D V and Rudenko V V 2009 *Engineering and Construction J.* **3** 38-41
[8] Chakrabarti S, Nayak G and Paul D 1988 *ACI Structural J.* 85
[9] SP 63.13330.2018 *Precast reinforced concrete structures. Design rules*
[10] SP 337.1325800.2017 *Concrete and reinforced concrete structures. Guidelines*
[11] SNiP 2.03.01-84 *Design of precast reinforced concrete structures*
[12] Guidelines to SNiP 2.03.01-84 *Design of precast reinforced concrete structures*
[13] EN 1992-1-1:2004 *Design of concrete structures. General rules and rules for buildings*
[14] SP 385.1325800.2018 *Protection of buildings and structures from progressive collapse. Design rules. Guidelines*
[15] EN 1991-1-7 *Action on structures. General actions – Accidental actions*
[16] Perelmuter A V, Kriksunov E Z and Mosin N V 2009 *Engineering and Construction J.* **2** 13-18

[17] Rastorguev B S 2003 *Seismic construction. Safety of Structures* **4** 45-48
[18] 2006 *Recommendations for protection of high-rise buildings from progressive collapse* (Moscow: Moskomarkhitektura)
[19] 2005 *Recommendations for protection of cast-in-situ residential buildings from progressive collapse*
[20] 1999 *Recommendations for preventing progressive collapse of large-panel buildings*
[21] Ramin Vagheia, Farzad Hejazia, Hafez Taheria, Mohd Saleh Jaafarb and Abang Abdullah Abang Alic 2014 *APCBEE Procedia* **9** 285-290
[22] Belov V V and Nikitin S E 2013 *Int. Offshore and Polar Engineering* **4** 142-147
[23] Ong K, Hao J and Paramasivam P 2006 *ACI Structural J.* **5** 103
[24] Frosch R J 1999 *ACI Structural J.* **4** 96

Modeling the warm-up process for hydraulic actuator components

E V Polovnikov, V V Konev and Sh M Merdanov

Institute of Transport, Tyumen Industrial University, 72, Melnikayte str., Tyumen, 625000, Russia

E-mail: polovnikovev@tyuiu.ru

Abstract. This paper discusses warm-up processes for the hydraulic system components of road construction machinery using mathematical and physical simulations. The efficiency of these processes is governed by warm-up and cool-down rates of the hydraulic actuator components. It studies how various hydraulic system components interface as they warm up. The general task is to estimate the dynamic behavior of the components and loops comprising such components and the entire warm-upsystem. The models of the transient heat flow process of the warm-up system are based on the heat balance equation of the flowing medium (of the heat transfer fluid – a hydraulic fluid) within the control volume. For experimental verification, the hydraulic actuator components warm-up modes using the warmed-up hydraulic fluid from the hydraulic tank have been tested on the hydraulic test bench that we developed. The example of consecutive warm-up of the hydraulic system components is given. Based on this, after the experimental findings were processed, correlation coefficients for different warm-up modes and types of relationships between the hydraulic actuator component temperatures and their warm-up times were determined.

1. Introduction

Most of the construction machinery used for building and operating facilities in northern regions of Russia is hydraulically driven; besides, these machines operate as part of packages or systems. This enhances work productivity and ensures a high level of work organization. Meanwhile, ifany one of the machines breaks down, it interrupts the entire construction and repair process onsite, causing other machines either to be downtime or operateless efficiently [1]. The problem of how to warm up internal combustion engines of hydraulically-actuated machines operating in the northern climate has mostly been solved, yet, the problem of how to warm up hydraulic actuators is still open. Different warm-up systems exist for warming up internal combustion engines. There are much fewer such systems for warming up hydraulic actuators. Construction machinery operators have to spend 30 to 90 minutes to warm up hydraulic actuators before starting the work. The hydraulic actuators are warmed up by way of no-load operation ('idle running').

Therefore, studying the heat exchange process in hydraulic actuators of construction machinery is of great relevance.

Warming up the hydraulic actuator components is necessary for stable and failure-free operation of hydraulically-actuated machinery. Reducing the hydraulic system warm-up time increases the machine operation time and operational availability.

2. Theoretical Rationale and Mathematical Description

The study assumes that the hydraulic fluid in the hydraulic tank is warmed up. Running through the hydraulic actuator components, the hydraulic fluid emits heat to the hydraulic actuator components.

Mathematical modeling of processes in the hydraulic actuator component warm-up system is based on the mathematical models of warm-up processes of its individual components: directional control valves, valves, a hydraulic tank with a heat exchanger.

The objective is to estimate the dynamic behavior of the components, loops comprising those components, and the entirewarm-up system. For this purpose, it is necessary to make assumptions that can simplify mathematical modeling of the warm-up system for the hydraulic actuator components.

The models of the transient heat flow process of the warm-up system are based on the heat balance equation of the flowing medium (of the heat transfer fluid – a hydraulic fluid) within the control volume$V$ [2]:

- $c_a m_a \frac{dT_a}{dt} - G_1 c_a T_1 + G_2 c_a T_2 + G' c_a T' - G'' c_a T'' = -Q_g + N_w;$

where $c_a m_a \frac{dT_a}{dt}$ is the rate of internal energy accumulation or degradation; $G_1 c_a T_1 + G_2 c_a T_2 + G' c_a T' - G'' c_a T''$ is the heat transfer fluid internal energy flow through the control volume surface, Q_g is the heat power supplied to $(Q_g < 0)$ or carried away$(Q_g > 0)$ from the heat transfer fluid to the hydraulic actuator component, N_w is the internal viscous forcepower, which can be neglected when designing the warm-up system [2].

The heat balance equation for a transient warm-up process of the hydraulic actuator component is written as follows:

- $c_a m_a \frac{dT_a}{dt} = Q_g - Q_b,$

whereQ_g is the heat power supplied to the hydraulic actuator component while warming up, kW; Q_b is the heat power transmitted from the hydraulic actuator component to the environment, kW; c_g is the heat absorption capacity of the hydraulic actuator component, $J/$°C; $m_g$is the hydraulic actuator component weight, kg.

The heat powers Q_g and Q_b are determined using the heat transfer equations:

- $Q_g = A_g(T_a - T_g),$
- $Q_b = A_b(T_g - T_b),$

where the heat transfer parameters$A_g = \alpha_a F_a$ and $A_b = \alpha_b F_b$ are derived from the consolidated experimental data.

3. Preparing for the Warm-Up Study with Regard to the Temperature of aHydraulic Actuator Components

In order to conduct experiments, a hydraulic test bench was engineered and manufactured. The hydraulic loop of the experimental test bench is shown in Figure 1.

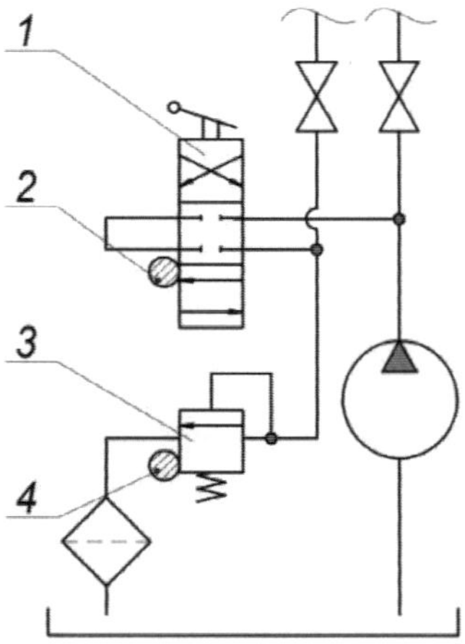

Figure 1. The experimental test bench hydraulic system schematic: 1 - directional control valve, 2 - directional control valve temperature sensor, 3 - hydraulic valve, 4 - hydraulic valve temperature sensor.

The following components of the hydraulic actuator were used: a hydraulic tank, a hydraulic pump, pipelines, a directional control valve, a hydraulic valve. The G 54-32 hydraulic valve and the VMMB 44 F directional control valve specifications are given in Table 1 [3].

The temperature sensors were positioned on the hydraulic valve body and on the directional control valve body. The sensors were positioned closer to the channel through which the hydraulic fluid flowed out of the hydraulic system component. This allowed evaluating the heat transfer from the hydraulic fluid to the hydraulic actuator component. The hydraulic fluid temperature in the hydraulic tank before the experiment started was 22°C above zero.

The directional control valve has three positions. The two end positions connect the pressure line with Line A or B, the central position shuts off the pressure line.

After the hydraulic system was started, it was determined on the hydraulic test bench that the temperature of the hydraulic actuator components (of the valve and of the directional control valve) was rapidly going up. The temperature in the hydraulic actuator while warming up changeddifferently between various components. As they cooled down, the irregularity in temperature changes was also found. The non-linear relationships were observedbetween the discussed processes. The main parameters of the VMMB 44 F directional control valve and those of the G54-32(M) hydraulic valve are shown in Table 1.

Table 1. Main parameters of the VMMB 44 F directional control valve and of the G54-32(M) hydraulic valve.

Parameters	Value	
Name	VMMB 44 F	G54-32(M)
Directional control reference tables	44	-
Nominal bore, mm	10	10
Hydraulic fluid rate, l/min	25-40	32
Hydraulic control pressure minimum (maximum), MPa (kgf/cm^2)	0.8-6.0 (8-60)	-
Pressure setting range, MPa	-	0.4-2.8
Weight, max, kg	6.45	2.4

4. Experiment

During the measurements, the test bench was used in the five following modes:

One –between 0 and 11 minutes, the hydraulic valve passes the hydraulic fluid through, the directional control valve is in the farthest position. The hydraulic fluid passes through the directional control valve.

The hydraulic actuator components are rapidly warming up, it can be seen that the warm-up is irregular. If the hydraulic actuator component is less heavy, the warm-up rate goes up. The correlation coefficient is 0.97.

Two –between 11 and 22 minutes, the directional control valve is in the central position. The hydraulic fluid does not pass through the directional control valve. This causes the warmed-up hydraulic valve to keep warming up. The correlation coefficient is minus 0.87.

Three –between 21 and 24 minutes, the mode is the same as between 0 and 11 minutes. The difference is that the warm-up takes place at different initial temperatures. The directional control valve warms up faster, even though it is heavy. The temperature of the hydraulic valve, which is lighter, does not change. The parameters correlate very little. The correlation coefficient is 0.44.

Four –between 24 and 27 minutes, there is no wind load on the hydraulic test bench. The temperature is rapidly growing. The heavy directional control valve with the lower initial temperature warms up at the same rate as the hydraulic valve that has the highest temperature at the beginning of Minute 24. The correlation coefficient is 0.98.

Five –starting from Minute 27, the hydraulic test bench shut down. It cools down. The lighter and hotter hydraulic valve cools down faster. Meanwhile, the heavier directional control valve cools down more slowly. The correlation coefficient is 0.89.

The crossplot of the temperatures of the hydraulic actuator components versus their warm-up times is shown in Figure 2.

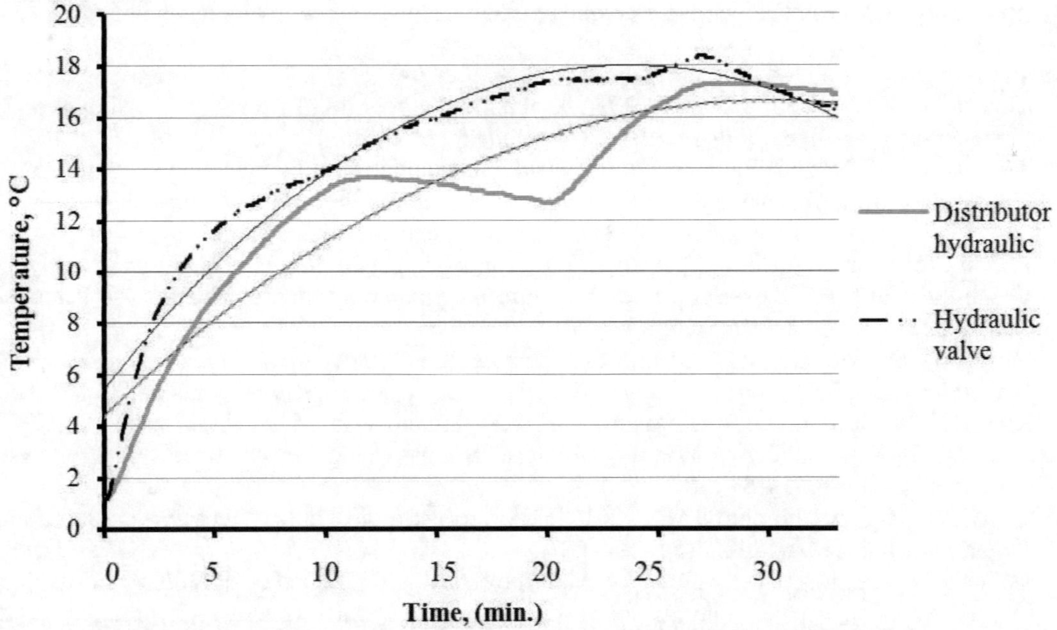

Figure 2. Hydraulic actuator components temperatures versus their warm-up times

The correlation of the hydraulic actuator components warm-up in different modes is shown in Table 2.

The relationship between these values is non-linear and can bemodeled as an exponential or polynomial relationship (Figure 2). The R-squared value R^2 is 0.88 - 0.95. The correlation values indicate that there is a close relationship between the output parameter of the hydraulic actuator component temperature and the factor of influence, which isthe warm-up time [3].

Table 2. Hydraulic actuator component warm-up temperature correlations

Mode	Duration (min)	Temperature variation		Correlation
		Directional control valve	Hydraulic valve	
1	11	Warm-up	Warm-up	+ 0.97
2	11	Cool-down	Warm-up	- 0.87
3	4	Warm-up	Unchanged	+ 0.44
4	3	Warm-up	Warm-up	+ 0.98
5	9	Cool-down	Cool-down	+ 0.89

5. Conclusion

The experiments demonstrated that various components of the hydraulic system require different time periods to warm up to a certain temperature. Their warm-up times are contingent on such factors, as the initial temperature of the components being warmedup (at the initial time point, it is equal to the ambient temperature) and their weights; the temperature of the hydraulic fluid (the heat transfer fluid); the rate of the hydraulic fluid, supplied to the component being warmedup; the hydraulic fluid flow direction; the number of the hydraulic actuator components being warmedup at the same time. In order to enhance the warm-up efficiency for the hydraulic actuator components, these factors must be taken into account. Meanwhile, the objective of the study is to consider each factor (its effect), both individually and as two-factor interface. Thisinterface process corresponds to a non-linear relationship. The experimental findings confirmed that the dynamic behavior of the components, loops comprising such components, and the entire warm-up system exists.

References

[1] Kuznetsova V N and Savinkin V V 2014 Analyzing the Working Efficiency of a Shovel *Bull. of the Siberian State Automobile and Highway Academy* **6(40)** 26-33

[2] Kuznetsova V N and Savinkin V V 2014 Substantiation of criteria for evaluating the effectiveness of the Komatsu PC300 excavator *Construction and road machines* **3** 9-12

[3] Savinkin V V 2013 Establishing dependencies between the main parameters of the Komatsu RS300 excavator *Architecture. Construction. Transport. Technologies. Innovations* 26-34

[4] Vashurkin I O 2005 *Warming up construction machinery in severe climates* (St Petersburg: Nauka) 238 p

[5] Merdanov Sh M, Konev V V and Zakirzakov G G 2016 *Road Construction Machinery Hydraulic Actuators for Operation at Low Temperatures: Monograph* (Tyumen: TIU) 160 p

[6] Konev V V, Borodin D M, Polovnikov E V, Merdanov Sh M and Zakirzakov G G 2019 Warming Up Hydraulic Actuators of Mobile Machines Using the Automatic Control System *Don Engineering Bull.* **6(57)** 38

[7] Sorokin E A and Khomutov M P 2011 Laboratory Studies of a Catalytic Heat Exchanger *Bulletin of Krasnoyarsk SAU* **3(54)** 154-157

[8] Velichko S A, Senin P V, Ivanov V I, Martynov A V and Chumakov P V 2017 Formation of thick layer electro-spark coatings for restoring worn-out parts of power hydraulic cylinders *Surface Engineering and Applied Electrochemistry* **53(2)** 116-123

[9] Velichko S A, Chumakov P V and Kolomeychenko A V 2019 Assessment of Technical Condition of C Series Power Hydraulic Cylinders of Mounted Hydraulic Systems of Tractors *Engineering technologies and systems* **29(3)** 396-413

[10] Karnaukhov N N, Yarkin A V and Kruk A R 2015 System for preheating of internal combustion engine operating in cold weather conditions *J. of Engineering and Applied Sciences* **10(1)** 7-10

Improvement of the technology for laying corrugated pipes for creating culverts under roads

A A Serebrennikov[1] and D A Serebrennikov[2]

[1] Tyumen industrial University, 38, Volodarsky St., Tyumen, 625000, Russia
[2] Tyumen Higher Military Engineering Command School, 1, L. Tolstoy St., Tyumen, 625001, Russia

E-mail: serebrennikovaa@tyuiu.ru

Abstract. The widespread use of polypropylene corrugated pipes for free-flow culverts requires that the construction technology ensure compliance with their physical and mechanical characteristics during their service. The use of polypropylene corrugated pipes for culverts in road construction poses greater demands on them. The paper considers the behavior of a buried pipe in interaction with the surrounding soil. The contradictions in the current building codes and regulations have been revealed. It has been proven that for polypropylene corrugated pipes, the resistance of the backfill soil, depending on the degree of compaction, is an essential factor in ensuring the long-term stability of the circular cross-section of the pipe. Theoretical calculations have confirmed that, depending on the backfill depth, the relative deformation of polypropylene corrugated pipes may exceed the maximum permissible values. An engineering solution has been proposed that provides an increase in the rigidity of the pipe bearing reaction under compressive forces and a decrease in the load from vehicles.

1. Introduction

Today, polypropylene corrugated pipes have almost everywhere come to replace pipes made of other materials (concrete, asbestos cement, metal, etc.) to create free-flow engineering systems. First of all, they are used in the newly constructed storm and waste sewers. However, the scope of their use is constantly expanding. Depending on their physical and technical characteristics and geometric parameters, corrugated pipes are systematically used to create culverts under roads, electrical channels for laying cables, ventilation systems and other engineering structures.

Initially, pipes from foreign companies (Pragma, Ecopal, DKS, etc.) were used on the Russian market. Later, due to high demand, European technologies were purchased and on their basis Russian production facilities were created (Korsis, Ikaplast, Politron, etc.), whose products are of comparable quality.

The widespread use of corrugated polypropylene pipes is explained by a number of their advantages: durability (estimated service life is at least 50 years); resistance to impacts from transported resources (corrosion, chemical, temperature); preservation of tightness during possible movements of the soil (deformation capacity due to the properties of flexibility).

An important indicator of drainage and sewerage networks is their resistance to wear from solid particles contained in the transported liquid. In accordance with the so-called Darmstadt Procedure polypropylene is the least susceptible to abrasive wear [1]. The results were obtained based on studies, the technique of which is that 1-meter long pipe sections (samples) are filled with an aqueous suspension with various additives of abrasive particles. Further, these samples on a special stand swing with a certain frequency. As a result of the movement of particles, real wear is simulated in an active pipeline. After a certain number of oscillations have elapsed, the change in the wall thickness of the samples is registered.

This assessment is carried out by Süddeutsche Kunststoffzentrum and other well-known companies (Borealis, Uponor, Wavin, et al.).

It has been experimentally proved that for polypropylene pipes such wear will not exceed 0.5 mm over the entire operating period (Fig. 1).

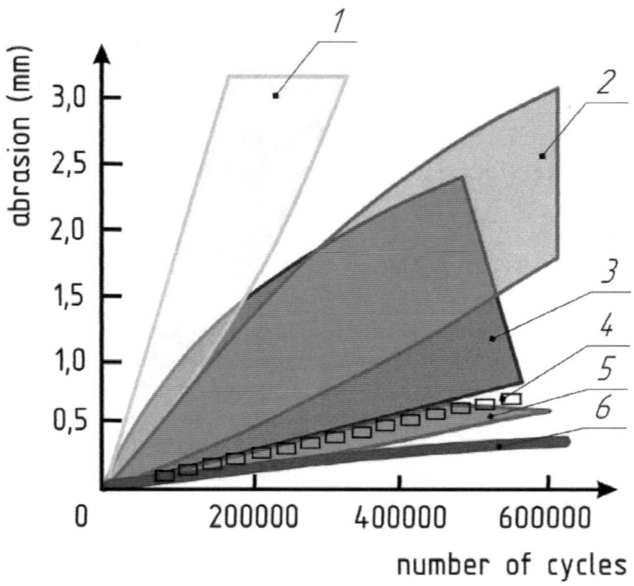

Figure 1. Abrasion curve of various pipe materials according to the Darmstadt procedure: 1 - concrete pipe; 2 - concrete pipe coated; 3 - GFK pipe; 4 - clay pipe; 5 - PVC pipe; 6 – PP pipe

The manufacturers claim that the use of polypropylene corrugated pipes is economically feasible, because not only their cost is lower in comparison with pipes made of other materials, but also there is lower costs of transportation, installation and maintenance.

However, the use of polypropylene corrugated pipes for culverts in road construction poses greater demands on them. In particular, water currents carrying significant amounts of sand, gravel, or stones can deform the material at a much greater rate than the results shown in Figure 1.

Also, the small wall thickness and, as a consequence, the relatively low ring stiffness of corrugated pipes (in comparison with concrete, metal, and even polyethylene) makes it necessary to strictly adhere to the requirements for laying technology.

2. Current situation and formulation of the problem

Culverts with corrugated polypropylene pipes are the most important structural elements of roads. The general condition of the road, ensuring the safety of traffic on it and protecting the environment, depends on their operation.

Currently, pipes of various types are used in the construction of highways of all categories. Polyethylene pipes are increasingly being laid by trenchless methods [2]. However, due to possible deformation when using known means of trenchless laying, the so-called cut-and-cover method continues to be used for polypropylene corrugated pipes, involving the creation of a trench.

After laying the pipe in a trench, the most stringent requirements are imposed on the process of compaction of the soil body surrounding the pipe (soil cage).

This process must be carried out in such a way that the pipe laid under the road either does not deform at all, or returns to its original position after the load from the vehicle. Such a return is possible due to elasticity and resilience of the pipe (Fig. 2).

(a) (b)

Figure 2. Diagram of deformation changes in a corrugated pipe: (a) deformation from loading from the side of the vehicle; (b) restoration of the shape upon termination of the load

The structural characteristics of a flexible pipe during its burial have been studied for a long time [3]. In a number of countries, the limiting state of the possibility of normal operation is established by comparing vertical deformations from loads with permissible deformations. As a rule, a series of experimental studies are carried out for this purpose (Fig. 3).

Figure 3. General view of laboratory test facilitie

As a result of such studies, the local maximum possible bending is assessed depending on the profile and thickness of the pipe wall [4, 5].

However, the real behavior of the buried pipe should be considered together with the surrounding soil cage, since they form a single structure that takes the loads acting on the structure [6, 7].

Thus, the rebound from the backfilled soil is an indicator that determines the behavior of the pipe (in conjunction with its own strength performance). Insufficient soil compaction can lead to lateral buckling and loss of pipe stability due to unacceptable deformations.

It is necessary to determine whether the standardized value of the degree of compaction ensures guaranteed preservation of the strength performance of the corrugated pipe material.

3. Analysis of reference documents and calculation methods

The organization of construction operations for the laying of water supply and sewerage networks is carried out in accordance with the regulated requirements.

The reference document of the Russian Federation on soil compaction around the pipe stipulates the following requirements: "When backfilling pipelines, a protective layer of sandy or soft local soil with a thickness of at least 30 cm, not containing solid inclusions (crushed stone, stones, bricks, etc.), must be installed above the top of the pipe. Tamping the pipeline with soil is carried out with a non-mechanized hand tools. Compaction of the soil in the cavities between the trench wall and the pipe, as well as the entire protective layer, must be carried out by manual mechanical ramming until the compaction coefficient established by the project is achieved. Compaction of the first protective layer with a thickness of 10 cm directly above the pipeline is carried out with hand tools".

For clarity, these requirements are shown in Figure 4.

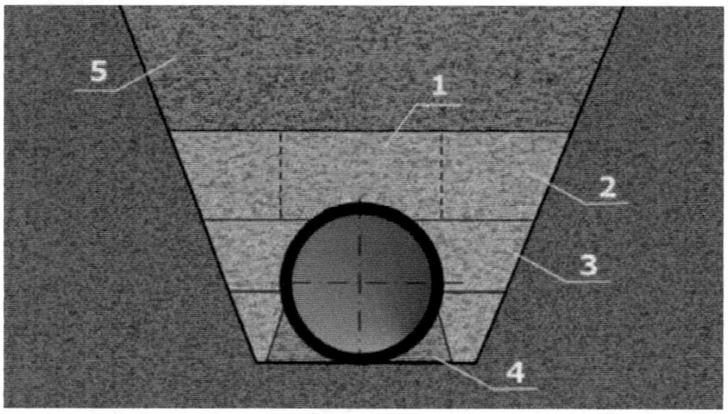

Figure 4. Soil compaction scheme when backfilling trenches: 1, 4 - layers of soil compacted with hand tools; 2, 3, 5 - layers of soil compacted with mechanized hand tools

From the information presented in the diagram, it follows that the labor intensity of earthworks is extremely high. According to the preliminary calculations, all the savings associated with the installation of a corrugated pipeline (versus, for example, a metal one) will be lost (and even exceeded) as a result of labor costs during layer-by-layer filling and compaction using manual labor (Fig. 5).

Figure 5. General view of a soil cage compaction

In addition, the outlined technology of soil compaction around a corrugated pipe conflicts with the requirements set out in the road construction codes and regulations, in accordance with which, the minimum coefficient of soil compaction in the working layer of the excavation (up to 6 meters) should not be lower than 0.95 (Table 1).

Table 1. Degree of Soil Compaction

Layer depth from the road surface (m)	Compaction coefficient under the capital road surface depending on the natural and climatic zone (I - V)		
	I	II, III	IV, V
< 1.5	0.98-0.96	1.0-0.98	0.98-0.96
1.5-6.0	0.95	0.96-0.95	0.95

The required compaction cannot be achieved using hand tools or mechanized hand tools even if the optimum moisture is maintained. Given the cramped conditions of the operations, it can be assumed that with a high probability the zones located directly under the pipe will remain not only non-compacted, but also not filled with sufficient quality.

Previous experience [8] was used in carrying out the necessary assessment of pipe deformations and a program of static calculations has been developed.

The serviceability of an underground pipeline (bearing capacity) can be determined by the condition of the maximum permissible value of the relative deformation of the pipe cross-section (shortening of the vertical diameter):

$$\frac{f}{D} \cdot 100\% \leq [\varepsilon], \tag{1}$$

where f - vertical deformation of the corrugated pipe (mm); D – nominal diameter (mm); $[\varepsilon]$ - maximum permissible relative deformation (%).

The soil cage should be considered as an elastoplastic medium, while it is necessary to assess the zones of maximum equilibrium in the body of the soil cage.

With an increase in deformation, the horizontal component of the soil pressure increases, gradually passing from a passive form to an active form. As a linearly deformable body, the soil cage resists the pipe pressure. Then, in accordance with the classic expression of Spengler, the deformation depends on the vertical load, the stiffness of the soil and the stiffness of the pipe itself:

$$\varepsilon = \frac{C_1 \cdot q}{C_2 \cdot SN + C_3 \cdot E_S}, \tag{2}$$

where q – intensity of vertical soil load (MPa); SN – ring stiffness of the pipe (MPa); E_S – secant soil modulus (MPa); C1 – coefficient of influence on the deformation of the intensity of the vertical load of the soil; C2 – coefficient of influence on deformation of pipe ring stiffness; C3 – coefficient of influence on the deformation of the secant soil modulus.

Distributed vertical load consists of two components:

$$q = q_1 + q_2\,, \tag{3}$$

Where q1 – distributed load from vehicle; q2 – distributed load from soil above the pipe.

The load from the vehicle side (according to the Boussinesq pressure distribution theory) depends on the pipe laying depth:

$$q_2 = 0{,}478 \cdot \frac{T}{H^2}, \tag{4}$$

where: T – vehicle axle load (kN); H – pipe backfill depth (m).

The density of the soil above the pipe (to determine q1) when compacted with power tools will not exceed 19 kN/m^3.

The standard ring stiffness for polypropylene corrugated pipes corresponding to the stiffness class SN 8 is 8 kN/m^2 (0.008 MPa).

The degree of compaction and the type of soil determine its secant modulus (ES). When compacting with power tools, its value cannot exceed 3.0 MPa.

The coefficients of influence on deformation (C1, C2, C3) are established by reference documents.

4. Results and Discussion

In accordance with the stated dependencies (1-4), statistical calculations have been performed. The secant modulus was set to the maximum possible for compacting the soil cage with power tool (ES = 3.0 MPa). The calculations have been carried out taking into account the movement of heavy land transport (T = 260 kN).

Table 2 shows, by the example, the results obtained for PP corrugated pipe "DN/ID 800 mm SN8".

Table 2. Calculated changes in relative deformation

Backfill depth (m)	1.0	1.5	2.0	2.5	3.0	3.5	4.0
Relative deformation (%)	8.7	6.3	5.2	5.2	5.8	5.9	6.3

The limiting relative deformations set by the manufacturers should not exceed 6%. Consequently, it turns out that the technology of laying polypropylene corrugated pipes described in the normative documents ensures the safety of their physical and mechanical characteristics only at depths ranging from 2.0 to 3.5 meters.

To address this contradiction, it is proposed to improve the technology of creating a soil cage around the pipe (Fig. 6).

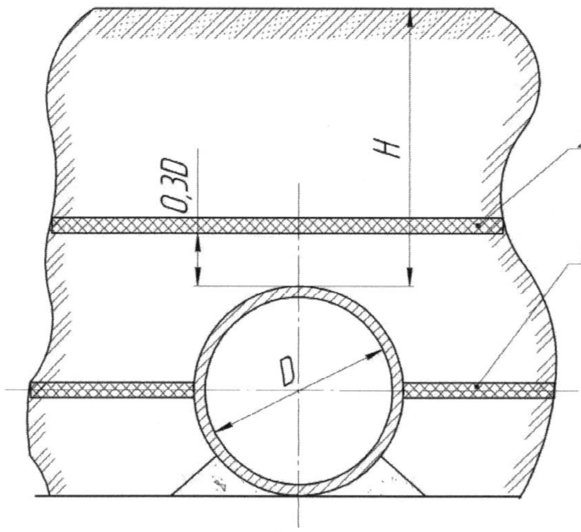

Figure 6. Laying scheme using reinforcing elements: 1 - horizontal stops; 2 - damping layer

The construction of additional membranes in a soil cage may be proposed as reinforcing elements. To create membranes, various non-woven and synthetic materials or a primer together with various binders can be used as structural and supporting materials. For example, it is proposed to make membranes from a geocell filled with sand-gravel or crushed stone, wrapped in geotextile.

The main task of the horizontal stops (1) is to increase the resistance to vertical loads acting on the pipe and to reduce its relative deformations. The damping layer (2) will allow the use of generally accepted mechanized technical means of compaction of the overlying soil. Both layers (1 and 2) will prevent possible settlement in the underlying soil mass, which is possible over time due to insufficient compaction.

5. Conclusion

The requirements for the technology of soil compaction around the corrugated pipe set out in the reference documents are in conflict with the requirements set out in the norms and rules for road construction.

It has been proven that for polypropylene corrugated pipes, the resistance of the backfill soil, depending on the degree of compaction, is an essential factor in ensuring the long-term stability of the circular cross-section of the pipe.

Theoretical calculations have confirmed that, depending on the backfill depth, the relative deformation of polypropylene corrugated pipes may exceed the maximum permissible values.

An engineering solution is proposed, which consists in the construction of additional membranes in a soil cage.

It is necessary to adjust building codes and regulations. These amendments should require the creation of additional reinforcing layers that provide an increase in the rigidity of the pipe resistance during deformation and a decrease in the load from the sealing means when creating a road and from vehicles during its operation.

References

[1] Marley Polyethylene Systems Solutions for a wide range of applications *Polyethylene Catalogue* p 79
[2] Serebrennikov A, Serebrennikov D and Hakimov Z 2016 Polyethylene Pipeline Bending Stresses at an Installation *American J. of Engineering and Applied Sciences* **9(2)** 350-355
[3] Hashash, N M A and Selig E T 1990 Analysis of the Performance of a Buried High Density Polyethylene Pipeî *Structural Performance of Flexible Pipe* (Sergand, Mitchell and Hurd edition Balkema, Rotterdam) pp 95-102
[4] Dhar A S and Moore I D 2006 Evaluation of local bending in profile-wall polyethylene pipes *J. of Transportation Engineering* **132(11)** 898-906
[5] Dhar A S and Moore I D 2001 Sensitivity of thermoplastic pipe behaviour to profile geometry Pipelines *Advances in Pipelines Engineering and Construction* 1-10
[6] Zhan C and Rajani B 1997 Load transfer analysis of buried pipe in different Backfills *J. of transportation engineering* **123(6)** 447-453
[7] Rajkumar R and Ilamparuthi K 2008 Experimental study on the behavior of buried flexible plastic pipe *Electronic J. of Geotechnical Engineering* **13** 1-10
[8] Serebrennikov A and Serebrennikov D 2018 Mathematical model of polyethylene pipe bending stress state *Int. Conf. on Mechanical Engineering, Automation and Control Systems* **327**

Behavior of asphalt concrete mixture during relaxation

E A Shishkin, S N Ivanchenko and S A Shemyakin

Pacific National University, 136, Tihookeanskaya St., Khabarovsk, 680035, Russia

E-mail: 004655@pnu.edu.ru

Abstract. The article discusses the asphalt concrete mixture compaction process with a road roller. The rheological model of a smooth drum and material interaction is adopted. The relaxation process that occur in the asphalt concrete mixture during compaction has been studied. During dynamic modeling, the physical and mechanical characteristics of the compacted material influence on the behavior of the latter during relaxation was established. The article reveals the relaxation process physical essence, which plays an important role in the material compaction. The law of asphalt concrete mixture behavior during relaxation was determined. The obtained dependencies make it possible to evaluate the effectiveness of the force effect of a smooth drum on each pass of the roller.

1. Introduction

A smooth drum roller is one of the main means to compact the asphalt concrete pavement of highways. Asphalt concrete mixture is an elastic-viscous-plastic material, therefore, a theoretical description of its behavior under load is a rather complicated task. Various processes take place in the material to be compacted under loading [1].

This paper presents an investigation of the stress relaxation process that occur in asphalt concrete mixture.

A common tool for the study of the process of material loading is rheological modeling [2]. We will use a rheological model, which includes elastic, viscous and plastic elements, to describe the asphalt concrete mixture (Figure 1).

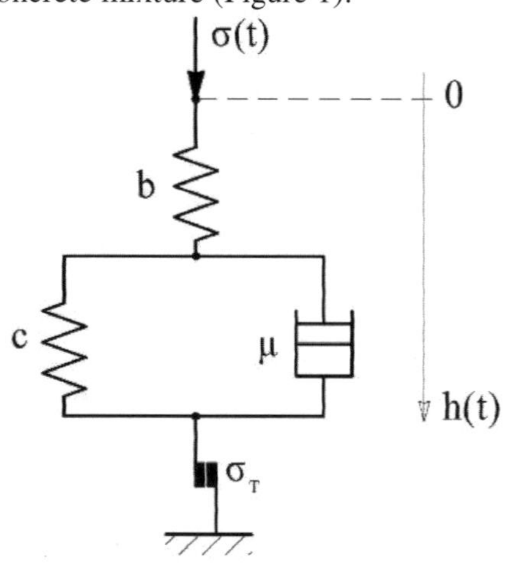

Figure 1. Rheological model of asphalt concrete mixture.

2. Study of the stress relaxation process

The force on the material is excessive due to the high loading speed when compacting the asphalt concrete mixture with a road roller drum [3, 4]. Therefore, we can talk about the relaxation phenomenon observed in the asphalt concrete mixture.

To study the relaxation process we exclude the dry friction element from consideration in the accepted rheological model (Figure 1), since it does not affect the specified process [5, 6]. The relaxation process is schematically shown in Figure 2.

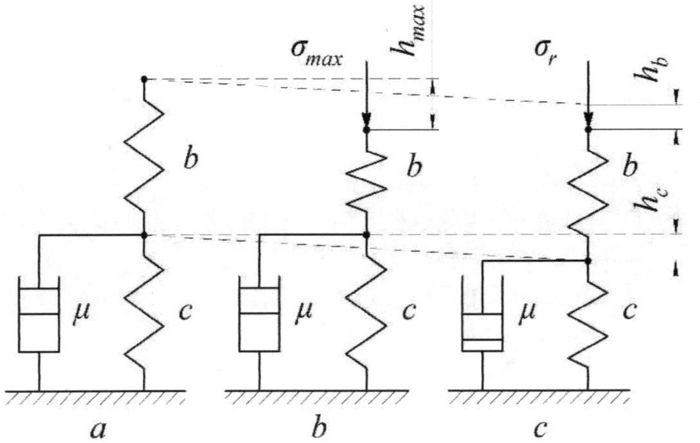

Figure 2. Scheme of the relaxation process: a – initial position; b – instant loading; c – relaxation.

The initial position of the considered model is shown in Figure 2, *a*. A load of σ_{max} value is instantly applied to the model at the moment of time $t = 0$, after which it is fixed. In this case, the deformation of the elastic element *b* is equal to h_{max}, and the deformation of the elastic element *c* is equal to zero. The following relation is true for the moment of time $t = 0$

$$\sigma_{max} = bh_{max}. \quad (1)$$

Over time, there will be a gradual tension of the elastic element *b* and compression of the elastic element *c*. At the moment $t \to \infty$, the deformation of the elastic element *b* will be h_b, and the deformation of the elastic element *c* will be h_c. For the moment of time $t \to \infty$ the following relation is true

$$\sigma_r = bh_b = ch_c. \quad (2)$$

where σ_r is the stress in the material after relaxation [7].

Since the total deformation of the considered model for the times $t = 0$ and $t \to \infty$ is invariable, then

$$h_{max} = h_b + h_c. \quad (3)$$

Taking into account (1) and (2), equation (3) takes the following form

$$\sigma_{max} b^{-1} = \sigma_r \left(b^{-1} + c^{-1}\right). \quad (4)$$

Expressing σ_r from (4), we obtain

$$\sigma_p = \sigma_{max} c\left(b + c\right)^{-1}. \quad (5)$$

Stress relaxation refers to the decrease over time of the stresses that initially arise in the body during its rapid deformation [8, 9]. In equation (5) the coefficient $c \cdot \left(b + c\right)^{-1}$ is always less than one, which means that $\sigma_p < \sigma_{max}$. Consequently, equation (5) describes the relaxation phenomenon.

In the model in Figure 2 elastic elements are connected in series (Figure 3, *a*). With such a connection, the following relation is valid

$$\sigma_{max} = bh_b = ch_c. \quad (6)$$

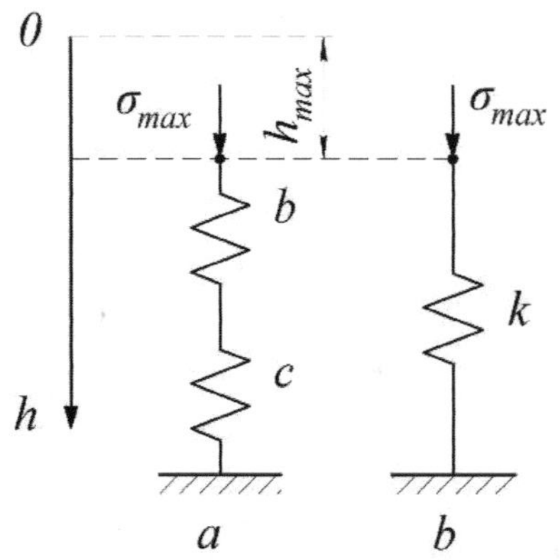

Figure 3. Connection of elastic elements: a - two elements; b - equivalent replacement by one element.

Taking into account (6), equation (3) takes the following form

$$h_{max} = \sigma_{max}\left(b^{-1} + c^{-1}\right). \tag{7}$$

We replace two series-connected elastic elements with one equivalent (Figure 3, *b*), for which the following equation is true

$$\sigma_{max} = kh_{max}. \tag{8}$$

Taking into account (8), equation (7) takes the following form

$$k^{-1} = b^{-1} + c^{-1}. \tag{9}$$

From the equation (9) analysis it follows that with a series connection of elastic elements, the total stiffness of the system will be less than the stiffness of any element of this system. This property is the cause of stress relaxation in an elastic-viscous body [10, 11].

It follows from the analysis of expression (5) that the larger the value of b in comparison with the value of c, the more pronounced the relaxation. We will establish the validity of this statement for the adopted rheological model. Since the material temperature and density during relaxation can be considered constant, the rheological model parameters *b*, *c* and μ are also assumed to be unchanged, and the equation of the rheological model (Figure 2) has the following form

$$\frac{d\sigma}{dt} + \frac{b+c}{\mu}\sigma = \frac{bc}{\mu}h_{max}, \tag{10}$$

where h_{max} is the model deformation achieved during loading.

Based on equation (10), a dynamic model was obtained, shown in Figure 4.

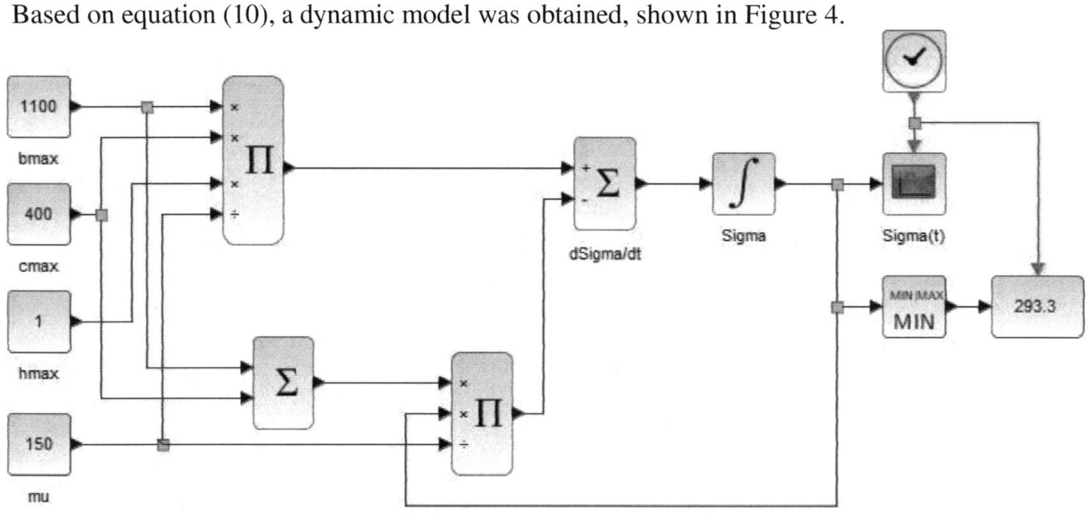

Figure 4. Dynamic model of the stress relaxation process.

IOP Conf. Series: Materials Science and Engineering 1103 (2021) 012026 doi:10.1088/1757-899X/1103/1/012026

The developed dynamic model makes it possible to obtain a stress-time diagram after fixing the model under instant loading (Figure 5).

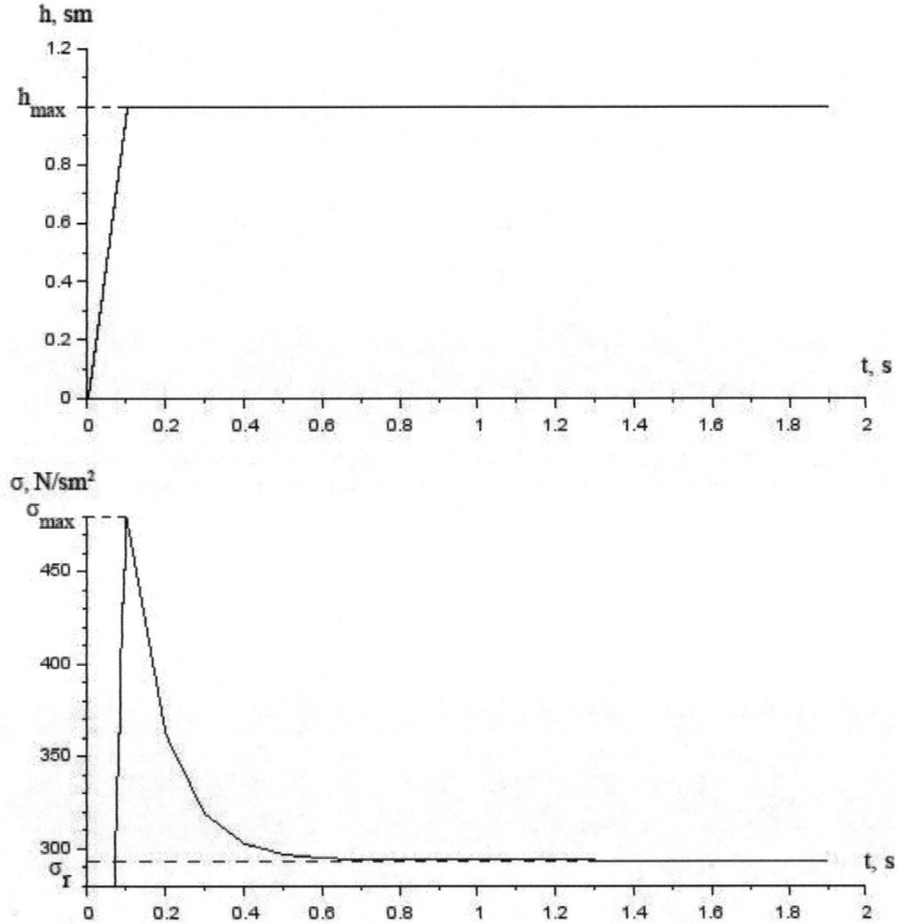

Figure 5. Strain-time and stress-time diagrams during relaxation.

Under the initial conditions $t = 0$, $\sigma = \sigma_{\max} = bh_{\max}$ the solution to the differential equation (10) will have the following form

$$\sigma = \frac{bc}{b+c} h_{max} + \left(\sigma_{max} - \frac{bc}{b+c} h_{max} \right) e^{-\frac{b+c}{\mu}t}. \tag{11}$$

As t → ∞, equation (11) takes the following form

$$\sigma = \sigma_r = \frac{bc}{b+c} h_{max}, \tag{12}$$

where σ_r is the stress value after the end of the relaxation process.

Taking into account (11), equation (12) takes the following form

$$\sigma = \sigma_r + \left(\sigma_{max} - \sigma_r \right) e^{-K_r t}, \tag{13}$$

where $K_r = (b+c)\mu^{-1}$ is the relaxation intensity factor.

Expression $\left(\sigma_{max} - \sigma_p \right)$ in equation (13) characterizes the relaxation resistance of the material.

The results of dynamic modeling are shown in Figures 6 and 7.

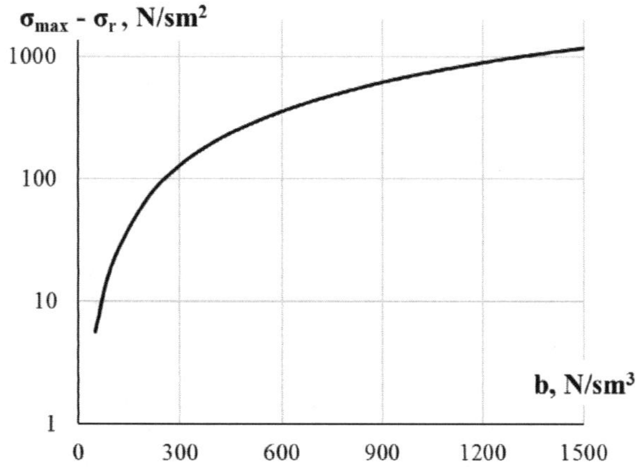

Figure 6. The relaxation resistance versus asphalt concrete mixture elasticity factor *b* graph

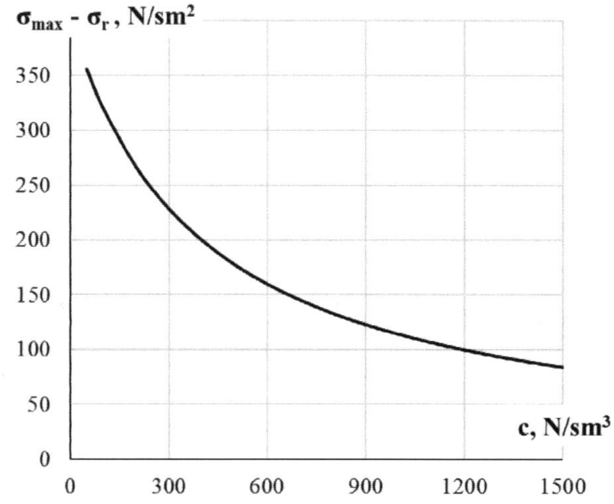

Figure 7. The relaxation resistance versus asphalt concrete mixture elasticity factor *c* graph

The analysis of the graphs in Figures 6 and 7 shows that with the increase in the difference between the values of b and c, all other things being equal, the difference between the initial and final stresses also changes. This confirms the adopted rheological model adequacy in describing the relaxation of the asphalt concrete mixture.

The equation (13) analysis allows concluding that the increase in the asphalt concrete mixture viscosity μ leads to the material relaxation time increase (Figure 8).

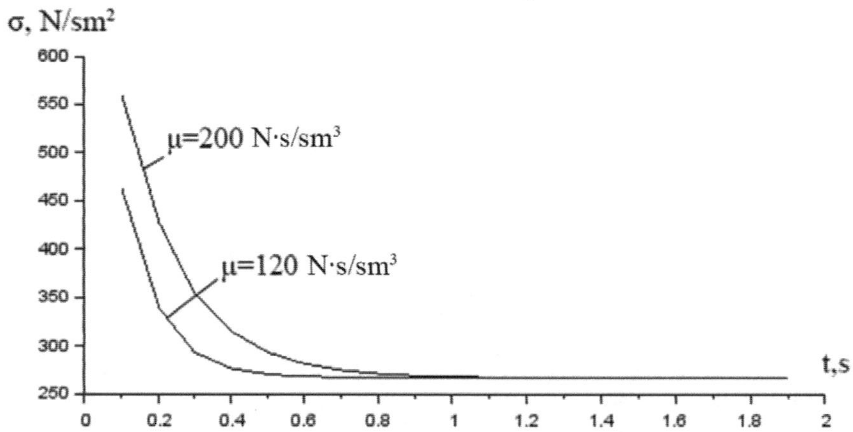

Figure 8. The relaxation process intensity graph at different values of μ.

3. Conclusion

It is found that for viscous-elastic bodies, not only the magnitude of changes in the attached load or deformations is important, but also the rate of these changes. Relaxation can proceed at different rates, and the material final stress can reach different values. The relaxation process intensity is influenced not only by the maximum value of the loading stress, but also by the mixture viscosity. The material elastic properties affect its relaxation resistance.

The results of the studies carried out can be used to describe the asphalt concrete mixture vibration compaction. In this case, it is necessary to determine relationship between the physical and mechanical characteristics of the material to be compacted and the drum vibrator parameters. This relationship can be determined in a laboratory experiment.

References

[1] Biggs W D 2013 *The mechanical behavior of engineering materials* (Amsterdam: Elsevier) p 156

[2] Belickij V D and Katunin A V 2013 The mathematical models thesaurus of asphalt concrete mixture compaction process *Omsk Scientific Bull.* (Russia: OmSTU) **3(123)** 18–21

[3] Ivanchenko S N 1996 The theory basic provisions of composite materials compaction under complex loading *Construction and road machines (Conference proceedings)* (Russia: KSTU) pp 100–107

[4] Zav'yalov M A and Zav'yalov A M 2002 Possible rheological model of asphalt-concrete mixture relaxation during compaction *Construction and road machines* (Russia: CRM Press) **7** 25–26

[5] Karimi M M, Tabatabaee N, Jahangiri B and Darabi M K 2017 Constitutive modeling of hardening-relaxation response of asphalt concrete in cyclic compressive loading *Construction and Building Materials* **137** 169–184

[6] Sun Y, Huang B and Chen J A 2015 Unified procedure for rapidly determining asphalt concrete discrete relaxation and retardation spectra *Construction and Building Materials* **93** 35–48

[7] Krishnan J M and Rajagopal K R 2004 Triaxial testing and stress relaxation of asphalt concrete *Mechanics of Materials* **36(9)** 849–864

[8] Krishnan J M and Rajagopal K R 2005 On the mechanical behavior of asphalt *Mechanics of Materials* **37(11)** 1085–1100

[9] Ren J, Liu Z, Xue J and Xu Y 2020 Influence of the mesoscopic viscoelastic contact model on characterizing the rheological behavior of asphalt concrete in the DEM simulation *Advances in civil engineering* **2020** 5248267

[10] Riccardi C, Falchetto A C, Losa M and Wistuba M P 2016 Rheological modeling of asphalt binder and asphalt mortar containing recycled asphalt material *Materials and Structures* **49** 4167–4183

[11] Zbiciak A, Grzesikiewicz W and Wakulicz A 2010 One-dimensional rheological models of asphalt-aggregate mixtures *Logistyka* **6** 3825–3832

IOP Conf. Series: Materials Science and Engineering 1103 (2021) 012027 doi:10.1088/1757-899X/1103/1/012027

One way to reduce the dynamic loads in the gearing engagement of construction machinery mechanisms

V I Skel

National Research Moscow State University of Civil Engineering, 26, Yaroslavskoe highway, Moscow, 129337, Russia

E-mail: skelvi@mgsu.ru

Abstract. In the gear mechanisms of the machine drive, vibrations occur mainly due to the difference in the circular pitches of a gear and a pinion, errors in the profile of the teeth, variable gearing stiffness and friction in the mesh. For involute gear transmissions of construction machinery, the difference between circular pitches of a gear and a pinion contributes significantly to the dynamic component of the load. The influence of the difference between the gear and pinion pitches on a change in the gear ratio of a gear pair with a gear frequency as a source of vibration is considered. It is known that the dynamic load is usually greater if the main pitch of the gear is greater than the main pitch of the pinion. The authors show that it is possible to choose such a gear displacement coefficient, with which the gear ratio error will not depend on the sign of the difference between the main gear and pinion pitches and will have the smallest possible value. Then, at low cost, it is possible to obtain the least probable dynamic effect of the difference in the main pitches of the meshing gears in the construction mechanisms.

1. Introduction

Involute gears are most common in drives of construction machinery. They have important advantages in terms of kinematics, technology and operation. At the same time, among the disadvantages of these gears, one can single out dynamic loads, which occur in engagement and in the drive mechanism as a whole. Dynamic loads in the engagement reduce the gear resource and lead to additional environmental pollution in the form of noise and product wear.

Among the reasons that cause the dynamics of gears, one can mention the difference in gear pitches, broken involute profile of the teeth due to manufacturing defects and wear of the side surfaces of the teeth, variable gearing rigidity due to the alternation of one-pair and two-pair engagement, sliding friction in engagement, etc. Some of these reasons can cause nonlinear vibrations.

In specific operating conditions and for specific designs of gear reducers, one or another reason for the vibration of the gear may prevail. The analysis showed that for involute gear transmissions of construction machinery, the difference between circular pitches of the gear and the pinion contributes significantly to the dynamic component of the load. Therefore, it is important to analyze the influence of the difference in pitches of the engaging gears on the dynamics of the gear transmission and ways to reduce this influence.

Many works from different fields of science and technology are devoted to the dynamics of gears. A number of works, for example [1], consider variable gearing stiffness and backlash of teeth as the cause of nonlinear oscillations of the gear transmission. Article [2] studies the effect of teeth modification on the dynamics of gear transmission, taking into account the damping ability of the lubricant film between the teeth, investigated by the finite element method. Work [3] considers the influence of stiffness and clearance in engagement, as well as profile errors, on the transmission dynamics in the presence of a compliant gear support (on a nonlinear model of engagement). The authors show for what conditions a positive effect of the compliant support of the gearbox on the transmission dynamics is possible. Many articles [1-5] pay attention to the choice of a dynamic model of a dynamic gear system. In these works, various sources of vibrations are considered with varying completeness, for example, variable stiffness in engagement, friction in engagement, errors in the profile of the teeth, modification of the profile of the teeth, errors in cutting the teeth of gear wheels and the difference in the main pitches of the meshing teeth for a gear and a pinion. It can also be noted that in many works attention is paid to methods for solving differential equations of dynamics.

Analysis of publications on the topic considered in this work showed that a decrease in the disturbing ability of the gear transmission could be obtained by improving the quality of cutting the teeth of the gear wheels, but not by the choice of the optimal gear displacement factor proposed here.

2. Results

The perturbing ability of the involute gearing due to the difference in the basic pitches of the gears can be described as follows. The advantage of the ideal involute gearing is the constancy of the gear ratio, when the contact point of the engaging teeth moves only along the straight line of the gearing. If there is a difference in pitches of the teeth of the engaging wheels, the contact point of the teeth digresses from the straight line of engagement in the initial or final period of engagement. As a consequence the gear ratio becomes variable with the frequency of teeth remating – the gear-mesh frequency. This frequency is equal to the product of the angular gear speed by the number of its teeth: $\upsilon = \omega z$. When the resonant frequencies of the gear-mesh (sometimes one of its harmonics) coincide with one of the natural frequencies of the system vibrations, as one knows, a sharp increase in the oscillation amplitudes occurs with known negative consequences for the oscillatory system of the mechanism.

If we take into account the difference between the main pitches of the gear and the pinion as:

$f_{pb} = p_{bt\,2} - p_{bt1} + \delta$

where p_{bt2} is the main circular pitch of the gear;

p_{bt1} is the main circular pitch of the pinion;

δ - deformation of the engagement, measured along the line of engagement,

then the graph of the gear ratio can be summarized in the form shown in Figure 1.

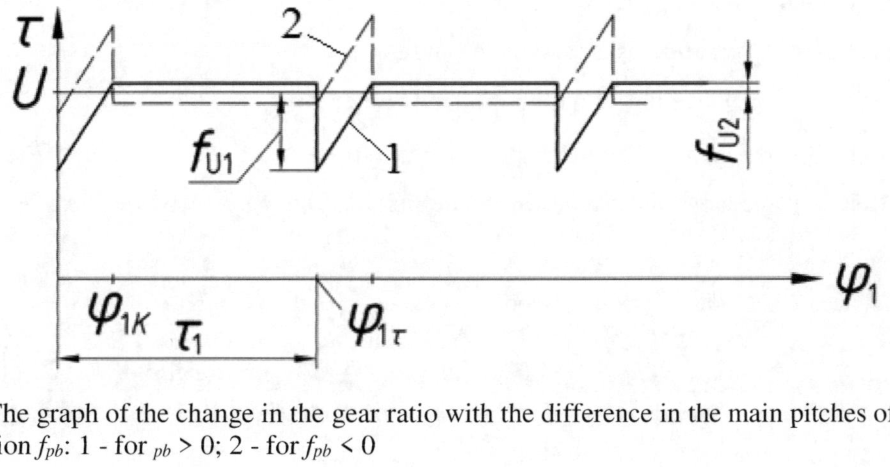

Figure 1. The graph of the change in the gear ratio with the difference in the main pitches of the gear and the pinion f_{pb}: 1 - for $_{pb} > 0$; 2 - for $f_{pb} < 0$

When $f_{pb} > 0$, the dynamic interaction of the teeth occurs when they enter engagement (edge engagement), and when $f_{pb} < 0$ – when the teeth come out of engagement. In the first case, the dynamic interaction of the teeth is greater than that in the second. This interaction in the equations of vibrations of a mechanical system can be taken into account by the magnitude of the change in the gear ratio and the time of its action. Let us represent the magnitude of the change in the gear ratio in the form:

$f_u = U - i$,

where U is the reduction ratio of the gear transmission;

i - gear ratio of the gear transmission.

The reduction ratio:

$U = z_2/z_1$,

where z_1 is the number of teeth of the gear,

z_2 is the number of teeth of the pinion.

The gear ratio:

$i = \omega_1/\omega_2$,

where ω_1 - angular speed of the gear.

ω_2 - angular speed of the pinion.

The figure also describes: f_{u1} - the largest error of the gear ratio at edge engagement; f_{u2} - error in the gear ratio when contacting involute sections of the tooth profiles; $\varphi_{1\tau}$ - angular gear pitch; φ_{1k} – angle of the edge engagement, measured by the angle of rotation of the gear φ_1; τ_1 – turning time of the gear corresponding to its angular pitch. In relation to the graph shown in the figure it can be noted that, taking into account the compliance of the teeth, sudden changes in the gear ratio are not instantaneous, but smoother, which is not of fundamental importance in our analysis. The authors also note that the average gear ratio is equal to the reduction ratio.

It is obvious that the smaller the error amplitude of the gear ratio, the lower the vibration activity of the gear transmission. In this work, the task is to reduce that part of the vibration activity of the involute gear transmission, which is associated with the difference in pitches of the gear and the pinion.

3. Discussion

The influence of the parameters of gear transmission on the error of the gear ratio f_{u1} with edge engagement was analyzed. For this, the value f_{u1} was expressed through the parameters of the gear transmission. As a result of the analysis, the following conclusions were made:

1) the higher $|f_{pb}|$ the higher $|f_{u1}|$;

2) f_{u1} has the same sign as f_{pb};

3) $|f_{u1}|$ increases with decreasing z_1 and increasing z_2;

4) the lower $|f_{u1}|$, the lower the ratio of the addendum h_a^*;

5) the lower f_{u1}, the greater the gear displacement coefficient x_1;

6) the lower f_{u1}, the larger the displacement coefficient колеса x_2, however, for $u \geq 5$ the influence x_2 is not enough;

7) dependence of $\frac{f_{u1}}{u}$ on $\frac{f_{pb}}{m}$ has a parabolic character.

Considering the degree of influence of the gear displacement factor x_1 on the gear ratio error f_{u1} you can see that there is such a value of the displacement coefficient $x_{1.0}$, at which $|f_{u1}|$ does not depend on the sign f_{pb}. If we denote the error of the gear ratio at $f_{pb} > 0$ through Δi_1, and at $f_{pb} < 0$ through Δi_2, we can state that:

$|\Delta i_1| > |\Delta i_2|$ for $x_1 < x_{1.0}$

$|\Delta i_1| = |\Delta i_2|$ for $x_1 = x_{1.0}$

$|\Delta i_1| < |\Delta i_2|$ for $x_1 > x_{1.0}$

Thus, by changing the gear displacement factor x_1, you can change $|f_{u1}|$ to a certain extent. Let us call the displacement $x_{1.0}$ optimal. It matters somewhat more than the displacement found from the condition that the tooth dedendum is not undercut.

The choice of the optimal displacement $x_{1.0}$ leads to the contact point of the teeth being closer to the engagement pole, thereby reducing the difference in the speeds of the contacting points of the teeth. When choosing the optimal displacement coefficient $x_{1.0}$, the line of engagement is shifted towards the symmetrical position relative to the pole of the engagement. An even greater effect is obtained by using an optimal displacement $x_{1.0}$ in combination with a reduced addendum coefficient, for example $h_a^* = 0{,}8$: the decrease in magnitude f_{u1} goes down by 10-50%, and hence a decrease in the shock load on the teeth occurs. The table shows an example of calculating the relative error of the gear ratio of an open gear pair of an excavator swing mechanism with the number of pinion teeth z_1 = 12 and the number of gear teeth z_2 = 60.

Table 1. Example of influence of $x_{1.0}$ and h_a^* on $\frac{f_{u1}}{u}$

No.	x_1	x_2	h_a^*	$\frac{f_{u1}}{u}$	
				$f_{pb} > 0$	$f_{pb} < 0$
1	0.3	0	1	0.1233	-0.1015
2	$x_{1.0}$ = 0.45	0	1	0.1045	-0.1045
3	$x_{1.0}$ = 0.338	0	0.8	0.0965	-0.0969

It should be noted, however, that in addition to a decrease in the probable value f_{u1}, applying $x_{1.0}$ and h_a^* = 0.8, can lead to a decrease in the theoretical overlap ratio ε_γ, an increase in the angle of pressure of the teeth and a thickening of the dedendum. At the same time, the influence of the displacement coefficient of the wheel x_2 has little effect on the relative error of the gear ratio and due to x_2 it is possible to compensate for the decrease in the overlap ratio of the open gear transmission of the construction machine.

The effect of the transmission parameters on the displacement coefficient $x_{1.0}$ was considered. As a result, the following conclusions were drawn.

1) $x_{1.0}$ changes little (up to 7%) with a changing u from 9 to 30; at $u < 9$ the change in $x_{1.0}$ is more noticeable;

2) $x_{1.0}$ depends on the number of gear teeth z_1: with increasing z_1 by one, $x_{1.0}$ decreases by ~ 2%;

3) $x_{1.0}$ decreases by ~ 15-17% with decreasing h_a^* by 0.1;

4) a change in x_2 has little effect on a change in $x_{1.0}$ and at $u > 9$ this influence can be neglected;

5) $x_{1.0}$ is almost independent of the difference in pitches of the engaging gears; so when changing f_{pb} from 0.002 m to 0.03 m $x_{1.0}$ decreases only 1.09 times;

6) in case of transmission operation in the multiplier mode (the driving of the gear instead of the pinion) $|f_{u1}|$ is much less, but the relative values are comparable.

This work is more concerned with the parameters of open gears of construction machinery mechanisms, for example, mechanisms for rotating the platform of an excavator or crane, or a mechanism for driving mechanical equipment, for example, a truck-mounted concrete drum. These transmissions usually have the number of gear teeth less than 17-15, a relatively large diametral pitch, low tooth cutting accuracy and the gear ratio is greater than that usually used in gearboxes with a closed housing.

4. Conclusion

The performed analysis showed that by choosing the recommended gear displacement coefficient, it is possible to reduce the probable dynamic effect of the difference in pitches of the meshing gears. In this case, the concept of probable impact implies the randomness of both the magnitude and the sign of the difference between the main tooth pitches. This recommendation can be especially useful for open gears on construction machinery, such as an open gear swing on a single bucket excavator or construction crane. In these construction mechanisms, the gear and the pinion do not have a common frame, which leads to the peculiarities of the gearing parameters. Such a simple approach is the optimal choice of the gear displacement factor from the point of view of the vibration activity of the gearing. It contributes to an increase in the quality of construction machines without significant costs involved.

References

[1] Litak G and Friswell M 2005 Dynamics of a Gear System with Faults in Meshing Stiffness *Nonlinear Dynamics* **41** 415-421
[2] Dimitrion K and Fokas N 2015 Calculation of Spur Gear Dynamic Transmission Error in Consideration of the Progressive Engagement of Compliant Profile-Modified Teeth *MATEC Web of Conf.* **28** 02004
[3] Zhang L, Chen C, Jie Liu J and Zhao S 2018 Dynamic Characteristic of Spur Gear with Flexible Support of Gearbox *J. of Vibroengineering* **20(3)** 1530-43
[4] Ozguvent H and Houser D 1988 Mathematical Models Used in Gear Dynamics: A Review *J. of Sound and Vibration* **121(3)** 383-411
[5] Wang Z, Zhang L, Luo Y Q and Chang-Zheng Chen C-Z 2017 Dynamic Model of Spur Gear Pair with Modulation Internal Excitation *Int. J. of Rotating Machinery* 1264904

Approaches to reduction of dynamic loads arising in truck-mounted hydraulic lifts

M A Stepanov and B A Kaitukov

National Research Moscow State University of Civil Engineering, 26, Yaroslavskoe highway, Moscow, 129337, Russia

E-mail: stepanov@mgsu.ru

Abstract. The article addresses approaches to reduction of dynamic loads arising in truck-mounted hydraulic lifts. Lifts are widely used to perform assembly and painting work in the course of construction. Dynamic loads, arising in the process of their operation, reduce performance and operational safety of lifts. The co-authors focused their research efforts on hinge mechanisms installed in truck-mounted hydraulic lifts. Hydraulic lift booms have hydraulic drives that ensure a uniformly continuous pace of hydraulic cylinder rod extension, but rotation speeds of the booms do not coincide, thus, causing them to move up by fits. This phenomenon triggers dynamic loads that arise in booms and make workers feel uncomfortable when on the platform. The co-authors provide motion equations for upper and lower booms of a lift. Having solved these equations, the co-authors obtained rational values of loads applied to the booms. Dependencies between varying values of the force in a hydraulic cylinder rod and the boom swing angle, as well as between the boom swing angle and the boom lifting angle are to be identified to reduce dynamic loads. These dependencies enable the co-authors to tailor the pump output. The co-authors suggest a rational system of proportional hydraulic cylinder control over acceleration or the force. Dependencies, obtained by the co-authors, can be applied to fine-tune the proportional distributor to make booms move smoothly, to reduce dynamic loads and to ensure a comfortable working environment for workers on the platform.

1. Introduction

Truck-mounted hydraulic lifts are construction machines that serve as mechanical aids in the process of construction. The number of accidents, involving hydraulic lifts, is driven by the growing volume of construction work. Worst accidents deal with the low reliability of hydraulic systems. Hydraulic lifts are equipped with one, two or more hinge booms. Levers and hydraulic cylinders are used to change the angle between them. The platform, attached to the top of the upper boom, is moved by changing the angle between top and bottom booms and by rotating the turntable [1]. Construction operations, involving hydraulic lifts, are performed by workers at some distance from the earth surface. Several accidents and injuries occurred due to the insufficient reliability of hydraulic systems [2]. Therefore, reliability improvement and development of recommendations for reduction of dynamic loads in truck-mounted hydraulic lifts have gained relevance [3, 4]. The following problems, dealing with the dynamic behaviour of hydraulic machines, arise: the temperature differential, demonstrated by the fluid inside different elements of the hydraulic system; the lack of an adequate description of the dynamic behaviour of hydraulic drive elements, which is needed to design a dynamic circuit composed of a hydraulic drive, a mechanical drive and metal constructions [5].

2. Results

Let us take a closer look at the behaviour of booms and hydraulic cylinders in motion. Research and analyses have shown that the rational choice of a swing angle, hydraulic cylinder rod pace, force and pressure values inside the hydraulic system can reduce dynamic loads. The resulting load, applied to the mechanism, is the total of dynamic and static loads. The hydraulic drive of a truck-mounted lift has a fixed capacity pump. Since the lifting speed depends on the swing angle of booms, it's important to identify the influence of speed variations on the dynamic behaviour in motion. Open-loop fluid power drive systems, installed in truck-mounted lifts, have a significant impact on changes in dynamic loads due to the irregular boom lifting pace.

The pump capacity is invariable; that's why the pace of the hydraulic cylinder extension in the course of boom lifting is constant; it depends on the diameter of the hydraulic cylinder responsible for lifting the boom, whereas the boom lifting (rotation) speed depends on the boom angle. Therefore, it's important to know the dependencies that describe the kinematics of booms.

We use the design, demonstrated in Figure 1, as the basic one, and determine the parameters of lift booms.

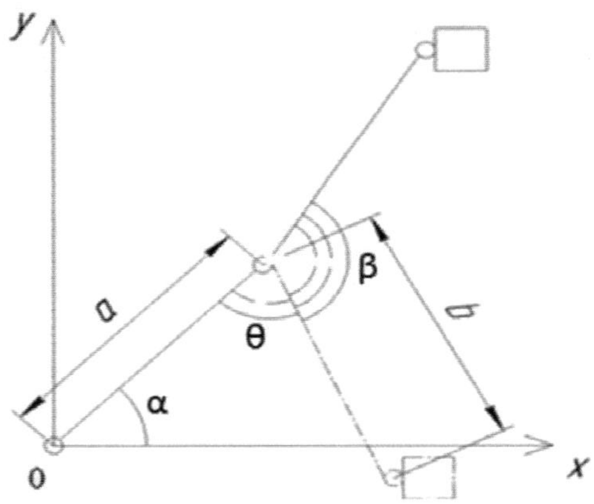

Figure 1. Design model.

Boom angle α (Figure 1) can vary from 0 to 90°, while β varies from 0 to 180°; therefore, the equation describing the motion of booms can be written as:

$$\begin{cases} x = a\cos\alpha + b\sin\theta \\ y = a\sin\alpha + b\cos\theta \end{cases}. \tag{1}$$

If $\beta > 180°$, then:

$$\begin{cases} \lambda+\beta = 90° + \upsilon \\ \lambda+\beta\text{-}90° = \upsilon \end{cases}. \tag{2}$$

Angle t=β-180°, therefore:

$$\begin{cases} x = a\cos\alpha + b\sin t \\ y = a\sin\alpha + b\cos t \end{cases}. \tag{3}$$

Let us identify dependence between the change in angle α and the change in the position of a hydraulic cylinder rod (Figure 2).

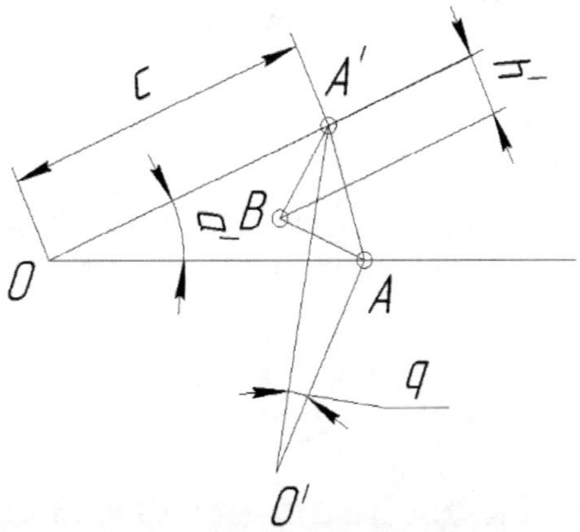

Figure 2. Dependence between the change in angle α and the change in the position of a hydraulic cylinder rod.

The rod pace is identified according to (4):

$$\upsilon = \frac{Q}{A} \tag{4}$$

where Q is the fluid flow rate inside the hydraulic cylinder in square meters per second; A is the piston surface area, identified according to (5):

$$A = \frac{\pi D^2}{4} \tag{5}$$

where D is the hydraulic cylinder diameter in square meters.

Let us identify the relation between the rod pace and the angular speed of the boom (Figure 3).

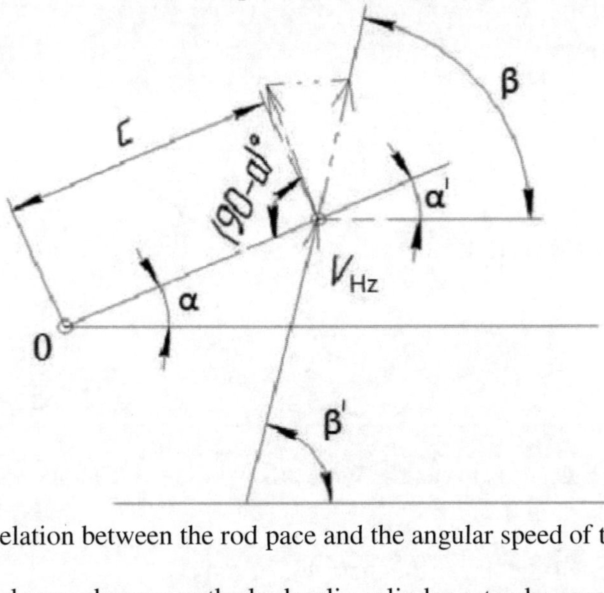

Figure 3. Relation between the rod pace and the angular speed of the boom.

The direction of the rod pace changes as the hydraulic cylinder extends; pace direction depends on the swing angle of the hydraulic cylinder (see Figure 3). Initially angle β', as demonstrated in Figure 3, is unaltered; however, further extension of the hydraulic cylinder causes the angle to change $\beta = \beta' + q$, while the speed of angular boom rotation can be calculated using formula (6):

IOP Conf. Series: Materials Science and Engineering 1103 (2021) 012028 doi:10.1088/1757-899X/1103/1/012028

$$\omega = \frac{\upsilon_0}{x}\frac{1}{c}. \quad (6)$$

Angle υ is the angle between speed v_{Hz} and speed v_0; it will be equal to $\upsilon = 90° + \alpha - \beta$, then speed v_0 can be calculated using the formula:

$$v_0 = v_{Hz} \cdot \cos \upsilon. \quad (7)$$

The speed of booms, installed in truck-mounted lifts, changes depending on the angle of swing, and speed v_{max} is attained if angle $\alpha = 80°$.

Let us analyze the motion of the first boom as the most heavily loaded element. We will consider two motion modes; they are acceleration and continuous motion:

$$M_d = J_{doom} \cdot \ddot{a} = P_д(t)h_0 - (m_1 h_1 + m_2 h_2)g - c\varphi \quad (8)$$

where J_{boom} is the moment of inertia, $m_1$is the mass of the first boom; m_2 is the mass of the second boom and the platform; $\ddot{a}, a$ stand for the boom acceleration and displacement; c is flexural rigidity of the boom, being a cantilevered beam resting on two supports.

The moment arising during the displacement of booms is calculated according to the following formula:

$$M = G_1 l_1 + G_2 l_2 + Qgl_3 \quad (9)$$

where G is the mass of the i-th boom in kilograms, l is the boom length in meters.

The force arising in the hydraulic cylinder is:

$$F_{Hz} = PA_п \quad (10)$$

where A_n is the piston area, $A_n = \frac{\pi D^2}{4}$.

Figure 4 shows dependence between the force changing in the hydraulic cylinder rod and the swing angle of a hinge boom.

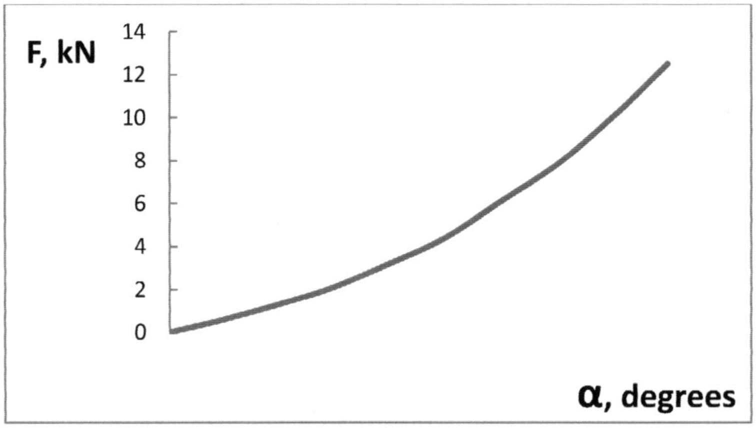

Figure 4. Dependence between the force arising in the rod and boom swing angle α.

The analysis has proven that the load change, caused by the changing swing angle, represents a parabola, if the load is applied to the hydraulic cylinder rod (11):

$$F = \alpha^2 + 48\alpha + 84. \quad (11)$$

Rod pace and angular speed are determined with regard for the swing angle (6), and ω is calculated according to the following formula:

$$\omega = \frac{v_{turn}}{R} \quad (12)$$

where ω is angular speed, $\frac{1}{c}$, R is rotation radius in meters.

Figure 5 shows dependence between angular speed ω and swing angle α.

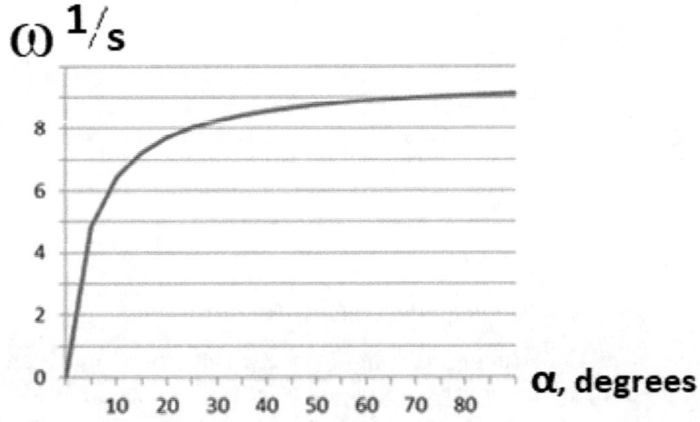

Figure 5. Dependence between angular speed ω and angle α.

3. Conclusion

The research results have evidenced dependence between the boom swing angle and the force in the rod which rises the hinge boom of a lift. Dependencies, obtained by the co-authors, will make it possible to rationally adjust the non-controllable pump output. It has been identified that any reduction of dynamic loads, arising in lifts, requires installation of controllable pumps or a system of proportional control over acceleration or forces in a hydraulic cylinder.

The analysis of various options has led the co-authors to the conclusion that the application of a system of proportional control over the force can ensure more reliable operation of a truck-mounted hinge lift and operational comfort on the platform. Besides, this measure can extend the service life of the metal construction of a lift boom and rationally fine-tune the extension pace of a hydraulic cylinder rod.

References

[1] Gudkov Yu I 1992 *Truck-mounted lifts and hydraulic platforms* (Moscow: Vysshaya Shkola)
[2] Stepanov M A and Tsibikov Ya I 2017 Analysis of operability of a truck-mounted hydraulic lift *Mechanization of construction* **78(11)** 9-12
[3] Inozemtsev A S and Rybalchenko D E 2007 *Pneumatic and hydraulic drives and systems* Part 2 *Hydraulic drives and systems. Fundamentals* (Moscow: Forum)
[4] Bashta T M 1971 *Fluid mechanics in machine building* (Moscow: Mashinostroenie)
[5] Gustov D Yu 2010 Problems of description of the dynamic behaviour of hydraulic construction machines *Vestnik MGSU* **4** 97-100

Garage complexes as the infrastructural basis of a modern city

S N Borychev[1], N A Suvorova[1] and E N Burmina[2]

[1] Department of Construction of Engineering Structures and Mechanics, Ryazan State Agrotechnological University, 1, Kostychev Str., Ryazan, 390044, Russia
[2] Autonomous non-profit organization of higher education Modern Technical University, 35A, Novoselov Str., Ryazan, 390048, Russia

E-mail: university@rgatu.ru

Abstract. The problem of the effective organization of parking spaces, the need for the construction of a multi-storey garage complex located in the transport-service system of the city investigated. The requirements of the existing federal law and regulations governing the conditions for placing the garage complex and car maintenance areas in relation to residential, public and business buildings located in the immediate vicinity have been worked out and considered. The construction site of Autohall garage complex, having weak, water-saturated clay soils, has been investigated. A constructive solution is considered and the advantage of using reinforced concrete column piles as supports reinforcing the soil is revealed. MONOMACH software complex solves a wide range of problems and has a number of advantages over other automated systems. The program implements a wide range of possibilities for inter-program communication. The construction of a monolithic frame of the complex, its volumetric model in COMPONOVKA program was consistently performed. Program GRUNT determined spatial modeling of the soil body according to the given engineering and geological surveys, the physical and mechanical characteristics of the soil, the bed coefficient. A monolithic foundation slab on a pile foundation was imported into PLITA program and the calculation of the optimal thickness of the slab, reinforcement of the lower and upper zones, numerical modelling of the work of the stress-strain state, i.e. an optimal design solution was chosen that excluded uneven deformations of the base, i.e. the design solution excludes ultimate deformations of the base.

1. Introduction

In modern urban conditions, the road transport network is often congested and the traffic is partially or completely halted. Among the problems generated by the high level of motorization of the urban population in large cities, architects and urban planners are aware of the urgent need to build parking lots, considering them to be the most important elements of the city's infrastructure. Optimizing the vehicle storage has changed enormously over the past century. Research of the problem of effective organization of parking spaces and the accumulated experience in design and construction makes it possible to distribute the space of urban areas effectively, to accommodate cars belonging to citizens has been investigated [1, 2]. The traffic performance of Ryazan is decreasing from year to year, due to cars parked along streets and squares, and the problem of the lack or insufficient number of organized parking spaces is actualized [3-5].

The construction site of the garage complex, the construction of which is directly associated with a decrease in the traffic performance of streets, has weak, water-saturated clay soils. The use of a foundation slab is not enough, since displacements obtained when calculating exceed the maximum permissible standard values. The calculation of the slab-pile foundation with the support of column piles on the rocky soil-limestone is carried out with the help of MONOMAH software package, which includes a wide range of functions agreed with regulatory documents. The software determines the bed coefficient by analyzing the stratification of the soil body, calculates deformations that occur both during construction and operation, taking into account the overall work of the complex frame.

2. Materials and methods

When speaking about multi-storey garage complexes, one means highly specialized structures designed for the organized parking of as many vehicles as possible. For this purpose, tiers located one above the other, various types of ramps, inclined floors, elevators, parking equipment and a complex of infrastructure (fire safety systems, ventilation, etc.), as well as maintenance workshops and other types of services are designed for the driveway. However, the purpose does not determine the shape of the transport structure, as well as the sum of the most important elements forming its basis. All the rest depend on the professionalism of design specialists: architects, designers and competent engineers. They conduct architectural and engineering research, choose the proportions and appearance of the building, its division, exterior, interior decoration and so on. Like any issue, when designing there should be the unity of form and content, including technical aspects [6, 7].

The garage complex is determined by some features and, first, by the multifunctionality of the use of the first floors. Garage users are divided into two large groups: those in need of short-term or long-term car storage. Storage areas with a relatively short duration of stay of vehicles should be located within walking distance. The needs of such users can be satisfied by low garages from one to four floors. In higher garages, parking spaces for them should be allocated on the lower floors. Usually the first floor is of multifunctional use and serves for the maintenance. Parking spaces are located on adjacent sites and on the ceilings above the first floor. In this case, the length of the path for everywhere or at the exit is small, the road in front of the car is free. There are no special difficulties in meeting the requirements of fire safety, ventilation, environmental friendliness and other requirements of Russian standards [8]. The structure of such a building is open to external observation from all sides. Therefore, its proportions, details, materials and even the color should be chosen with great care. Those who need long-term storage of cars usually use large multi-storey car parks. In both cases, the main factor influencing the choice of a garage complex is the attractiveness of the construction site and its involvement in the city's infrastructure.

The urban development situation in Ryazan determined the choice of the construction site for the Autohall garage complex (Fig. 1), being an area free from buildings and located in Kalnaya Street, 3A (Fig. 2). Optimization of design and architectural solutions took place at the stage of developing the design assignment and influenced the appearance of the entire structure.

Figure 1. Model of garage complex Autohall

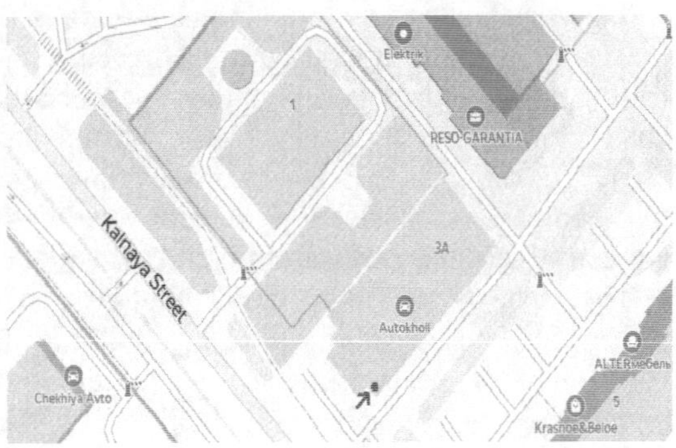

Figure 2. Microdistrict Kalnoe, Ryazan region (garage complex Autohall)

The important role in design and construction is played by the foundation, as one of the main elements of the building. The longevity of the entire structure will depend on the quality and reliability of it, since it perceives the loads from the walls of the building and overlying structures and transfers them to the base soils, counteracting the lift forces of the soil.

3. Results and discussion

A monolithic reinforced concrete slab was designed for the entire building on a solid pile foundation in garage complex Autohall to redistribute loads, eliminate uneven deformations of the base and ensure joint work. On the basis of the calculation, reinforced concrete piles 6 m long with a section of 30x30 cm were taken as supports reinforcing the soil. The depth of the foundation was 1.350 m from the ground surface.

The program interface MONOMACH is intuitively accessible, since when creating a design model, work is carried out with familiar objects: axes, beams, ceilings, columns, floors and other structural reinforced concrete elements. The basis for calculating building structures is the finite element method.

A volumetric model for the garage complex was created with the help of KOMPONOVKA program, which is included in the main module of MONOMACH software package (Fig. 3). The software has a wide range of functions and allows predicting the influence of design factors. The load-bearing structures of the complex are monolithic reinforced concrete columns on which the monolithic reinforced concrete floor slabs are supported.

According to Table 8.3 [9] the dead load of all structures, obtained in automatic mode, when forming a volumetric model of the complex and setting temporary loads (for multi-storey garages - 4 kPa) were taken as constant loads (Fig. 3)

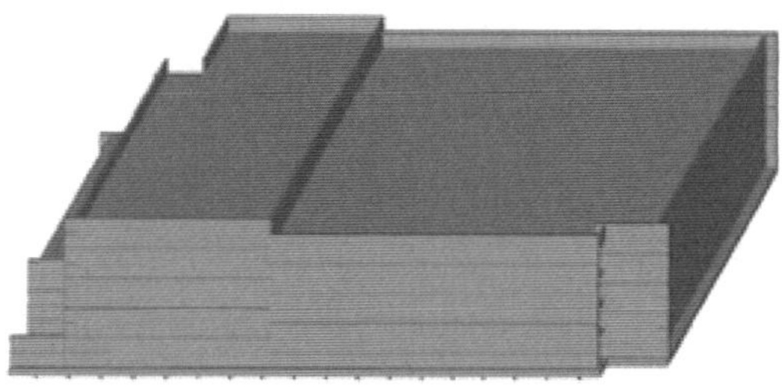

Figure 3. Volumetric design model of the garage complex

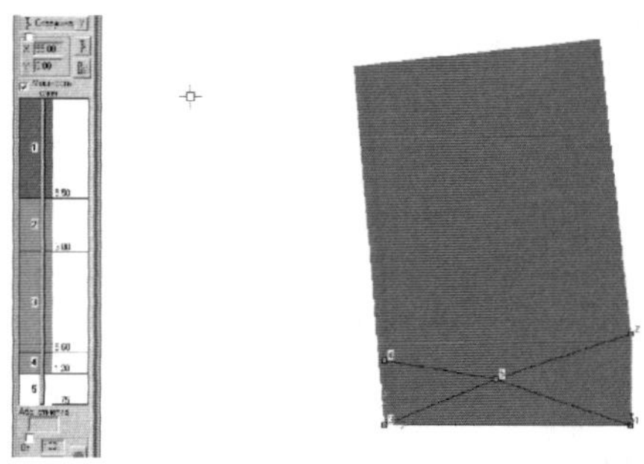

Figure 4. Stages of building a 3D model of a soil body for engineering and geological surveys

The issues of engineering and geological phenomena on the territory of the building, the relief and the results of the physical and mechanical characteristics of soils were determined by the results of drilling wells and their desk study by specialists of RYAZANGRAZHDANPROEKT CJSC. Weak, water-saturated soils were exposed under the base of the foundation slab.

Hydrogeological conditions of the construction site are as follows: soils to a depth of 8.0 m have a high degree of corrosiveness towards steel and soils to a depth of 2 m are moderately aggressive to concrete of normal permeability.

The next step was the spatial modeling of the soil body according to the data of engineering-geological wells located within the building area. Having set the absolute elevation of the mouth, using GRUNT program, a spatial model of the soil base was formed according to the specified engineering and geological conditions of the construction site, vertically reinforced with rigid elements (Fig. 4).

After placing the piles in the plan, their total number was 250 (Fig. 5). To check this technical solution, a numerical simulation of the work of the stress-strain state was carried out in universal software package MONOMAH, certified in the Russian Federation.

Based on the simulation results, it has been established that the use of a monolithic foundation slab on a solid pile foundation, on weak clay soils, provides regulatory requirements for the deformability of the foundation (Smax<Su). One of the most important tasks of calculating a pile-slab foundation is to find the dependence of the zones of influence for the second group of limiting states.

A slab on a pile foundation was imported into PLITA program. Column piles were used to support on the rocky limestone soil, which was load-bearing. The material for the foundation slab is heavy concrete of class B25 and reinforcement of class A 400 [10]. The slab thickness depended on the loads and the pile pitch (Fig. 5-6). The foundation slab was at the same time the floor of the garage complex.

The analysis of the calculation results in the software and hardware, made it possible to draw the following conclusions: the maximum displacement was 24.5 mm, which was less than the permissible value of 180 mm according to Appendix D, SP 22.13330.2016 Foundations of buildings and structures. Updated edition of SNiP 2.02.01-83* (Fig. 6); the optimal slab thickness was 500 mm (Fig. 8); the foundation slab was reinforced in the lower zone with separate rods with a diameter of 20 mm and a pitch of 200 mm and in the upper zone the ones with a diameter of 16 mm and a pitch of 200 mm, reinforcement of class A400 (Fig. 7-10). Technically, exceeding the permissible yielding (according to the second group of limiting states) led to a complete or partial loss of serviceability.

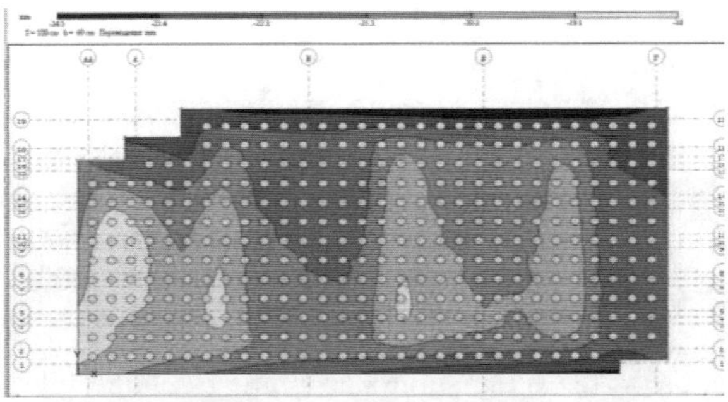

Figure 5. Distribution of displacements (yielding) in the foundation slab, S=100cm, h=60cm

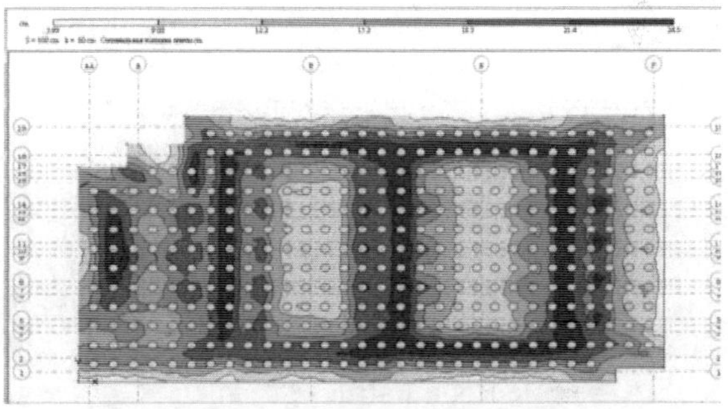

Figure 6. Results of calculating the optimal slab thickness, S=100cm, h=60cm

Isofields of reinforcement in PLITA program are presented in Figures 7-10. At the end of the design process, a complete model of the complex is obtained. The results of the calculation made it possible to substantiate a variant of the optimal design solution for a multi-storey garage complex, considering the complex engineering and geological conditions of the construction site.

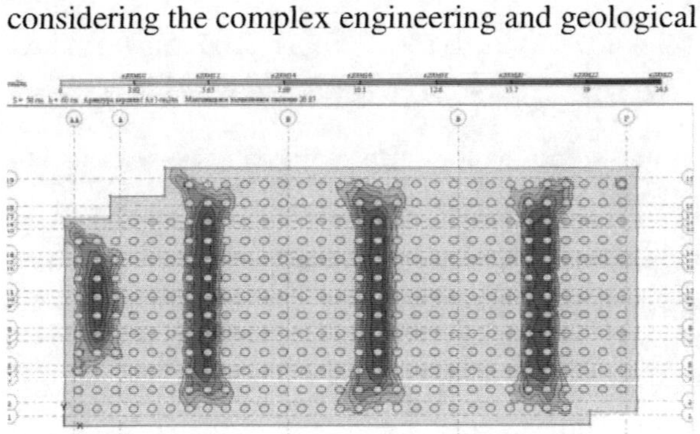

Figure 7. Top reinforcement along X axis, S=50cm, h=60cm

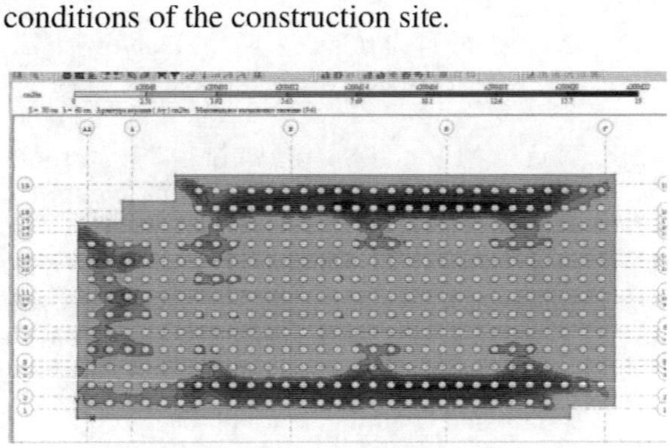

Figure 8. Top reinforcement along Y axis, S=50cm, h=60cm

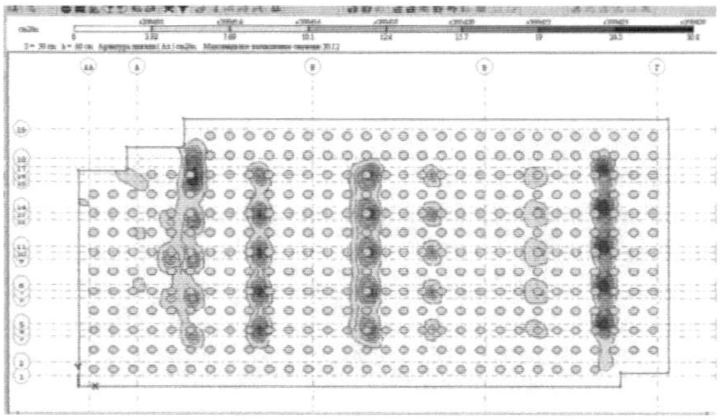

Figure 9. Bottom reinforcement along X axis, S=50cm, h=60cm

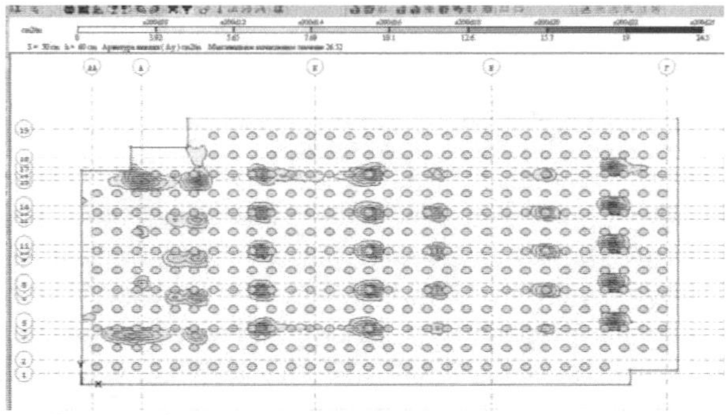

Figure 10. Bottom reinforcement along Y axis S=50cm, h=60cm

4. Conclusion

The foregoing allows concluding that universal program complex MONOMACH is aimed at sequential computer-aided design of buildings in conjunction with the created 3D model of the soil body based on engineering and geological data.

The program requires not only time to build a computational model of the inter-program data transfer and the calculation itself, as well as the availability of professional, construction knowledge that allows interpreting the results obtained and qualitatively prepares a reasonable choice of design

As a result of the analysis of calculations carried out at the modern technological level and their interpretation, the following design decisions were made: the thickness of the foundation slab is 500 mm with heavy concrete of class B25; reinforced concrete piles have a section of 30x30cm and are 6m long; slab foundation is reinforced in the lower zone with separate rods with a diameter of 20 mm and a pitch of 200 mm and the rods in the upper zone with a diameter of 16 mm and a pitch of 200 mm, reinforcement of class A400.

The selected optimal design solution is the erection of the foundation slab uniting column piles resting on rocky soil that is the most technologically advanced and reliable when weak, water-saturated clay soils. These findings can be used when designing similar buildings.

This direction of design is relevant and in demand. The construction of multi-level garages in densely built-up Ryazan will help solve the problem associated with storage of vehicles, traffic and expansion of the road network and a high level of motorization of the urban population.

References

[1] Kodish E N, Trekin N N, Kelasiyev N G, Terekhov I A and Soskov A A 2017 *Steel framed multilevel parking structure design: Steel Construction Development Association* (Moscow) 176 p

[2] Shalin M V and Suvorova N A 2017 Problems and solutions for car parking in large settlements *Student research to science and education of the XXI-st century. Mater. of the IX-th Int. student sci.-pract. conf.* (Ryazan) pp 101-105

[3] Terentyev V V, Andreev K P and Shemyakin A V 2018 Application of road energy-absorbing barriers to improve road safety *Transport, Transport Facilities* **1** 5-12

[4] Suvorova N A and Burmina E N 2020 Choosing a contractor for construction work *Student scientific research - to science and education of the XII century. Mater. of the XII-th Int. student sci.-pract. conf.* (Ryazan) pp 68-72

[5] Belikova T S, Tkach T S and Sheremet I V 2020 Technology of concreting in winter conditions *Scientific and practical aspects of the innovative development of transport systems and engineering structures* (Ryazan) pp 340-348

[6] SP 42.13330.2011 *Urban planning. Planning and development of urban and rural settlements.* Updated edition of SNiP 2.07.01-89 * (with Amendment)
[7] Federal Law No. 190-FZ of December 29, 2004 (as amended on July 3, 2016) *Urban Development Code of the Russian Federation* 2017 (Moscow: Eksmo-Press)
[8] SP 113.13330.2016 *Car parking.* Updated edition of SNiP 21-02-99 * (with Amendment No. 1).
[9] SP 20.13330.2016 *Loads and impacts.* Updated edition of SNiP 2.01.07-85*.
[10] SP 63.13330.2012 *Concrete and reinforced concrete structures. Basic provisions.* Updated edition of SNiP 52-01-2003.

IOP Conf. Series: Materials Science and Engineering 1103 (2021) 012030 doi:10.1088/1757-899X/1103/1/012030

Choosing the effective modes of compaction of hard concrete mixtures

V G Vasiliev

Moscow State University of Civil Engineering, 26, Yaroslavskoe highway, Moscow, 129337, Russia

E-mail: VasilievVG@mgsu.ru

Abstract. In the manufacture of reinforced concrete products and structures, one of the main labor-intensive operations is the compaction of concrete mixtures. All the basic physical and mechanical properties of the molded products, such as density, strength, frost resistance and others, as well as the specified shapes and high-quality surfaces, depend on the quality of mixture compaction. A decrease in the density of a product after compacting concrete mixes by 1% leads to a decrease in the strength of hardened concrete by 5-7%, and when molding products from hard mixtures – up to 13%. The effectiveness of vibration molding of concrete and reinforced concrete products is largely determined by the methods and modes of compacting concrete mixtures in molds. These questions are especially relevant when compacting hard and very hard concrete mixes. The analysis of the existing modes of molding reinforced concrete products from hard concrete mixtures is carried out in this paper. The equipment used for molding reinforced concrete products applies different types of compaction: vibration, shock-vibration and shock. The main parameters of vibroforming machinery are frequency, amplitude and acceleration of the working body, which can be symmetric and asymmetric. The advantages of using asymmetric shock-vibration modes are shown, and a method for creating shock-free asymmetric vibrations is proposed. The results of theoretical and experimental studies are presented, which have shown the high efficiency of compaction of hard concrete mixtures in shock-free asymmetric modes.

1. Introduction

In the manufacture of reinforced concrete products and structures, one of the main labor-intensive operations is the compaction of concrete mixtures. All the basic physical and mechanical properties of the molded products, such as density, strength, frost resistance and others, as well as the specified shapes and high-quality surfaces, depend on the quality of mixture compaction. A decrease in the density of a product after compacting concrete mixes by 1% leads to a decrease in the strength of hardened concrete by 5-7%, and when molding products from hard mixtures – up to 13% [1, 2].

To increase the efficiency of shaping reinforced concrete products, especially from hard concrete mixtures, it is very important to rationally approach the choice of molding equipment and use the optimal modes of vibration on the compacted concrete mixture.

2. Materials and methods

The equipment used for molding reinforced concrete products applies different types of compaction: vibration, shock-vibration and shock. The main parameters of vibroforming machinery are frequency, amplitude and acceleration of the working body, which can be symmetric and asymmetric [1-4].

Evaluation of the effectiveness of vibration impact on the concrete mix during the molding process is most often carried out according to the intensity I, which is determined by the acceleration of vibrations of the screed vibrator [5]:

$$I = A \cdot \omega^2, \tag{1}$$

where A - amplitude of displacement; ω - circular vibration frequency.

3. Results and Discussion

Numerous works on the study of the processes of molding concrete products have convincingly shown that asymmetric shock-vibration modes are most effective for compacting hard concrete mixtures [1, 2, 4]. Comparison of the specific power of the screed vibrators during the compaction of concrete mixtures showed that the power of the vibrators functioning in asymmetric modes are about eight times higher than the capacities of the vibrators operating in the symmetric vibration mode [2].

When implementing asymmetric shock-vibration modes, two accelerations are recorded (depending on the position of the screed vibrator): top A_{g_h} and bottom acceleration A_{g_l}. Moreover, A_{g_l} plays a primary role in molding, since maximum compaction forces develop in the concrete mixture during this period of time [1, 3].

The results of comparative experimental studies, showing the effectiveness of various modes of molding products from hard concrete mixtures, are presented in Table 1 [6]. In this case, the compaction time of concrete mixtures was used as a criterion.

Table 1. Compaction time of concrete mixtures under various vibration modes

Concrete mixture stiffness (s)	Compaction mode		Compression ratio
	Symmetric $f = 50$ (Hz) $A_g = 3.5g$	Shock-vibration $f = 10$ (Hz) $A_{g_h} = 2\text{-}2.5\ g;\ A_{g_l} = 5\text{-}7g$	
20	30	15	0.98
50	150	50	0.98
70	300	120	0.95

The data in Table 1 convincingly show that the asymmetric shock-vibration mode of compaction of concrete mixtures is more preferable in comparison with the standard mode of harmonic vibrations.

Theory substantiating the effectiveness of using shock-vibration while molding reinforced concrete products in comparison with symmetric harmonic vibrations is presented in [7]. It was found that in the case of impact of harmonic vibrations on the concrete mixture, the particles of the mixture vibrate with the frequency of the generated vibration action. In the case of shock-vibration impact, the displacement of concrete particles during the propagation of a shock wave is directed against the direction of wave propagation, thereby providing a compaction effect. The intensity of compaction process for concrete mixtures in real conditions is also influenced by the shape and spectrum of impulses of shock-vibration effects.

One of the ways to reduce the time of vibration compaction of concrete mixtures is to use screed vibrators with a free (without fixation) form installed on the vibration table, which creates conditions for the form to collide with the table surface [2, 4]. As a result, the processes of vibration compaction of mixtures occur more intensively, since at the moment of impact a wide spectrum of frequencies, containing a large number of high energy components, is generated [3].

In shock-vibration modes, the maximum value of the bottom acceleration A_{g_l} (when the working body collides with the limiter) largely determines the magnitude of the dynamic stresses and the strength characteristics of the screed vibrator. Since the value A_{g_l} reaches 100 m/s^2 and more, this leads to frequent crashes and breakdowns of vibration equipment [2, 4]. In addition to reducing the reliability and durability of equipment, the permissible noise and vibration levels are also exceeded.

As shown in [8], one of the effective modes of compaction of hard concrete mixtures is the use of the so-called asymmetric shock-free vibrations, to a large extent imitating the shock-vibration mode and differing from the latter in the absence of collisions of the vibrator with the balancing frame.

Asymmetric shock-free vibrations are achieved by combining two harmonic vibrations of the same amplitude, differing in frequency by two times. For the practical implementation of asymmetric vibrations, a system of inertial vibrators is required, containing two pairs of eccentric weights, synchronously rotating in opposite directions, and the circular frequency of rotation of the extreme weights ω_1 should be twice the rotation frequency of internal weights ω_2 (Figure 1).

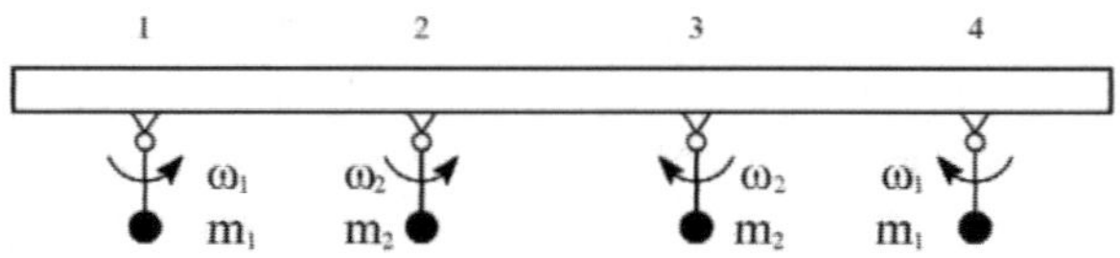

Figure 1. Inertial vibrator system. 1, 2, 3, 4 - eccentric shafts; $\omega_1 = 2\ \omega_2$.

To obtain the maximum asymmetry of the resulting vibration, it is necessary that, with equal eccentricities of the weights, the ratio of the eccentric mass of the external weights m_1 to the internal m_2 was 1:4 [8]. Taking into account these recommendations, the screed vibrator was developed and created with the following technical characteristics: lifting capacity - 1.5 t, vibration frequency - 7...32 Hz, vibration amplitude - 0.3...3.5 mm, electric motor power – 11 kW.

Researching the process of vibration compaction of hard concrete mixtures by asymmetric vibrations showed that the efficiency of compaction is determined not only by the absolute values of accelerations A_{gl} and A_{gh}, but also depends on the duration of action of the bottom compaction acceleration A_{gl} [8].

Figure 2 shows oscillograms of displacements of a mold with a concrete mixture under shock-vibration and in asymmetric shock-free modes of molding with the same values of the frequency f and the amplitude of the lower acceleration A_{gl}.

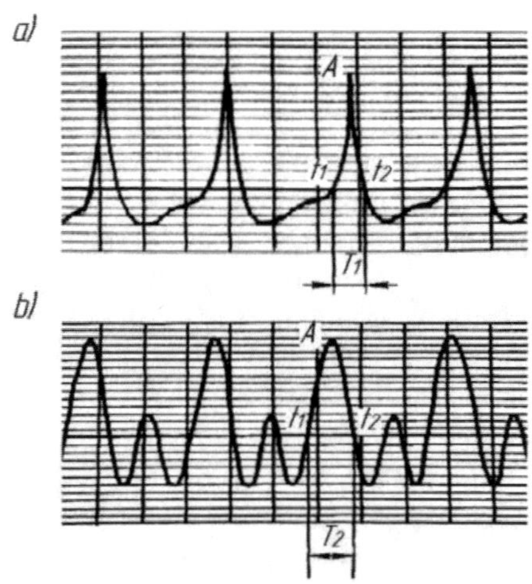

Figure 2. Oscillograms of concrete mixture mold displacements in the process of shaping by shock-vibration (a) and in asymmetric shock-free mode (b)

Let us determine the work performed by vibrators at the time interval during which the concrete mixture is affected by the compaction accelerations A_{gl}. The work done by the screed vibrator with shock-vibration (Figure 2*a*) is simplified by the area of a triangle, limited by the moments of time t_1, t_2 and the amplitude of the vibration A.

Work A_p, performed by a vibrator with asymmetric shock-free vibrations over a period of time from t_1 to t_2, that is, over a period of T_2 (Figure 2*b*) can be represented as:

$$A_p = \int_{x_1}^{x_2} F dx = \int_{t_1}^{t_2} F v dt, \tag{2}$$

where F and v are the force and speed of the impact of the screed vibrator working body;

x_1, x_2 and t_1, t_2 are the start and end coordinates of the point and their time values.

Specifying the trajectory of motion of a point on the vibrating table surface by the expression $x = Asin\omega t$, taking the change in force over time as a function $F = fcos\omega t$, a speed change $v = \frac{dx}{dt} = A\omega cos\omega t$, work force for the period of vibration from t_1 to t_2 will have the form:

$$A_p = fA\omega \int_{t_1}^{t_2} cos^2\omega t dt. \tag{3}$$

Having chosen the initial moment $t_1 = 0$, $x_1 = 0$ and the final moment $t_2 = \pi/2$, $x_2 = 0$ and substituting these values into the equation (3), after some transformations we get:

$$A_p = \pi/2 \cdot A^2 f^2. \tag{4}$$

These results are in good agreement with the criterion for evaluating the effectiveness of vibration impact on the mixture, which is determined by the acceleration of vibrations of the screed vibrator [5].

Comparing the operation of vibrators during compaction of concrete mixtures by shock-vibration (Figure 2*a*) and asymmetric shock-free mode (Figure 2*b*) with the same values of amplitude, vibration frequency and accelerations, it can be seen that the exposure time of asymmetric shock-free vibrations T_2 is longer than shock T_1 value. Therefore, the work performed by asymmetric shock-free vibrations is greater than the work in the shock-vibration mode.

The results of experimental studies of the kinetics of compaction of hard concrete mixtures, specifically by lowering the surface of the mixture in the process of vibration by various vibration effects, are shown in Figure 3 [8].

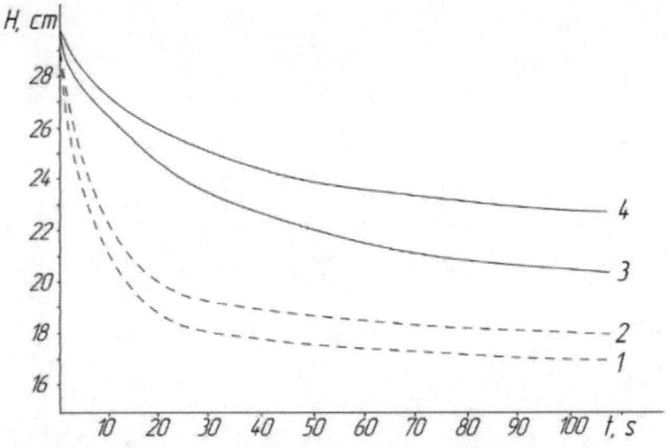

Figure 3. Dependence of the height of ***H*** layer on the compaction time ***t*** for hard concrete mixtures with hardness values of **1, 3** - 35s; **2, 4** - 60s at different vibration modes:
\- - - - - - asymmetric shock-free mode;
_______ shock-vibration mode

As graphs in Figure 3 show, the time of compaction of concrete mixtures in the asymmetric shock-free mode is approximately 3.5-5 times less compared to the shock-vibration mode with a higher degree of compaction. Consequently, the asymmetric shock-free vibration mode is more effective than the shock-vibration mode.

IOP Conf. Series: Materials Science and Engineering 1103 (2021) 012030 doi:10.1088/1757-899X/1103/1/012030

4. Conclusion

The analysis of the existing modes of molding reinforced concrete products from hard concrete mixtures is carried out. The performed set of experimental studies has shown the high efficiency of compaction of hard concrete mixtures (hardness index of 30...80s according to a technical viscometer) when using asymmetric shock-free modes in comparison with shock-vibration and symmetrical vibrations. This makes it possible to reduce the overall molding cycle to 60...100s and increase the strength of concrete by 12...20%, depending on the vibration frequency, acceleration and hardness of the concrete mixture.

References

[1] Gusev B V and Zazimko V G 1991 *Vibration Technology of Concrete* (Kiev: Budivelnik)
[2] Savinov O A and Lavrinovich E V 1986 *Vibration Technology of Compaction and Formation of Concrete Mixtures* (Moscow: Striyizdat)
[3] Gusev B V, Deminov A D, Kryukov B I et al 1982 *Shock-Vibration Technology of Compaction of Concrete Mixes* (Moscow: Striyizdat)
[4] Shmigalsky V N 1968 *Forming Products on Vibroplatforms* (Moscow: Striyizdat)
[5] Gusev B V, Axelrod E Z, Zvezdov A I et al 1988 *Manual on the Technology of Molding Reinforced Concrete Products* (Moscow: Striyizdat)
[6] Sinyaeva E A 1982 *Influence of the Parameters of Vertical-Directional Symmetric and Asymmetric Vibrations on the Compaction of the Concrete Mix:* dissertation abstract (Moscow: NIIZHB)
[7] Vasiliev V G 2017 The Choice of Parameters of Vibration Effects in the Molding of Concrete Products *J. Mech. of const.* **11** 21-25
[8] Vasiliev V G 2017 Block Vibration Platforms with Shockless Asymmetric Vibrations *J. Mech. of const.* **7** 14-16

IOP Conf. Series: Materials Science and Engineering 1103 (2021) 012031 doi:10.1088/1757-899X/1103/1/012031

Geoecological screening of the state of small and medium-sized river basins to assess the impact of measures for the operation of hydraulic structures

Y D Yurova[1] and V A Shirokova[1,2]

[1] State University of Land Use Planning, 15, Kazakova Str., Moscow, 105064, Russia
[2] S.I. Vavilov Institute for the History of Science and Technology of the Russian Academy of Sciences, 14, Baltiyskaya Str., Moscow, 125315, Russia

E-mail: Yuliya.yurova.1996@mail.ru

Abstract. The relevance of the work is due to the annual increase in anthropogenic load on components of geosystems of basins of small and medium rivers. The paper proposes an integrated approach to solving the main problem using a combination of classical and modern methods of analyzing natural systems - the screening method. This paper notes the need for screening control during the implementation of measures for the operation of hydraulic structures to assess the current state of catchment basins of small and medium-sized rivers, which includes the forecast of possible changes in the components of geosystems under the influence of anthropogenic load in order to prevent, minimize or eliminate negative environmental and related socio-economic and other consequences and maintain optimal living conditions for the population. Solving the problem using the screening method will make it possible to better analyze the state of catchment areas, in particular, the basins of small and medium rivers with a high anthropogenic load, which will make it possible to promptly identify environmental risks that have a negative impact on the components of geosystems and take more effective measures to eliminate them. The development of a methodology for screening control will make it possible to comprehensively assess the impact of anthropogenic factors on the geoecological state of the catchment area of the Osetr river, to study the current ecological state of the study area, identify the main factors of impact, as well as assess the degree of pollution, zone the study area according to the degree of anthropogenic load, and develop a geoecological passport for the catchment area of the Osetr river.

1. Introduction

A catchment basin is a system of landscape formations that differ in structure and function and are combined with a landscape aquatic ecosystem. There is a close relationship between the geographic structure of the catchment area and the functioning of the hydrographic network.

Small and medium-sized rivers reflect the ecological state of the environment and determine the hydrological and hydrochemical characteristics of large rivers, which makes them vulnerable to overexploitation of both water resources and drainage basins.

When developing a methodology for assessing the geoecological situation of small river basins, which significantly react to any anthropogenic impacts within the catchment area, N. P. Karpenko noted that the catchment areas of medium-sized and small rivers had become the subject of special attention due to the high degree of their transformation as a result of anthropogenic activity [1].

In their works on the geoecological analysis of the catchment area of a small river as exemplified by the Ugra basin (Levin, 2007); geoecological assessment of the state of small watercourses in the Kaliningrad region (Nagornova, 2012); assessment of the impact of the development of the catchment area of the Verkhnyaya Sukhona River (Vologda region) on the chemical composition of the waters of its tributaries (Ivicheva, Filonenko, 2017); the influence of diffuse pollution on water bodies (Ratkovich, Markin et al., 2016), the authors indicate that medium-sized and small rivers have a lower capacity for self-purification and are more prone to pollution, as a result of which their ecological state deteriorates. At present, the anthropogenic impact on river basins, catchment areas and directly on water bodies is growing very rapidly, which is associated with the high growth rates of cities and the developing industry, agriculture, an increase in transport networks and transport itself, which leads to a change in the quality and functioning of aquatic ecosystems.

Long-term operation of surface and underwater structures of hydraulic structures on small and medium rivers has a significant impact on the catchment basin and the natural environment as a whole, with subsequent positive or negative effects.

I. S. Moiseev, V. V. Malakhanov, Yu. P. Lyapichev note in their works on studying the experience of designing and building earth dams, classifying the states and criteria of the operational reliability of hydraulic structures and hydrological and technical safety of hydraulic structures, that the safety of hydrosystems, especially small ones, decreases due to unreliable initial data for assessing the environmental situation at the site of work. One of the significant results of the influence of hydraulic structures on the catchment basin is a change in the hydrological regime of rivers and formation of conditions for the qualitative composition of water [2–4].

In this regard, it becomes necessary to carry out activities for integrated environmental monitoring and geoecological assessment, which will make it possible to trace changes in water quality in the basins of small and medium-sized rivers and determine their impact on larger water bodies and draw up a picture of anthropogenic impact in general. For geoecological assessment of the catchment areas of small and medium-sized river basins, an integrated approach to solving the main problem using a combination of classic and modern methods of analyzing natural systems is proposed.

River catchment area screening is a comprehensive method for assessing the components of geosystems. The application of the "screening" method will allow one to obtain an objective assessment of the geoecological situation in the studied catchment area, as it will provide an opportunity to better analyze the state of the catchment areas, in particular the basins of small and medium rivers with a high anthropogenic load, which will make it possible to quickly identify environmental risks that have a negative impact on the components of geosystems, and take more effective measures to eliminate them [5, 6].

Adam, Begun, Dmitriev et al. (2016); Dmitriev, Kupressova, Ledovskaya, Romanenko (2018) note in their works on substantiating the methodology of screening control of hydrosphere objects to detect accidental oil spills and the use of screening methods for environmental monitoring of hydrosphere objects, that this approach will reveal the excess of a given normative (or background) determined parameter of a water body, information about which is possible to obtain remotely in real time, as well as through additional research of selected samples in the laboratory, to analyze data on the state of water bodies for a certain period; identify the causes of pollution and factors affecting its distribution, and predict the state of water bodies in different phases of the water regime, before and after various emergencies.

The main goals of geoecological screening are: to classify (zone) the sites according to the degree of anthropogenic load on the environment of the water catchment, taking into account the indicators of the state of the environment in the territory of the method implementation; to determine the contribution of anthropogenic sources of pollution; to find out how pollutants enter the water body; to describe environmental problems, at solving of which the project is aimed; to determine characteristics of possible changes in the state of the components of geosystems, etc. [5, 6].

2. Materials and methods

The development of a screening method for obtaining an objective assessment of the geoecological situation in the studied catchment area of the Osetr river basin is based on long-term monitoring data for the period 2015-2020, scientific concepts, regulatory documents, statistical and cartographic materials of the Federal Service for Hydrometeorology and Environmental Monitoring, the Ministry of Ecology and Nature Management of the Moscow region, the Administration of the Zaraysky district, Ryazanproekt LLC, RusHydro, the company "Shooting from the air", "MOSEKOMONITORING", analytical laboratories "Meshchersky Scientific and Technical Center" and "EcoCenter", Federal Agency for Water Resources, Federal Forestry Agency, etc.

The work is based on methods for calculating complex indicators (water pollution index, specific combinatorial index of water pollution); analysis of the environmental risk of anthropogenic impact; assessing the anthropogenic load on river ecosystems, taking into account their regional characteristics (point-rating method); methodological recommendations for the development of unified integrated survey programs for catchment areas, waterworks and drinking water MR 01-19 / 52-17 and Earth remote sensing data [7-13].

Geoecological studies will be carried out on the basis of analysis of samples of water, soil, bottom sediments and bacterioplankton taken from different parts of the Osetr River catchment area.

The territory of the catchment basin of the Osetr river within the Moscow region from Serebryanye Prudy settlement to Akatevo settlement was taken as the studied area. The middle and lower section of the Osetr river from Zaraysk settlement to the mouth zone of the river (the area where it flows into the Oka River – Akatyevo settlement) was taken as a screening site.

3. Results and discussion

Osetr is a river in the European part of Russia, in the Moscow, Ryazan and Tula regions, that is classified as an average river in length and catchment area and is one of the largest southern tributaries of the Oka river.

Studies of the current state of the territory of the catchment area of the Osetr river are of great importance, both in theoretical and practical aspects, and are becoming more relevant due to the growth of anthropogenic load. Due to the fact that in the study area in the middle and lower river course of Osetr (from Zaraysk settlement - Vlasyevo settlement - Akatyevo settlement) there are no gauging stations or there is the least number of gauging stations operating year-round, a significant problem arises for monitoring surface waters. Early studies led to the conclusion that, like many other small and medium-sized rivers in the region, the Osetr river is highly sensitive to anthropogenic impact, which is due to a number of reasons associated with hydrological and natural-climatic characteristics, namely: relatively small volumes of runoff, low self-cleaning ability, dependence of the water regime on climatic and weather conditions, etc. [7,10-13].

The main sources of pollution in the river basin are industrial enterprises, utilities, agricultural land, transport, whose effluents are insufficiently treated and contain a large amount of pollutants. In addition, industrial enterprises dump garbage and household waste in water protection zones, ravines and gullies.

Despite the lack of study of the river Osetr basin, some issues require complex generalization and further solutions. The most problematic for Osetr is the content of organic pollution (according to BOD_5), biogenic elements (nitrite ion), suspended solids and some metals. A significant concentration of chemical pollutants falls on the lateral inlet. The degree of water pollution of the Osetr river is characterized as high, which is due to the violation of the existing standards for 5 ingredients. Long-term operation of the surface and underwater components of the structure of the Zaraisk dam has no less significant impact on the natural environment [2-4, 7, 12, 13]. Due to the absence or gross violations of the rules for the operation of the dam, the low efficiency of measures to ensure public safety, an incompetent assessment of the expected flooding zones, poor engineering training during the construction of the dam, etc., the long-term operation of the structure led to the wear of the dam structures by more than 70% [7, 8]. An equally important aspect is the lack of design documentation for the majority of owners, which determines an increased interest in the problem and makes it difficult for specialists to assess the safety of hydraulic structures according to the established class of structures. Measures for the reconstruction and operation of hydraulic structures have a significant impact on the components of geosystems with subsequent positive or negative consequences [2-4,7, 8].

According to the Administration of the Zaraysk District and Ryazanproekt LLC, during the reconnaissance survey, it can be noted that no critical violations in terms of environmental indicators, including assessment of radiation, chemical, sanitary-epidemiological and physical risk factors, were found. Monitoring of interaction of the hydraulic structures with the environment is of great importance for taking urgent measures, if necessary, to prevent undesirable effects or to mitigate them. The composition of monitoring studies is determined on the basis of an analysis of specific aspects of the impact of hydraulic structures on the environment [7, 8].

The study is aimed at a comprehensive geoecological study and assessment of anthropogenic impact on the catchment area of the Osetr river. The development of a methodology for screening control will make it possible to comprehensively assess the impact of anthropogenic factors on the geoecological state of the catchment area of the Osetr river, study the current ecological state of the study area, identify the main factors of impact, as well as assess the degree of pollution, zone the study area according to the degree of anthropogenic load and develop a geoecological passport for the catchment area of the Osetr river [5, 6].

Complex geoecological monitoring, analysis and assessment of the catchment area, changes in the hydrographic network and water content of the Osetr river will be conducted using Earth remote sensing data, modeling data and ground measurements.

To assess the quality of water in the screening and study areas, the following parameters will be determined: water temperature, pH value, electrical conductivity, turbidity and color, dissolved oxygen content, ionic composition, mineral and general forms of nitrogen and phosphorus, concentrations of heavy metals, oil products and organic substances. The measurements will be carried out by the following laboratory-field methods: high performance liquid chromatography (HPLC), atomic absorption spectrometry, potentiometry, fluorimetry, spectrometry, gas chromatography with mass spectrometric and flame ionization detection [9].

To assess the level of eutrophication of a water body, to determine the ability to self-purification and, as a consequence, to calculate the proportion of anthropogenic load, in accordance with the methods used in hydrobiological studies, a laboratory study of the main significant representatives of the phytoplankton community (Bacillariophyceae, Cyanobacteria) will be carried out. It is planned to determine the concentration of heavy metals and oil products in the bottom sediments [14].

Sampling and assessment of the qualitative state of soils are planned to be carried out at the screening site at the boundaries of the zones of sanitary protection zones, taking into account local topographic conditions and wind rose for non-specific, specific and microbiological indicators in accordance with MR 01-19 / 52-17.

The resulting geoecological models (schematic maps) of changes in the natural environment are supposed to be implemented in the form of GIS-oriented maps.

Based on the proposed methods and approaches for processing and analyzing the data obtained, a geoecological passport of the catchment area of the Osetr river basin will be created. A geoecological passport of a water body is a document reflecting the ecological state of a water body in accordance with generally recognized water protection standards and requirements for recreation areas to ensure the functioning of ecological well-being, forecasting eco-crisis situations and early warning of environmental hazards when assessing the state of a water body [15, 16]. A geoecological passport of the catchment areas of small and medium-sized rivers contains comprehensive information on the current state of nature and anthropogenic impact and can be used to assess and compare the characteristics of different basins. Studies on the creation of a geoecological passport include complex hydrological-hydrochemical, hydrobiological, ecogeochemical, radiation, landscape-ecological and sanitary-bacteriological surveys, as well as data from long-term field measurements using GIS technologies and mathematical modeling [15, 16].

4. Conclusion

Studies of the current state of the catchment areas of small and medium-sized rivers are of great importance, both in theoretical and practical aspects, and are becoming more relevant due to the increase in anthropogenic load.

Small and medium-sized rivers, as close as possible to consumers, reflect the ecological state of the environment, determine the hydrological and hydrochemical characteristics of large rivers, which makes them vulnerable to excessive use of both water resources and catchment basins.

This research is aimed at a comprehensive geoecological study and assessment of anthropogenic impact on the catchment area of the Osetr river.

Early studies led to the conclusion that, like many other small and medium-sized rivers in the region, the Osetr River is highly sensitive to anthropogenic impact, which is due to a number of reasons associated with hydrological and natural-climatic characteristics, namely: relatively small volumes of runoff, low self-cleaning ability, dependence of the water regime on climatic and weather conditions, etc. For example, the long-term operation of the surface and underwater components of the Zaraisk dam structure led to the wear of the structure of the structure by more than 70%, which significantly affected the natural environment [2-4,7,8].

In this regard, the development and application of the screening method will make it possible to obtain an objective assessment of the geoecological situation in the studied area of the catchment area of the Osetr river basin, since it will provide an opportunity to better analyze the state of the catchment areas, in particular, the basins of small and medium rivers with a high anthropogenic load, which will make it possible to promptly identify the occurrence of environmental risks that have a negative impact on the components of geosystems, and take more effective measures to eliminate them.

With an annual increase in anthropogenic load on the components of the geosystems of small and medium-sized rivers basins, the creation of a geoecological passport for a catchment basin using a screening approach that takes into account a comprehensive geoecological assessment of anthropogenic impact on the catchment area of the Osetr river basin is an urgent land management and water management task in solving issues of sustainable development of the territory.

Acknowledgments
The reported study was funded by RFBR, project number № 20-35-90019.

References

[1] Karpenko N P 2018 Assessment of the geoecological situation in river basins based on attributive indicators and generalized geoecological risks *Environmental Engineering* **2.** Retrieved from: https://cyberleninka.ru/article/n/otsenka-geoekologicheskoy-situatsii-rechnyh-basseynov-na-osnove-atributivnyh-pokazateley-i-obobschennyh-geoekologicheskih-riskov
[2] Moiseev I S 2000 Experience in the design and construction of embankment dams *Hydraulic Engineering* **8-9** 41-44
[3] Malakhanov V V 2000 Classification of states and criteria for the operational reliability of hydraulic structures *Hydraulic Engineering* **11** 8-14
[4] Lyapichev Yu P 2008 *Hydrological and technical safety of hydraulic structures: Study guide* (Moscow: RUDN) 222 p
[5] Adam A M, Begun V M, Dmitriev A V, Kupressova E A, Ledovskaya A M, Romanenko S V and Savichev O G 2016 Screening control of hydrosphere objects for detection of emergency oil spills *Izvestiya of Tomsk Polytechnic University. Georesources engineering* **327(9)** 29–38
[6] Dmitriev A V, Kupressova E A, Ledovskaya A M and Romanenko S V 2018 The use of screening methods for environmental monitoring of objects in the hydrosphere *Energy and resource efficiency for sustainable development: collection of scientific papers of the int. sci. conf.* (Tomsk: Tomsk Polytechnic University) pp 254-255

[7] Yurova Y D, Shirokova V A and Khutorova A O 2018 Assessment of the safety of hydraulic structures as part of environmental monitoring of small rivers (as exemplified by the Osetr river in the Moscow region): environmental aspects *Collection of scientific works of young scientists, graduate students, students and specialists following the results of the IX Youth Environmental Conf. "Northern Palmyra"* pp 109-113

[8] Overhaul of the dam on the Osetr River in the city of Zaraysk, Moscow Region 2017 *Technical report on the results of environmental engineering surveys* (Ryazan: Ryazanproekt LLC) **3 (17/84 IEI)** 173 p

[9] Karaushev A V 1981 *Methodological framework for assessing anthropogenic impact on surface water quality* (Leningrad: Gidrometeoizdat) 174 p

[10] Aminnikova V V and Arustamov E A 2017 Ecological and social problems of the Zaraisk district of the Moscow region in the year of ecology of Russia *Waste and Resources* **2** 1-10. Retrieved from: https://resources.today/PDF/04RRO217.pdf

[11] Reshetnyak O S, Nikanorov A M, Trofimchuk M M and Grishanova Yu S 2017 Assessment of hydroecological risk in the Oka river basin *Water and Ecology* **3** 158–170

[12] Yurova Y and Shirokova V 2020 Geoecological Assessment of Anthropogenic Impacts on the Osetr River Basin *Geosciences* **10(4)** 121-133

[13] Shirokova V A, Khutorova A O, Gurov A F, Shirokov R S and Yurova Yu D 2019 Integrated environmental monitoring of Osetr river *Advances in Engineering Research* **182(3)** 310–315

[14] Lukianova T S, Shirokov R S, Shirokova V A and Khutorova A O 2020 Energy resources of the world ocean ecosystem: study and problems *IOP Conf. Ser.: Earth Env.* **579** 012043

[15] Lamkov I M 2015 On the content of environmental passports *Interexpo Geo-Siberia* **2.** Retrieved from: https://cyberleninka.ru/article/n/o-soderzhanii-ekologicheskih-pasportov

[16] Zinchenko T D, Saksonov S V, Senator S A et al 2019 Environmental passport of the Usa River (Right-bank tributary of Volga) *Samarskaya Luka: Problems of regional and global ecology* **28(2)** 156-188

IOP Conf. Series: Materials Science and Engineering 1103 (2021) 012032 doi:10.1088/1757-899X/1103/1/012032

Study of the tension of a heterogeneous hydroelastic system

M V Zaretskaya

Department of Computer Technologies and Applied Mathematics, Kuban State University, 149, Stavropolskaya str., Krasnodar, 350040, Russia

E-mail: zaretskaya.m@kubsu.ru

Abstract. The purpose of the article is to analyze seismic-acoustic wave fields and stress-strain state in a complex heterogeneous seismogenic structure of the contact area of the continental and oceanic crust, taking into account two types of impact: technogenic and pressure of the oceanic water column. The study uses the theory of block structures and the block element method, developed on the basis of a topological approach. It is shown that the contact stress can be considered as a superposition of stress waves occurring at the interface between the liquid and elastic media. The number of waves, their amplitude and frequency are determined by the physical and geometric characteristics of the volume of the liquid, the elastic semi-bounded layer, the elastic foundation, as well as the oscillation frequency of the external harmonic load. The model can be useful in the design of a transport network, production facilities on the coastal territory, since it allows one to assess seismic risks before the start of the process, as well as the risks of associated hazardous exogenous phenomena - landslides, tsunamis, and volcanism activation.

1. Introduction

The article presents new results that can be useful for studying dangerous endogenous processes in the densely populated and actively developed territory of the North Caucasus, in particular, the Azov-Black Sea region of the Krasnodar Territory. The manifestations of endogenous activity include the high seismic hazard of the region, Late Cenozoic volcanism and mud volcanism.

Natural disasters in the densely populated, nationally and socially problematic territory of the North Caucasus are fraught with serious environmental, political and economic consequences. In addition, the region is currently being intensively developed. Numerous critical structures (roads, pipelines, bridges, unique high-rise buildings, Olympic facilities) are under construction and have already been built [1]. Therefore, the study of the causes of the occurrence of strong earthquakes, other dangerous endogenous processes and accompanying exogenous catastrophic phenomena is an urgent scientific task.

The high seismic hazard in the North Caucasus is obvious [2, 3]. In addition, the region is characterized by a high population density and mainly low quality of civil buildings.

The study of catastrophic natural phenomena, primarily strong earthquakes, is an important part of a more general and very urgent problem of ensuring environmental safety and security of the population, especially in view of a noticeable increase in damage from strong earthquakes and associated landslides, landslides, tsunamis, etc. A reliable prediction of strong earthquakes in a practical aspect in the world has not yet been resolved, although these phenomena have become the objects of serious priority scientific research and protective measures, both in Russia and abroad.

There are several approaches to assessing seismicity [4-12].

Studies carried out by traditional seismic-tectonic methods are aimed at finding precursors, that is, geological criteria for seismicity. They are based on a comprehensive comparative analysis of the geological structure and the latest structure, using seismic statistics methods. The areas of increased seismic hazard identified in this case are located mainly where relatively strong earthquakes have already occurred.

The second approach, the extra-regional seismotectonic method, is the result of many years of research in this direction. The essence of the method is to conduct geodynamic zoning of the earth's crust, independent of seismological data, using features that characterize its current structure and state, using cluster analysis of a complex of geological and geophysical data.

The third approach is to analyze the tension of seismogenic structures responsible for the realization of the seismic potential of the region [13, 14]. According to the results of the analysis of the locations of earthquake foci and their intensity, one of such geological structures is the contact area of the continental and oceanic crust. Such structures make mutual slow horizontal movement. This process is accompanied by the gradual subsidence of the oceanic plate under the continental plate (plate subduction). It is obvious that the state of such a mechanical system is significantly influenced by a layer of oceanic or sea waters. Structures of this type are widespread in the Krasnodar Territory, the North Caucasus and Russia as a whole. Below, we propose a method for studying seismic-acoustic wave fields in a complex heterogeneous seismogenic structure of the contact area of the continental and oceanic crust, taking into account two types of impact: technogenic and pressure of the oceanic water column.

2. Method for studying the stresses of seismogenic structures of complex structure

Dynamic processes in the structures of the earth's crust, despite the large number of studies devoted to them, are far from complete description today. A number of approaches to predicting earthquakes requires studying the stress-strain state of lithospheric structures.

The complex structure of geological structures, the presence of inhomogeneities (faults of different thickness, inclusions of media of different rheology) significantly affect the propagation of wave fields generated by various sources.

The block structures of the geophysical medium, considered in the work, are distinguished by complex geometry, the materials of individual blocks have significantly different physical and mechanical properties. This makes it practically impossible to apply the methods of solid mechanics to describe wave processes.

In modern geophysics, direct numerical methods (finite difference, finite element method), the method of fundamental solutions, variants of the boundary element method, as well as methods based on variational approaches have become widely used.

None of these approaches can be considered universal. The classical schemes of the indicated methods are applicable only in limited regions. To estimate the energy carried away to infinity from the source, the questions arise of conjugation of the obtained solutions with the asymptotic behavior of the far zone. In addition, to estimate the energy carried by individual modes, it is necessary to use special signal processing methods.

The described problems, generated by the formulation of boundary value problems that simulate the behavior of wave fields in the block structures of the geophysical medium, are quite complex and require the development and application of a set of new methods. The topological approach used in this work - the block element method - is an effective tool for studying the dynamic behavior of geological structures modeled by a block-layered elastic medium in contact with blocks of a different rheology. The block element method has several advantages over the existing computational finite and boundary element methods. It serves as a qualitatively new addition to existing methods, allows one to study boundary value problems that simulate natural and man-made processes, for which direct numerical ones are ineffective [14, 15].

The use of the integral approach and its generalization - the differential factorization method - makes it possible to find the optimal representation of the solution, which is convenient for the analytical study

IOP Conf. Series: Materials Science and Engineering 1103 (2021) 012032 doi:10.1088/1757-899X/1103/1/012032

of the problem and numerical analysis, as well as to describe the behavior of solutions at infinity and to isolate the waves carrying away energy.

The method is implemented in three stages. At the first stage, the investigated geological fault-block structure is associated with the block structure as a set of separate blocks, the material of which has different mechanical, physical chemical properties and its own internal structure. Further, in each block, its own boundary problem is set, corresponding to the process under study, the connection of the blocks is carried out through interblock boundary conditions. Further, the boundary value problem is reduced to a scalar or matrix functional equation. At the second stage, the obtained functional equation is investigated, and an automorphism is realized on the block element, which leads to a system of pseudodifferential equations. Their solution gives us the opportunity to get a solution to the problem in each block. At the third stage, applying homeomorphisms and constructing factor-topological spaces, we sew the solutions for individual blocks into a solution for the block structure

3. Mathematical formulation of the problem

The problem of assessing the stress-strain state and seismic-acoustic wave fields in a complex heterogeneous seismogenic structure of the contact area of the continental and oceanic crust is mathematically reduced to studying the distribution of contact stresses at the interface between an elastic foundation and a semi-bounded fluid basin. We consider a block structure (Figure 1).

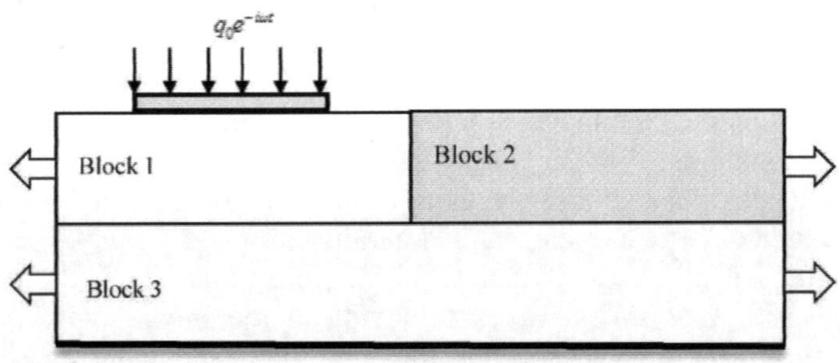

Figure 1. Seismogenic structure of the contact area of the continental and oceanic crust

Block 1 is a semi-bounded elastic layer simulating a section of the continental crust.

Block 2 is a semi-bounded layer that models a fluid mass (ocean, sea).

Block 3 - non-deformable elastic foundation, simulates the oceanic crust. A homogeneous elastic half-space or a homogeneous elastic layer rigidly adhered to a non-deformable base can be considered as a base.

Time-harmonic oscillations in the system under study are modeled by a distributed load $\{0, 0, q_0\}$ and correspond to a certain technogenic impact (production process, the impact of transport infrastructure, extraction of minerals). A steady-state oscillation process is considered, i.e. the dependence of all unknown and given functions on time is determined by a factor $e^{-i\omega t}$.

The motion of points in the liquid, described by the velocity potential, satisfies the wave equation:

$$\Delta\varphi(x,z) = -\frac{\omega^2}{c^2}\varphi(x,z) .$$

The displacements of the constituent elastic media satisfy the Lamé equations, represented for steady-state wave processes in the form:

$$(\lambda+\mu)\,\mathrm{graddiv}\,\mathbf{u} + \mu\Delta\mathbf{u} + \rho\omega^2\mathbf{u} = 0 .$$

The material of elastic blocks has the following characteristics: ρ – density; λ , μ – Lame constants. c – the speed of sound in the liquid; ω – frequency of harmonic vibrations.

On the vertical boundary of the liquid and the elastic semi-bounded layer, the impermeability conditions apply.

The conditions of continuity of displacements and stresses are taken as the conditions for conjugation at the boundary of the elastic semi-bounded layer and the base.

The interaction of the liquid medium and the elastic foundation is determined by the equality of the vertical components of the velocities of the points of the liquid and the elastic medium in the contact zone.

If an elastic half-space is considered as a base, the condition decreasing displacements during infinity and the radiation condition.

As a result of the implementation of the algorithm of the block element method [15] for the investigated system "Liquid - Elastic base", systems of pseudodifferential equations were obtained. Their transformation made it possible to construct and solve integral equations of the first kind with a kernel depending on the difference and the sum of the arguments, and to obtain expressions for determining the contact stresses at the interface between a liquid and an elastic medium. Then integral representations of solutions for displacements of points of an elastic foundation are calculated, as well as an explicit expression for the potential of velocities in a fluid. The inverse of the integral for the displacements of the points of the elastic foundation is performed numerically.

4. Results and Discussion

Numerical calculations were performed for a model structure. Water was chosen as a liquid. The following parameters were chosen as variable parameters: the thickness of the liquid layer, its density, the velocities of the longitudinal and transverse waves, and the frequency of oscillations of the external source. For various combinations of these parameters, the values of contact stresses at the interface between the liquid and elastic medium for the layer and half-space are obtained. Analysis is also possible for other combinations of geometric and mechanical parameters.

The study showed that in all cases of the initial variable parameters, the contact stress can be considered as a superposition of stress waves occurring at the block interface. The physical and mechanical characteristics of these waves, such as the amplitude, frequency, and their number, depend on the specified variable properties of the material in the sides of the structure under study and its geometric parameters.

After conducting the amplitude-frequency analysis of the system under study, it was concluded that, depending on the vibration frequency of the external source, it is possible to obtain different dependences for the blocks of an elastic semi-bounded layer and an elastic foundation in the form of a layer or half-space:

1) In the case of a half-space, the critical frequencies are those at which the wave number of the longitudinal vibrations of the elastic foundation coincides with the wave numbers of the natural vibrations of the fluid. In this case, there is a sharp, almost exponential increase in the amplitude of contact stresses.

2) There is a dependence of the nature of the growth of the amplitude of contact stresses on the vertical size of the liquid block. With an increase in the vertical dimension, a sharp, almost exponential increase in the amplitude of contact stresses corresponds to lower values of the frequencies of the external load.

3) There is a dependence on the stiffness of the base: a more rigid base corresponds to lower extreme values of the amplitude of contact stresses while maintaining other mechanical and geometric characteristics of the structure blocks and frequencies of external harmonic action.

4) In the case of a block of an elastic foundation in the form of a layer, a different dependence of the amplitude of contact stresses on the frequency of the external harmonic load is observed. Under harmonic influences of low intensity, the contact stresses are distributed uniformly over the interface between the liquid and elastic blocks. With an increase in the intensity of external influence, this pattern is violated.

5) The dependence of the nature of the growth of the amplitude of contact stresses on the vertical size of the liquid block was not noted.

5. Conclusion

Analysis of the results of computational experiments made it possible to establish the dependences of the stresses in the contact zone of the media of a block structure on the physical, geometric and frequency parameters of the problems under consideration for two models of an elastic foundation: a half-space and a layer with a restrained bottom edge.

The novelty of the presented study lies in the analysis of the dynamic behavior of a structure, including blocks of different types, characterized by different mechanical, physical and chemical properties, based on the use of the block element method, the solution of pseudodifferential equations in the form of integral equations, rather than using the averaged characteristics of contact stresses.

This approach to solving the problem makes it possible to increase the accuracy of the description of real processes and the reliability of assessments of the possible consequences of technogenic impact. This is especially important in regions where production, transport and mining infrastructure facilities are located near settlements or a coastal recreational zone. This study can be useful in the design of the transport network, production facilities on the coastal area, since it allows to assess seismic risks before the start of the process, as well as the risks of associated hazardous exogenous phenomena - landslides, tsunamis, and volcanism activation.

Acknowledgments

Some of the research results were obtained with the support of the Russian Foundation for Basic Research (project No. 19-08-00145), the Russian Foundation for Basic Research and the Administration of the Krasnodar Region (project No. 19-41-230002).

References

[1] Osipov V I 2016 *Ecology and Industry of Russia* **20** 4
[2] Yetirmishli G J, Mammadli T Y, Rogozhin E A and Sysolin A I 2019 *Izvestiya - Atmospheric and Ocean Physics* **55** 870
[3] Rogozhin E A 2019 *Izvestiya, Physics of the Solid Earth* **55** 111
[4] Anbazhagan P 2013 *Disaster Advances* **6** 66
[5] Cheng G 2017 *Mechanics and Mining Sciences* **94** 123
[6] Dziewonski A M and Chou T-A Woodhouse J H 1981 *J. of Geophysical Research* **86** 2825
[7] Rogozhin E A, Gorbatikov A V, Zaalishvili V B, Andreeva N V and Kharazova Y V 2015 *Doklady Earth Sciences* **462** 543
[8] Arefiev S S, Rogozhin E A, Bykova V V and Dorbath C 2006 *Izvestiya, Physics of the Solid Earth* **42** 27
[9] Gorbatikov A V, Stepanova M Yu and Korablev G E 2008 *Izvestiya, Physics of the Solid Earth* **44** 577
[10] Rogozhin E A, Gorbatikov A V, Stepanova M Y, Ovsyuchenko A N, Andreeva N V and Kharazova Y V 2015 *Geotectonics* **49** 123
[11] Gorbatikov A V and Tsukanov A A 2011 *Izvestiya, Physics of the Solid Earth* **47** 354
[12] Prilepin M T, Balasanyan S, Baranova S M, Guseva T V, Mishin A V, Nadariya M, Rogozhin E A, Rozenberg N K, Skovorodkin Yu P, Hamburger M, King R and Reilinger R 1997 *Izvestiya, Physics of the Solid Earth* **33** 486
[13] Morozov V N, Manevitch A I and Tatarinov V N 2018 *Mining Informational and Analytical Bull.* **2018** 123
[14] Babeshko V A, Evdokimova O V and Babeshko O M 2018 *Acta Mechanica* **229** 2163
[15] Babeshko V A, Evdokimova O V, Babeshko O M, Gorshkova E M, Zaretskaya M V, Mukhin A C and Pavlova A V 2015 *Doklady Physics* **60** 515

Influence of intermediate layer on the strength of freezing of soil with the metal surface of machines at negative temperatures

S A Zenkov, P Yu Dryupin and I S Bondalet

Bratsk State University, 40, Makarenko St., Bratsk, 665709, Russia

E-mail: mf@brstu.ru

Abstract. The development of moist cohesive soils in conditions of negative temperatures leads to a decrease in the productivity of earth-moving machines due to the increase in adhesion and friction. When wet soil and a metal bucket interact, the soil freezes to the sides of the bucket, which leads to a decrease in the useful volume of the bucket and, therefore, to a decrease in the productivity of the machine. The experimental studies have been carried out on the effect of the intermediate layer at the interface between the soil and the working body of the machine, on adhesion. The prophylactic lubricant "Niogrin" was used as an intermediate layer. As a result of processing the obtained experimental data, a regression equation was obtained; the dependence of the soil shear stress on various factors is presented and the influence of external factors on the adhesion of soils at negative temperatures is estimated. The comparison of the obtained dependencies showed that the strength of soil freezing when using a preventive liquid decreases 1.5-2 times.

1. Introduction

To date, there are various ways to combat soil freezing to the working bodies of earth-moving machines when operating in freezing temperatures. These methods can be divided into four groups: the formation of an intermediate layer at the contact boundary; external influence (for example, thermal effect) on the working body in the contact zone; design and technological method; a combination of two or more of the previously listed methods.

2. Formulation of the problem and method of solution

The experimental studies of the effect of the intermediate layer on adhesion to the surface of excavator buckets at negative temperatures have been carried out using the methods of physical modeling of systems and the mathematical theory of planning experiments.

The research data on the establishment of the dependence of the shear stress on the main factors that determine the adfreezing strength of soils with a metal sliding surface were carried out using a matrix of rotatable central compositional planning, the basis for which was the analysis of a priori information and the conditions for conducting experiments.

The parameters of the equipment for creating an intermediate layer (their rational values) did not change during the experiment. The prophylactic lubricant "Niogrin" was used as a preventive liquid.

Factor levels and variation intervals are presented in table 1.

Table 1. Factor levels and variation intervals for preventive exposure

Factor	Factor levels					Variation intervals
	-2	-1	0	1	2	
X_1 – dispersion of soil, D, mm	$9\cdot10^{-3}$	$7\cdot10^{-3}$	$5\cdot10^{-3}$	$3\cdot10^{-3}$	$1\cdot10^{-3}$	$2\cdot10^{-3}$
X_2 – normal ground pressure, P, kPa	0	10	20	30	40	10
X_3 – weight soil moisture, W, %	7.5	12.5	17.5	22.5	27.5	5.0
X_4 – ambient temperature, T, °C	5	-5	-15	-25	-35	10
X_5 – duration of soil contact with metal, t, min	0.5	10.5	20.5	30.5	40.5	10

The planning matrix and the results of a multifactorial experiment with preventive exposure are presented in table 2.

Table 2. The planning matrix and the results of a multifactorial experiment with preventive exposure

Experience number	Factors					Response functions		
	X_1	X_2	X_3	X_5	X_4	Y_{av}, kPa	Y_{prev}, kPa	K_{ef}
1	-	-	-	-	-	29.15	16.03	1.84
2	+	-	-	-	+	42.26	25.39	1.69
3	-	+	-	-	+	62.65	28.22	2.24
4	+	+	-	-	-	40.22	12.08	3.35
5	-	-	+	-	+	105.23	36.84	2.88
6	+	-	+	-	-	53.94	21.59	2.52
7	-	+	+	-	-	48.36	19.36	2.52
8	+	+	+	-	+	150.65	78.35	1.94
9	-	-	-	+	+	128.06	61.47	2.06
10	+	-	-	+	-	68.54	39.76	1.74
11	-	+	-	+	-	82.08	35.26	2.35
12	+	+	-	+	+	212.38	78.59	2.72
13	-	-	+	+	-	124.37	41.06	3.05
14	+	-	+	+	+	275.67	104.78	2.65
15	-	+	+	+	+	294.96	106.17	2.76
16	+	+	+	+	-	269.38	80.83	3.35
17	-2	0	0	0	0	50.26	16.06	3.14
18	2	0	0	0	0	120.37	58.97	2.06
19	0	-2	0	0	0	69.88	35.65	1.98
20	0	2	0	0	0	164.17	77.19	2.15
21	0	0	-2	0	0	68.56	25.39	2.72
22	0	0	2	0	0	292.37	96.47	3.05
23	0	0	0	-2	0	40.14	19.28	2.06
24	0	0	0	2	0	305.13	100.67	3.05
25	0	0	0	0	-2	59.55	27.96	2.15
26	0	0	0	0	2	239.47	90.97	2.65
27	0	0	0	0	0	185.76	59.46	3.14
28	0	0	0	0	0	179.35	55.62	3.25
29	0	0	0	0	0	192.05	74.87	2.58
30	0	0	0	0	0	187.38	56.23	3.35
31	0	0	0	0	0	184.55	92.28	2.02
32	0	0	0	0	0	184.43	88.54	2.06

As a result of the implementation of the experiment planning matrix (table 2) with varying the main factors (table 1), causing the adhesion of soils to the metal surface of the working body at negative temperatures, and processing the results of experimental studies using the "MODEL N" program, a regression equation was obtained that approximates experimental data:

- with preventive exposure in coded form:

$$Y_{prev} = 71.46 + 7.61X_1 + 7.29X_2 + 13.93X_3 + 19.71X_4 + 15.83X_5 - 8.71X_1^2 - 3.99X_2^2 - \\ - 2.86X_3^2 - 3.1X_4^2 - 3.23X_5^2 + 1.54X_1 \cdot X_2 + 4.21X_1 \cdot X_3 + 1.44X_1 \cdot X_4 + 0.74X_1 \cdot X_5 + \\ + 4.31X_2 \cdot X_3 + 0.98X_2 \cdot X_4 + 2.1X_2X_5 + 2.71X_3 \cdot X_4 + 4.55X_3 \cdot X_5 + 3.4X_4 \cdot X_5; \quad (1)$$

- with preventive exposure in natural form:

$$\tau_{prev} = 6227.75D + 0.34P + 2.73W - 0.85T + 0.8t - 5.4 \cdot 10^5 D^2 - 9.97P^2 - \\ - 2.86 \cdot 10^{-2} W^2 - 7.77 \cdot 10^{-3} T^2 - 8.07 \cdot 10^{-3} t^2 - 19.25D \cdot P - 100.25D \cdot W + 18D \cdot T - \\ - 9.25D \cdot t + 2.15 \cdot 10^{-2} P \cdot W - 2.45 \cdot 10^{-3} P \cdot T + 5.27 \cdot 10^{-3} P \cdot t - 1.35 \cdot 10^{-2} W \cdot T + \\ + 7.7 \cdot 10^{-3} W \cdot t - 8.5 \cdot 10^{-3} T \cdot t - 17.7; \quad (2)$$

3. Results and discussion

The analysis of equation (1) for the optimum showed that such a point is in the negative response region. All factors in the investigated range mainly contribute to an increase in the freezing strength. In ascending order of increasing adhesion, they make up the series: D, P, t, T, W.

The increase in the influence of pressure during preventive exposure can be explained by the disappearance of the cementation bonds of ice (its melting) under the influence of thermal energy and the transition of bound water into loose and free water.

The quasi-one-factor dependences of the shear stress under preventive action on external factors are shown in figures 1-5.

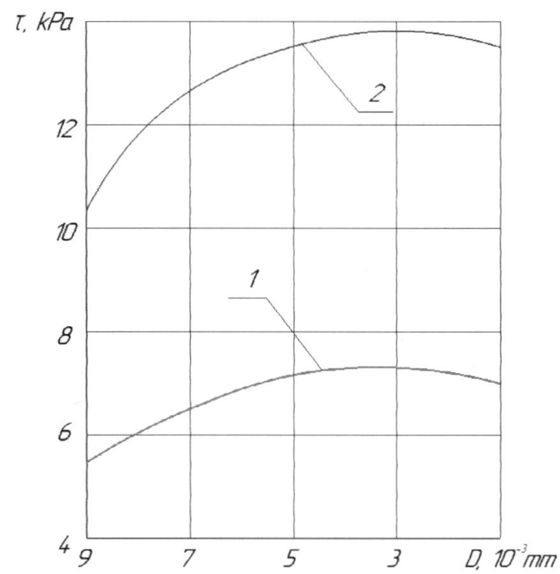

1 – with preventive exposure; 2 – without impact;

Figure 1. Dependence of the shear stress of the soil on its dispersion at P=20 kPa; W=17%; T=-15 °C; t=20.5 min

Figure 1 shows that with an effective particle diameter D = $3 \cdot 10^{-3}$ mm, the maximum freezing strength is observed.

The resulting dependence of the effect of soil dispersion on the strength of soil freezing to a metal surface during the formation of an intermediate layer can be described by the empirical formula:

$$\tau_{prev} = 67{,}41 + 3541{,}25D - 544375D^2 \quad (3)$$

where τ_{prev} - resistivity to shear during preventive treatment, kPa; D – soil dispersion, mm.

With an increase in pressure, the shear stress increases, which is explained by an increase in the true area of contact between the soil and the surface, after which the pressure affects the shear stress insignificantly (figure 2).

The resulting dependence of the influence of the initial pressing pressure on the strength of freezing of the soil to the metal surface during the formation of an intermediate layer can be described by the empirical formula:

$$\tau_{prev} = 60{,}2 + 0{,}76P - 9{,}97 \cdot 10^{-3} P^2 \quad (4)$$

where τ_{prev} - resistivity to shear during preventive treatment, kPa; P – initial contact pressure, kPa.

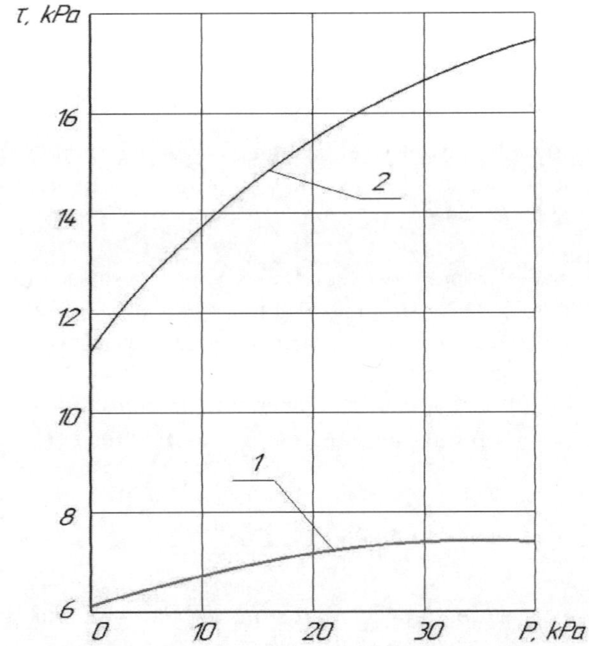

1 – with preventive exposure; 2 – without impact;

Figure 2. Dependence of the soil shear stress on the initial pressing pressure at D = $5 \cdot 10^{-3}$ mm; W = 20%; T = –15 °C; t = 20.5 min.

The analysis of the dependence of soil shear stress on moisture (figure 3) shows that with increasing moisture, the shear stress increases.

The obtained dependence of the influence of the weight moisture content of the soil on the strength of freezing of the soil to the metal surface during the formation of an intermediate layer can be described by the empirical formula:

$$\tau_{prev} = 2{,}33W + 27{,}87 - 2{,}86 \cdot 10^{-2} W^2 \quad (5)$$

where τ_{prev} - resistivity to shear during preventive treatment, kPa; W - weight soil moisture, %;

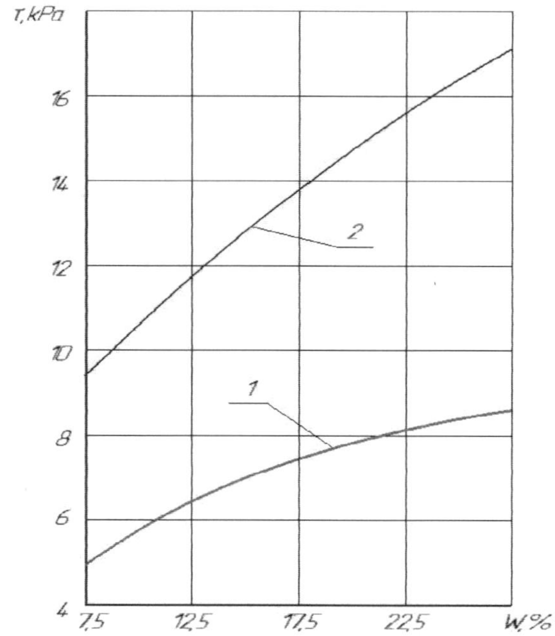

1 – with preventive exposure; 2 – without impact;

Figure 3. Dependence of the soil shear stress on the weight moisture content of the soil at D = $5 \cdot 10^{-3}$ mm; P = 20 kPa; T = –15 °C; t = 20.5 min.

The temperature in the zone of contact between the soil and the surface of the working body is one of the main factors affecting the adhesion of the soil (Figure 4).

The dependence of the shear stress in the investigated range of variation of the factors is practically directly proportional to the change in temperature.

The resulting dependence of the effect of temperature in the contact zone on the strength of freezing of soil to a metal surface during the formation of an intermediate layer can be described by the empirical formula:

$$\tau_{prev} = 54{,}96 - 1{,}22T - 7{,}77 \cdot 10^{-3} T^2 \quad (6)$$

where τ_{prev} - resistivity to shear during preventive treatment, kPa; T - contact zone temperature, ºC.

The quasi-one-factor dependence of the shear stress on the contact time (figure 5) shows that with an increase in the time of contact between the soil and the metal surface, the shear stress increases.

The obtained dependence of the influence of the contact time on the strength of freezing of the soil to the metal surface during the formation of the intermediate layer can be described by the empirical formula:

$$\tau_{prev} = 51{,}91 - 1{,}12t - 8{,}07 \cdot 10^{-3} t^2 \quad (7)$$

where τ_{prev} - resistivity to shear during preventive treatment, kPa; t - duration of soil contact with a metal surface, min.

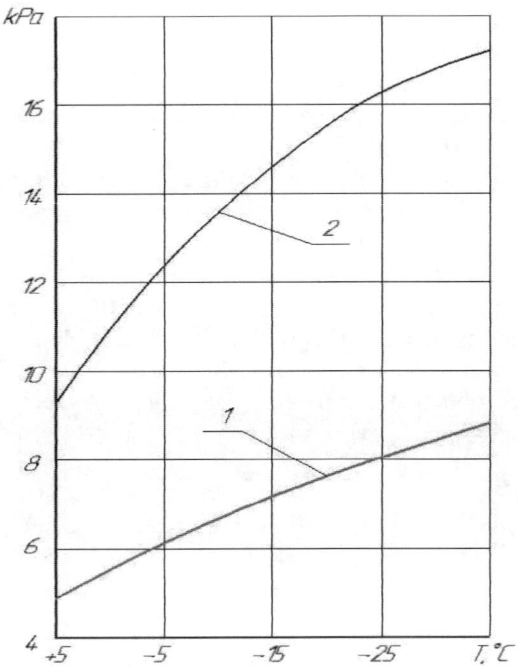

1 – with preventive exposure; 2 – without impact;

Figure 4. Dependence of the soil shear stress on the temperature in the contact zone at D = $5 \cdot 10^{-3}$ mm; P = 20 kPa; W = 17%; t = 20.5 min.

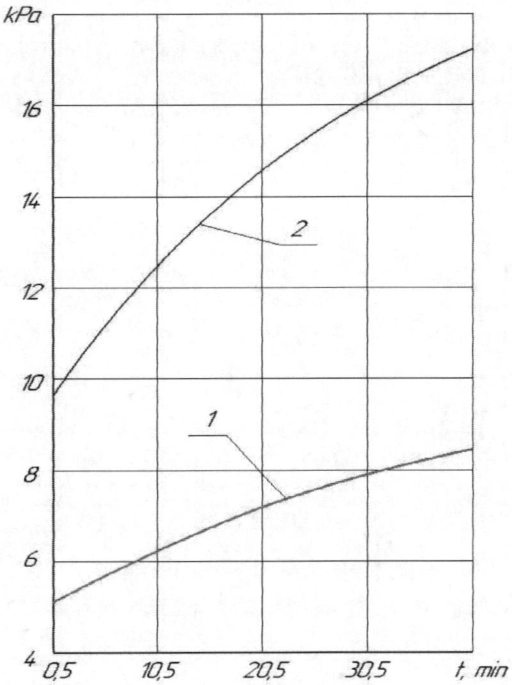

1 – with preventive exposure; 2 – without impact;

Figure 5. Dependence of soil shear stress on the contact time at D = $5 \cdot 10^{-3}$ mm; P = 20 kPa; W = 17%; T = –15 °C.

4. Conclusion

1. It was experimentally established that of the variety of external factors affecting the adhesion of soils at negative temperatures, the main ones (in decreasing order of the degree of influence of the factor) were: weight moisture content of the soil, temperature in the contact zone, contact time, pressure of initial pressing, dispersion of the soil.

The equations (1.2) were obtained that connected these factors with the freezing force. It was found that in the selected range of changes in external factors, all factors contributed to the increase in freezing strength.

2. As a result of studying the preventive effect on the adhesion of the soil at negative temperatures, equations (3-7) were obtained, connecting external factors with the freezing force.

All the factors in the investigated range contribute to the increase in the freezing strength. In an ascending order of increasing influence on adhesion, they made up the series: D, P, t, T, W.

3. The comparison of the obtained dependences showed that the strength of soil freezing when using a preventive liquid decreases 1.5-2 times.

References

[1] Zadneprovskiy R P 2005 *Theory of Sliding Friction* (Volgograd: Ofset) p 51
[2] Rajaram G and Erbach D C 1999 *J. of Terramechanics* **36** 39–49
[3] Zenkov S A, Balahonov N A and Kirichenko O P 2018 Decrease of soil adhesion to working bodies of earth-moving machines *Advances in Engineering Research* **158** 468–73
[4] Zenkov S A, Balahonov N A and Kirichenko O P 2019 Improving the efficiency of technological machines in the interaction with adhesive materials at negative temperatures *IOP Conf. Ser.: Mater. Sci. Eng.* **560** 012097
[5] Sharma V K, Drew L O and Nelson L 1977 *Transactions of the ASAE* **20** 46–51
[6] Tong J, Ren L, Yan J, Ma Y and Chen B 1999 *Int. Agricultural Eng. J.* **8** 1–22
[7] Wang X L, Ito N, Kito K and Garcia P P 1998 *J. of Terramech.* **35** 87–101
[8] Azadegan B and Massah J 2012 Effect of temperature on adgesion of clay soil to steel *Cercetiiri Agronomice in Moldova* **XLV 2(150)** 21-7
[9] Zadneprovsky R P 1992 *Working Bodies Earthwork and Reclamation Machines and Equipment for Soil and High Humidity Materials* (Moscow: Mashinostroenie) p 176

IOP Conf. Series: Materials Science and Engineering 1103 (2021) 012034 doi:10.1088/1757-899X/1103/1/012034

Quality of construction materials as an indicator of the design of railway transport facilities in Russia

A Korchagin[1] and D Zaycev[2]

[1] Moscow State University of Civil Engineering (NRU MGSU), Department of Economics and management in construction, 26, Yaroslavskoe shosse, Moscow, 129337, Russia

[2] Moscow Automobile State Road Transport Technical University, Department of Road Economy, 64, Leningradsky pr., Moscow, 125319, Russia

E-mail: zaycevdv77@mail.ru, korchaginap@mgsu.ru

Abstract. The purpose of the research is to assess the global quality of project design documentation for capital construction at the preliminary stage of investment projects in the context of the properties and qualities of applied construction materials. The problem of the assessment of the quality of construction and finishing materials in the subdivisions of Russian Railways, the largest Russian company, raises fundamental issues of regulatory and methodological support during the choice of engineering solutions and reasonability grounds of investments in the construction and reconstruction of transport infrastructure facilities. According to the formal indicators for the assessment of the properties of materials in the company, the authors used a method based on a separate analysis of the regulatory and methodological support of the construction preparation process. The main quality criterion was adopted in the form of deviations in documentation by the internal examination service of the company projects due to the unreliability in the determination of the physical properties of material resources. The main tool to influence the design process is the formalization of requirements for accounting construction materials, especially innovative ones, in project design documentation. These requirements are reflected in the proposed methodology for the generation of documentation at the early stages of investment project. A retrospective analysis of the verification results of infrastructure projects in the company showed that after the introduction of uniform requirements for the applied materials and the introduction of the basic provisions of materials science in the methodological documents, a positive trend has occurred during internal verification. The decrease in the number of incorrectly determined values of the need for construction materials indicate that cost certification has become more explicit and accessible to developers. It is proved that the tool for the management of the quality of construction materials by specifying the structure of documents justifying the main design decisions is the most effective. The idea is confirmed that the greatest responsibility for the efficiency of the applied material resources falls on the early stages of project development. The efforts made by the Company in this direction lead to positive changes.

1. Introduction

The priority task in the improvement of the efficiency of Russian Railways investments in the context of modernization of its own infrastructure is to conduct a technological and economic examination of the investment projects being implemented in terms of the applied material resources. At the same time, two main industry-specific requirements should be presented to the expert activity. The first is the need for the initial technological examination of the project and comparison of it with the main directions of development [1]. Then it becomes reasonable to carry out a financial and economic assessment of the project and manage its cost, as it is noted in the work [2]. The second requirement is the existence of its own system of criteria for the assessment of the technological [3] and economic [4, 5] project efficiency, which would allow recording and evaluating the specific features of infrastructure projects on the railway network already at the stage of the investment concept. A number of authors note the predominant dependence of efficiency on the level of cost norms [6] and methods of their application [7].

During the implementation of investment programs, the important issue is the efficiency of application of funds invested in the creation or maintenance of railway infrastructure facilities. Every construction or reconstruction project must be justified from the standpoint of investment efficiency even before the start of design work [8]. However, the situation that has developed over the past two decades in the field of technical regulation in Russia has led to the absence of a specialized industry methodology for the development of the main document of investment project - investment explanation.

2. Materials and methods

In the investment process of Russian Railways, pre-project economic studies must include investment explanation taking into account the current legislation. The purpose of the work performed at this stage is, first of all, to optimize the future costs of the Company choosing the best option for the project implementation. The most frequently used in this case is the factorial model [9, 10].

The authors proposed the system of requirements for pre-project documentation which is set out in the Methodological Recommendations on the composition of investment explanation sections and requirements for their content (including the calculation of economic efficiency) for investment projects of Russian Railways. This document is intended to improve the methodology for the development of pre-project documentation - explanation of investments in the construction of enterprises, buildings and structures of railway transport in Russia. The analysis of regulatory documents on the explanation of investments, preparation of primary material for a preliminary assessment of the estimated cost, analysis of the impact of decisions taken at the pre-design stage on the organization of construction on the investment value of the object was carried out in order to make the main provisions of the methodological recommendations as accessible as possible. In this case, the consolidation of problems corresponds to the idea outlined by Yu Zhang et al. in the publication [11].

During the preparation for the development, the normative and methodological literature, as well as orders, decrees and documents of both federal and industry levels, were previously studied [12]. The degree of impact on the activities of construction organizations was taken into account in accordance with the works of Silka [13, 14]. According to the requirements of the existing documents at Russian Railways, considered during the development of the methodology, the investment justification document should have the following structure.

The 1st section "Initial data" contains the materials of the investment application developed by the initiator of the investment and materials of the pre-design engineering surveys carried out after the approval of the investment project by the Investment Committee of the Company.

The 2nd section "Market environment of the project" describes the economic, marketing, social, transport and infrastructure environment in which the project is supposed to be implemented. The main markets for the sale of products, tariff policy, plans for the implementation of transport services are determined. According to the work [15], the factors of success in the implementation of investment projects were taken into account.

In the 3rd section "Basic technological solutions", the main production indicators of the facility, the technology of the transportation process and the technical solutions used for various farms affected by the project should be specified.

During the development of the 4th section "Provision of the project with resources", the need for basic resources for the implementation of the investment project is determined. The possibilities to obtain them are studied, as well as the prospects to obtain necessary construction resources as a result of tendering [16].

In the 5th section "Location of the object" the possibility of the territorial location of the construction object is substantiated, the basic needs of land reservation and restrictions in their use are determined.

For the development of the 6th section "Basic construction solutions", the main structural and technological solutions of buildings and structures are selected as a part of the investment project.

In the 7th section "Environmental Impact Assessment" the analysis of the initial state of the environment is carried out, the calculations of man-made impact as a result of the project implementation and the comparison of the existing exposure limits with the calculated ones are carried out.

The 8th section "Enterprise structure. Personnel" defines the possibilities and needs of staffing at the stage of creation and operation of the project.

In the 9th section "Project Implementation Schedule", the main timing of the investment project and its parts is calculated. Moreover the basis for scheduling works, commissioning of capacities and financial management is created. The calculation of the construction time, along with the normative method can be assessed using statistical methods [17].

The part of expenses on the project is determined in the 10th section "Determination of investment expenses". It contains the calculations of the amount of costs for the implementation of an investment project based on the use of enlarged indicators or analogous objects.

The 11th section "Evaluation of the effectiveness of an investment project" defines the indicators of the effectiveness of investment in the project, the main economic consequences of the project for its participants, the state and society. It is necessary to mention the possibility of the assessment of the effects using the mathematical models proposed by Lie-Jane Kao [18].

The 12th section "Conclusions and proposals" completes the development of the investment case. It contains a comparison of the results of the calculation of economic efficiency for the developed alternative project options, proposals for choosing the best option for further development of project documentation, similar to the work [19].

Such a structure for the explanation of investments is primarily reasoned by the needs of the subsequent design process, i.e., a detailed study of the main decisions made in investment explanation.

3. Results

According to the practical results of departmental examination of projects, most technological errors in projects are laid down at the stage of the development of investment application. Table 1 shows the statistical distribution of the dependence of the values not confirmed by the calculation of the explanation of the effectiveness of investments or investment costs from various factors of pre-project and project activities increasing the cost of the project over the parameters of this calculation.

Table 1. Distribution of reasons for overinvesting

The reasons for the decrease in quality during the determination of the project cost	Share of unreasonable investment costs, %
Inaccuracy in the collection of initial data on the economic state of an object and its surroundings	3
Incorrect definition of the organizational pattern of the object construction, stages and terms	26
Wrong choice of structural solutions for buildings and technological equipment	12
Incorrect determination of the cost of work at the pre-project stage	15
Errors in the calculation of the economic efficiency of investments	8
Overestimation of the estimated cost of design work completion	14
Lack of coordination of the goals of construction (reconstruction) of the facility with the development strategy of the landfill and the technology of transportation process	7
Inaccuracy in the collection of initial data on the economic state of an object and its surroundings	18

The significant contribution to the problem of unjustified investments is caused precisely by the defects at the stage of the emergence of an investment concept and the development of explanations for upcoming investment project. The correlation between the economic and financial results of the implementation of an investment project is a separate problem with an ambiguous solution [20], since the overestimation of costs can be interpreted in different ways. The prevention of inaccuracies in the determination of costs at the early stages of a project, rather than elimination of them later, is the task of the specialists of functional customer and investment departments of roads, who develop investment applications and explain investments.

4. Conclusion

The monitoring of investment projects shows that pre-project activities cover a significant part of the responsibility for ineffective investments. The contribution to this problem at the design and costing stage is not as great as it is commonly believed. The proposals for the systematization of the content of design materials at the pre-design stage are aimed at the increase of the validity of the calculations, therefore, at the improvement of the quality of the project as a whole. The allocation of sections of resource provision and descriptions of the construction time are obligatory. The commercial environment of the project provides the basis for detailed calculation of cost elements in a future construction project. The improvement of the quality of the explanation of investment projects should be considered a guarantee of future success in the improvement of investment efficiency.

References

[1] Baranovskaya N I 2004 Problems of transition to a new budget and regulatory framework for pricing in construction *Bulletin of civil engineers* **1** 174-178

[2] Bashirov A E, Rubchenko D S and Kurbackaya E P 2012 On pricing in the construction of railway transport projects *Economy of Railways* **6** 34

[3] Solov'ev V V 2019 Economic indicators of the use of machines in construction *Bulletin of the Moscow international higher school of business. MIRBIS* **1(17)** 184-194

[4] Solov'ev V V and Kuznecova A E 2016 Modeling of cost standards for railway construction objects *Economy of Railways* **12** 44-51

[5] Solov'ev V V 2016 Branch features of definition of integrated indicators of construction cost *Economy of Railways* **6** 46-55

[6] Vaganova M D and Shcherbakova N A 2019 The process of transition to a new budget and regulatory framework in construction *Young scientists - development of the National technology initiative (Search)* **1-2** 158-161

[7] Poltava A V and Korchagin A P 2019 Problems of determining the reliable cost of construction in the framework of project management, *In the collection: Development of the methodology of modern economic science, management and education in the context of information and digital trends: Materials of the III Interdisciplinary all-Russian scientific conference* pp 208-213

[8] Mishchenko A S 2019 Problems of budget and regulatory framework and pricing in construction, *In the collection: in the world of scientific discoveries: Materials of the III international student scientific conference* pp 87-89

[9] Mitchell Dayton C 2020 Logistic Regression Analysis. Retrieved from: https://www.researchgate.net/publication/268416984_Logistic_Regression_Analysis

[10] Yalu Ping 2020 Analysis of the Influence of Multiple Linear Regression on Construction Price *Statistics and Application* **09(01)** 56-60

[11] Yu Zhang, Dangqiang Yin and Yating Zhu 2019 Method for Construction Progress Measurement for EPC Oil and Gas Pipeline Projects *Journal of Oil and Gas Technology* **41(06)** 39-47

[12] Solov'eva E V, Teunaeva F S and Hapaev T I 2019 Modern approaches to the development of budget documentation based on the new budget and regulatory documentation and the Federal state information base *Economy and entrepreneurship* **2(103)** 905-913

[13] Silka D N and Babaeva M S 2018 Ways to improve the quality and efficiency of relationships between participants in investment and construction activities *Economy and entrepreneurship* **1(90)** 682-685

[14] Silka D N 2011 Key factors for identifying the phases of the business cycle in construction *Bulletin of the University* **21** 210-212

[15] Hasan Mathar, Sadi Assaf, Mohammad A Hassanain, Abdullatif Abdallah and Ahmed M Z Sayed 2020 Critical success factors for large building construction projects: Perception of consultants and contractors *Built Environment Project and Asset Management* **9(5)**

[16] Tihonova E A 2019 Organization of contract auctions (competitions) for objects and services in construction *Young scientist* **4(242)** 261-264

[17] Chistyakova A P 2019 Value of factor analysis in assessing the financial position of a commercial enterprise *Young scientist* **6(244)** 156-158

[18] Lie-Jane Kao, Cheng-Few Lee and Tzu Tai 2014 Discriminant Analysis and Factor Analysis: Theory and Method, in: *Handbook of Financial Econometrics and Statistics* pp 2461-76. DOI: 10.1007/978-1-4614-7750-1_89

[19] Saransh Khandelwal and Dhananjay Kumar 2020 Computational Fact Validation from Knowledge Graph using Structured and Unstructured Information Publication:*CoDS Proceedings of the 7th ACM IKDD CoDS and 25th COMAD* pp 204–208. DOI:10.1145/3371158.3371187

[20] Niu Zilong 2020 *Essays in Empirical Asset Pricing and International Finance CentER*: Dissertation Series, Research Group: Finance, *Center Ph. D*, Vol **618**

Transport hubs as experimental ground for new construction materials

S Guzhov[1] and A Korchagin[2]

[1] Department of Road Economy, Moscow Automobile State Road Transport Technical University, 64, Leningradsky Ave., Moscow, 125319, Russia
[2] Moscow State University of Civil Engineering (NRU MGSU), Department of Economics and management in construction, 26, Yaroslavskoe shosse, Moscow, 129337, Russia

E-mail: korchaginap@mgsu.ru

Abstract. The research purpose is to clarify the role, significance and problems inherent in transport hubs in the implementation of engineering solutions using new construction materials. The traditional scheme of commutation of passenger traffic between different modes of transport allows building physical models. However, the efficiency of materials in structures and the study of their operational properties are additional tasks for the calculation in new conditions. Historically, many large cities are characterized by a lag in the development of transport infrastructure in comparison with the growth rates of cities. The material and technical supply of construction inevitably faces the problem of the depletion of materials from their operational properties. Massive daily movement of the population to places of job and high operational loads reveal disadvantages primarily in terms of the traditional materials. The way out of this situation is the development of transport interchange hubs, while the use of traditional constructive and technological solutions causes additional difficulties for transport companies and businesses. The presentation of the key elements of the transport infrastructure as a testing ground for testing the properties of new materials such as high-alloy steels, composite materials, superplasticized concretes allows solving the problems of construction engineering in a new way. The constructed transport hubs rarely comply with the loading standards, which is caused by the incorrect approach to their design: the existing construction rules are not comparable with the functional needs of new type facilities. For the assessment, such factors as material consumption, creation of regulatory conditions, quality and safety of materials, reduction in construction time, stability of functioning and others are used. The risks associated with the creation of transport interchange hub (TIH) are noted. The development of transport hubs in a megalopolis can be implemented during simultaneous improvement of the quality of applied construction materials and satisfaction of the commercial interest of investors. In addition to the direct technological effects, there are accompanying operational consequences in various areas of urban life.

1. Introduction

The purpose of conducted research is to analyze the interaction of the investment and the social spheres in cities based on the study of investment policy in transport construction area.

The methodology is based on dual system of views on the social-economic model [1]. On the one hand there is a use of factor analysis of different processes – construction, exploitation of transport objects, enterprises activity. On the other hand, the authors [2] approached social economic of a city as a synergetic system, creating additional effects thanks to interaction of separate elements.

This issue is covered by researches of many authors - A.I Popov (1998), A.A. Gavrilenkov (2011), A.V. Zvyagintsev (2018), L.A. Melnik (2018), Alfonso Valenzuela (2013), Irena Mocanu, Paul-Razvan Serban (2015).

Big cities in modern society acquired new qualities, that complicated systems have – multi-level management, lack of availability of resources, structural properties. Several decades ago there were two basic ways of a city development – either according to a centralized plan (from the centre to outskirts) or stochastic development of outskirts with their integration into cities. Nowadays all the cities in the countries of former USSR find new ways of growth in their development based on a combination of interests of the government, business and society [3]. The key part of the scheme of development is transport infrastructure, created by different types of transport (main, city and private). The success of transport system defines the quality of living standards and economic development rate. That is why the issue of investment management in transport construction is getting basic in the system of the social-economic development of the city.

During the process of the study of transport construction development the following tasks were stated and described:

1. Connection between gross figures of investment activity and macroeconomic indicators of economic and social development [4]
2. Social performance assessment of current construction projects in the passenger transport sphere, introduction of the project classification and substantiation of their selection criterion.
3. Building of social effect models accompanying urban development, determination of the proportion of the effect in technological development of construction and transport industries.

According to the authors, the above mentioned tasks suggest consideration of the important problem of calculation accuracy of investment needs. As the comparative analysis shows, this indicator in Russian investment complex is less 2 – 3 times than in American and European ones [5].

The interest towards the economic assessment of urban growth in Russia is increasing annually. The uncontrolled increase in the population in urban agglomerations can lead to negative consequences for the state and society, as it is noted in the work [6]. The authors made attempts to create scientific approaches to determine the limits for a city by the criterion of the equipment of its transport infrastructure by analogy with the methods proposed in the work [7]. The evolution of this idea over time should lead to the emergence of new scientific assessment tools. In addition, the useful consequences of these studies will be in the field of legislation, in the development of urban planning policies and urban development plans [8].

2. Materials and methods

According to the research work [8], one of the main problems in the development of urban areas is the inconsistency of the transport component with the rest of the components of any investment project. Thus, for Moscow urban agglomeration, the parameters of transport accessibility of a residential or industrial facility have no less significant impact on the return on investment than land lease rates. At the same time, the differentiation of this influence and its accounting in the investment budget of the project are significant. Other modern researchers come to similar conclusions [9]. For example, for housing construction on the scale of a micro-district at a distance of more than 30 kilometers from the administrative boundaries of the city, the costs of the provision of local transport infrastructure (in-yard passages, exits to the existing street-road network, urban passenger transport facilities and parking lots) comprise to 10-15% of the volume investment.

For the same type of construction within a 30-kilometer zone, the share of costs for transport infrastructure is reduced to 5-10%, and within the city it does not exceed a few percent. This is due to the increase in the value of land, as well as a high saturation of local roads. In such conditions, the problem of the provision of the population with a local transport infrastructure is solved quite successfully, which is confirmed in the work [10]. Such a conclusion could become in-depth only with the traditional local distribution of the population in relation to the objects of employment: when the majority of workers at the enterprise live in a settlement or town for workers.

The complexity of the determination of large construction projects costs is explained by multifunctionality of the objects in specific cities conditions. There is an imbalance between the growth of demand for transport services and the carrying capacity of the city passenger transport and the capacity of road system. According to the population census data the Moscow population rose over to more than 12 million in 2010. At the same time the speed of traffic flows in the city decreases and as a consequence the duration of the population movement rises. Due to this productive time decreases and economic development of the city is getting slow and as a result investment flows are decreasing. However, the city is supposed to be developed improving living standards of its habitants and getting economic capacity better. It causes new standard requirements for all spheres of the city life, including the passenger transportation.

With the modern structure of the housing and labor market, specialists deal with significant daily movements of several tens of kilometers in different directions using various types of transport. This is typical for the modern stage of development of society [11]. This requires the development of not a local, but a unified infrastructure of suburban and urban passenger transport, as presented in the work [12]. The strategy for the development of passenger transport that existed in the twentieth century had an adequate supply of passenger traffic, but the population was determined by a target of 6 million people with sufficient employment in the suburbs. The sharp population growth that has taken place in recent decades, labor migration and job losses in small settlements have determined the need to strengthen the existing directions of passenger traffic. A completely similar situation is currently emerging in the countries of Asia [13, 14] and Europe [15, 16].

At the same time, some disadvantages were revealed such transfer hubs between railway stations, stations of underground and street passenger transport. If the throughput of suburban railways has enormous potential due to the creation of additional tracks, organization of clock traffic, introduction of new rolling stock (including two-story rolling stock), and increasing traffic speeds, then the connection between the station and the city is still organized according to the old, centuries-used scheme i.e. through the universal station square.

This brings to nothing the efforts of the relevant departments of the railways that are implementing investment projects in the field of passenger transportation. The actual efficiency of investments is also decreasing, since the significant time spent on transferring reduces the competitiveness of suburban public transport in favor of personal.

In order to get rid of bottlenecks in public transport system, the state budget remains the main source of investment. The participation of business and private capital in this area is limited only by the part of deductions from investment projects centralized in the budget in the form of rent, targeted fees and taxes. The second most important aspect is the participation of the transport operators themselves. However, even the railway industry does not have the opportunity to invest on its own because the limited investment budget and the actual dependence of any transport project on the interests of other participants play its role [17]. To determine the possibilities of engagement of various budgets in the implementation of the project for the construction of a network of transport interchange hubs (TIH), it is necessary to consider the management of economic and legal risks in this segment of investment sphere.

The main reason for the lack of investment initiatives in the development of THI is the lack of comprehensive legal and economic mechanisms for interaction between public-private investors within the framework of the agreement of public-private partnership (PPP). This raises questions about the way how to manage this type of project. Today, TIHs are considered only for transferring passengers from ground passenger transport to high-speed off-street (metro, railroad, etc.), parking vehicles, providing related services and regulation of ground transport routes, in other words, for extended social functions, as noted in the work [18]. The safety aspect in the development of transport acts as a prerequisite that affects the design and cost [19].

In a scientific point context, TIH is a nodal infrastructure element, as opposed to linear objects. In large countries, the concept of infrastructure security has always meant the development of linear facilities. In the city, an interchange hub is extremely important and occupies a significant share in local traveling and has a significant impact on the natural and social environment [20].

3. Results

The studies show that in the most loaded nodes, where modern design solutions are applied, the density of passenger traffic reaches its limit values - up to 8-10 people / m^2, in less loaded nodes during peak hours the maximum density is 3-6 people / m^2. Only two TIHs out of 37 in Moscow meet the specified requirements. The operation of other transfer hubs does not meet the specified requirements as there is no optimal scheme for the movement of transport and passengers, intercepting parking lots, waiting pavilions, covered passages and information on the movement of ground transport. Almost all transport hubs in Moscow have a deficit in the length of the loading-unloading line by about 30–70%. The analysis, carried out through the examples of Moscow railway stations, showed the discrepancy between the construction codes and the commercial performance of the operators of suburban passenger transportation [5].

Thus, the growth of the general passenger traffic for all directions of TIH does not guarantee the provision of the increase in revenue for suburban traffic. Therefore, the participation of railways in investment projects for the implementation of TIHs is based on the risks determined by the imperfection of the used methodology. Having spent financial resources on the integration of station complexes into TIHs, there is a possibility not to get the result on passenger traffic in the zone of particular economic interests.

The social efficiency of construction is reasoned by the factors of the improvement of the transport accessibility of urban facilities. In this research on the determination of the economic efficiency of the development of TIH system, the following factors were determined and quantified:

1. The formation of a single system of transport hubs and new modern bus stations in the city and suburbs in order to increase the comfort and attractiveness of the use of transport system, which reduces the risk of underutilization of the transport line. The current state and problems of the transport industry in Moscow are characterized by the following parameters: Moscow in terms of population density (100.2 people / ha) is 1.5–2 times higher than similar indicators of the largest megacities in the world. The density of the road network is 3.3 km / km^2, which is 2–4 times lower than similar indicators in the world's largest megacities. The volume of metro traffic is 1.5–2 times higher than those of the largest megacities in the world. The share of public transport in the total traffic (metro, railways and surface urban passenger transport) is about 74%.

The forecast of the changes in traffic volumes was performed, in accordance with which it was planned to increase the volume of traffic by public transport by 40%. At the same time, the maximum growth - one and a half times was predicted precisely on the railway.

2. The creation of regulatory conditions to take passengers and rolling stock of suburban, intercity and international traffic. The planning structure of mass passenger transport networks is characterized by the geometric features of the construction of communications in the city as a whole (for communications connecting the planning zones of the city with each other and with the suburban area) and in individual planning zones, the density and straightness of communications and the branching of the route network.

3. The improvement of the quality and safety of transport services, reducing the time for transferring at large transfer hubs will entail additional jobs. The quality of passenger service provided by the rolling stock of urban transport is checked on the basis of the calculation of the standard number of vehicles W_t on the network of this type of transport:

$$W_t = \frac{120 L_c \mu z}{V_m i}$$

where i – standard traffic interval on passenger transport networks, min.

The following is taken $i \leq 5$ min – to connect the main residential areas with the city center and the largest industrial zones, $i \leq 10$ min – for other networks;

V_m – route speed of this type of transport;

z – coefficient of reliability to ensure the normative interval of movement, equal to 1,6;

L_c – length of the network of lines of this type of transport, km;

μ – route factor, taken in the range of 1.5–4.

For TIH, this formula is of particular importance as the need for mutual coordination of transportation schedules for all types of transport will require changes in route coefficients, primarily for street public transport.

4. The elimination of unauthorized points of arrival and departure of buses. Such points, although insignificantly, have a negative effect on the economic development of transport.

5. Reduction of the time of communication between the city and the region.

The purpose of TIH is determined by the following tasks:

- to provide a safe and comfortable transfer of passengers from stopping points at the Moscow Railway Station to stopping points of other types of transport;
- to minimize the time spent on the transfer;
- to provide pedestrian communication of the stopping point with the adjacent territory, including ensuring transit.

6. The increase of the attractiveness of public transport. Transport hubs should consist of technological facilities that ensure the transit of passenger traffic from terminals to other facilities, including facilities of other types of transport.

7. The reduction of the load on the main road network due to the formation of a system of intercepting parking lots as part of transport hubs.

8. The provision of the conditions for the sustainable functioning of ground passenger transport systems through the implementation of special measures at the approaches to the nodes. The principles of TIH reconstruction are determined by the time spent on travel to the place of work for 90% of working citizens (60 minutes). However, this figure does not take into account the traffic flow from nearby cities, as for settlements in other large countries (the USA and China).

The reconstruction of transport hubs should be carried out taking into account their role in the system of transport services in Moscow, based on the following principles: provision of comfort and convenience of passengers, functionality in the placement of node elements and the complexity of the formation of nodes.

Nowadays there are two main ways of reconstruction of transport interchange hubs: placement in the hubs of multi-storey interchange complexes (as in the USA and Japan) and the implementation of local plane measures (as in busy cities in China). Taking into account the significant cost of multi-storey transport hubs, as the first stage, it is advisable to implement a set of local measures prior to the construction of large stationary facilities. In order to implement this project, it is necessary to analyze the risks that would affect the economic efficiency, which is important during the implementation of the project in the format of public-private partnership (PPP) with the prevalence of private investor investments.

It is necessary to note that in addition to positive effects, the project of TIH also carries a number of potential risks and negative consequences, such as:

1. Risk of underloading. The discrepancy between real traffic and projected indicators can lead to the decrease in the expected socio-economic effects and at the same time negatively affect the financial component of the project. The underloading factor is calculated by the formula: C_{ul} = Tariff × Ls × Length of the section × N, where Tariff - the adjusted level of nominal tariffs (RUB / vehicle-km); Ps - losses during collection of fares (number of cars, pairs of trains) (per day); Section length, km; N - days in a year;

2. Risk of public resistance. The deterioration of the environmental situation and low awareness of the population about the purpose of the construction of a transport facility can cause discontent among people with the start of construction.

4. Conclusion

The basic conclusion of the authors is the quantitative determination of the investment process impact on social sphere of any city. It is determined that the biggest correlation has factors dependent on the quality of the pricing system. The results of the research are significant for the process of the pricing improvement in construction and urban planning regulation.

The main conclusion of this research is that the project of TIH can actually have a significant positive impact on the process of the use of new construction materials, without the reduction of the economic effect for railway operators. The direct beneficiaries from the construction and operation of TIH will be shopping centers, parks, hotels and business centers [8]. From a socio-economic point of view, new jobs will be created for the population of TIH-based region.

The direct beneficiaries of the project will be construction companies and private investors among the enterprises at the construction stage and due to the multiplier effects, economic benefits will be reflected in almost all sectors of the economy of Moscow. A separate effect directly related to the emergence of new transport services will be a reduction in the costs of transport companies, which ultimately will have a positive effect for most trade and industrial organizations due to the decrease in the cost of transport component.

References

[1] Kötter T 2019 Urban Development, in: *The Bonn Handbook of Globality* pp 701-712. DOI: 10.1007/978-3-319-90377-4_61
[2] Bryant C and White L G 2019 Managing Urban Development, in: *Managing Development in the Third World*, 253-273. DOI: 10.4324/9780429052125-15
[3] Anon. 2019 Urban Development and the New Elites, in: *Class and Power in Roman Palestine* pp 16-70. DOI: 10.1017/9781108656757.002
[4] Li Qiang 2019 The Lessons and Experiences of Urbanization and Urban Development at Home and Abroad *Research Series on the Chinese Dream and China's Development Path* 39-102. DOI: 10.1007/978-981-13-9451-5_3
[5] Kleniewski N and Thomas A R 2019 Urban Development in the United States *Cities, Change, and Conflict* 66-86. DOI: 10.4324/9780429022920-4
[6] Harris M 2019 Population Change and Urban Development *A Future for Planning* 26-30. DOI: 10.4324/9781315201092-3
[7] Muresan, Georgiana Silvia, Carmen Gheorghe and Nicolae Filip 2019 Road Traffic Studies for Urban Development *30th SIAR Int. Congress of Automotive and Transport Engineering* pp 304-311. DOI: 10.1007/978-3-030-32564-0_36
[8] Bzhasso A M, Maslensky V V and Teryukova L I 2019 The impact of urban transport systems on sustainable urban development *Scientific Development Trends and Education* 4. DOI: 10.18411/lj-05-2019-69
[9] Dörry Sabine 2019 Finance as technocratic agent in urban development *Planning and Knowledge* 115-126. DOI: 10.1332/policypress/9781447345244.003.0009
[10] Yep Ray, June Wang and Thomas Johnson 2019 *Handbook on Urban Development in China*. DOI: 10.4337/9781786431639
[11] Sorokin G V, Eroshenko T I, Fedoseenkov A V and Malyshev A V 2018 Postmodern and Social Problems of Cities *Materials Science Forum* **931** 765-769. DOI: 10.4028/www.scientific.net/MSF.931.765
[12] Dedík Milan, Tomáš Čechovič, Jozef Gašparík and Jozef Majerčák 2019 Rationalization of the passenger transport system as an important transport system *Transportation Research Procedia* **40** 193-200. DOI: 10.1016/j.trpro.2019.07.030
[13] Chen Zhao and Ming Lu 2019 Urban System and Urban Development in the People's Republic of China *Cities of Dragons and Elephants* 81-102. DOI: 10.1093/oso/9780198829225.003.0004

[14] Jacquemin Alain R A 2019 Urban Development in India *Urban Development And New Towns in the Third World* 48-68. DOI: 10.4324/9780429433771-4
[15] Bajnai László and Attila Józsa 2018 An Insight into Operational Urban Development in Hungary in the Light of Regulation-Based Urban Development *Acta Universitatis Sapientiae, European and Regional Studies* **13** 39-64. DOI: 10.2478/auseur-2018-0004
[16] Salgado Canseco and María José 2019 Urban Development and Human Health *Encyclopedia of the UN Sustainable Development Goals* 759-767. DOI: 10.1007/978-3-319-95681-7_103
[17] Solov'ev V V and Korchagin A P 2014 Ways to improve the quality of pre-project documentation in Russian Railways *Economy of Railways* **3** 21-26
[18] Asgharian Mehran and Nazanin Asgharian 2019 Transportation and Sustainable Urban Development *Int. J. of Civil Engineering* **6** 17-22. DOI: 10.14445/23488352/IJCE-V6I4P105
[19] Berg Leo Van Den, Peter M J Pol, Giuliano Mingardo and Carolien J M Speller 2019 Security and Urban Development *The Safe City* 7-32. DOI: 10.4324/9780429060625-2
[20] Abu-Lebdeh Ghassan 2017 Smart Transport Systems: Planning and Designing Transport Systems to Support Public Health *J. of Transport & Health* **5** 34. DOI: 10.1016/j.jth.2017.05.319

Reasonable choice of materials for project management in transport construction

V Solovev

Moscow State University of Civil Engineering (NRU MGSU), Department of Economics and management in construction, 26, Yaroslavskoe shosse, Moscow, 129337, Russia

E-mail: solovevvv@mgsu.ru

Abstract. Material and technical supply is traditionally considered as a set of activities united by one project purpose. As a rule, a hierarchical approach to management functions is not applied and the links between management functions are determined by the classical form of the implementation of a construction project. The selection of the best types of material resources and its component that is the costs determined by the properties of the applied materials are becoming a key link in project management in modern Russia. This is an important scientific phenomenon. The research of this phenomenon allows finding the problem points in costing and the solution to the problems of the increase in the efficiency of building materials assembly through them. The relevance of this research is important for those countries where there is strong regulation of the estimated cost. The authors of the paper propose the measures to improve the efficiency of the choice of construction materials due to a more reliable determination of the estimated cost. For this, there are two ways to improve the project bill of materials: through the improvement of the regulatory framework and management of the process of the application of this framework. The quality level of resource and cost norms is a paramount problem in conditions of state regulation. This is intensified by the use of the basic costing method. The lag of norms calculated at the basic cost level can not be compensated for the use of conversion indices due to territorial and industrial classification. The attempts to create in Russia a unified information system for setting costs for construction products have not yet had measurable results. This results in a large percentage of outdated regulations that take into account materials with insufficient usability. The participants of the construction process have to spend large human resources to manage material resources in projects. The process of management in terms of material and technical equipment of construction is on top in terms of importance in the field of project management in transport construction. The quality of the projects being implemented largely depends on the success in the improvement of the regulatory framework.

1. Introduction

The purpose of the paper is to study the problem of the choice of construction materials that correspond to the goals of the project through the determination of the reliable estimated cost of construction of transport facilities associated with the shortcomings of the estimated regulatory framework and the search for the causes of errors. The existing possible ways to improve the accuracy of economic calculations is a well studied area of economics. At the same time, the performance properties of materials that make up a significant part of project costs are studied separately. Thus, materials science and costing are not related areas in the actual engineering design process. According to previous studies [1], a qualitative system of norms is the basis of the final cost of an object. This is most reasonable for the creation of high-speed passenger transport [2], including the creation of tourism infrastructure [3].

Nowadays, there is a steady growth in the construction sector of economy, accompanied by a significant amount of work on the design, construction and commissioning of both new construction and reconstruction projects. This is most reasonable for high-speed rail lines, built using new structures and extended overpasses [4]. The choice of the best design of the railway track becomes directly related to the degree of accuracy of the cost estimates [5]. As it is noted in the work [6], the issues of the formation of construction project management systems are becoming relevant. The project management system allows overcoming the difficulties associated with the uncertainty of both the external and internal environment and the chronic lack of resources necessary for the production of construction products.

Under these conditions, high requirements for the quality of project planning are formed, which allows organizing and controlling the processes of project implementation as efficiently as possible, despite external constraints and risks for participants [7]. The criteria for the success of a project are its timely completion with a given level of quality and within the allocated budget. Therefore, it is not surprising that the issue of the determination of the project budget in conditions of a permanent shortage of investment resources is becoming increasingly important.

The issues of the management of the cost of construction projects are widely reflected in the works of different authors. The determination of the estimated cost of construction and the development of estimated standards at the present stage in Russia, Europe and China are more and more focused on filling the gaps in the construction of high-speed transport [8, 9], while in Russia there is an additional tendency to develop consolidated cost standards [10].

The features of construction projects include non-stationarity, non-uniformity of construction production and the nature of final product, the participation of a large number of organizations, which determines a high degree of cooperation in construction production, the role of climatic and natural conditions. Construction projects, especially in the construction of atypical facilities, have many risks and problems in the field of organization and communication. Design features also greatly complicate the organization of construction production [11]. They require significant capital investment and tight coordination of realization, financing and quality control processes. This is compounded by the presence of several actors (contractors, designers, customers) who are forced to work closely together.

Nowadays one of the most relevant areas of the construction industry is transport construction, since through the provision of transport accessibility to regions, uniting them into a single economic space and creating the basis of national economy, it contributes to the provision of the strategic and economic security of the state [12]. Large-scale construction of transport infrastructure facilities requires increased attention to project management issues. Apart from all the above-mentioned difficulties in the management of construction projects, the implementation of road construction projects is complicated by the linear nature of the work, the territorial remoteness of objects from production bases and head offices, which complicates the speed and accuracy of decision-making on the project, requires consideration of the lag of managerial decisions, leads to significant restrictions on logistics providing the project with material and technical resources and equipment, technical limitations and inaccessibility of materials at operating units. An interesting example in this aspect is the example of construction in Singapore, when climatic risks influenced the construction management model [13].

The management of any project has similar tasks: implementation schedule, creation of conditions for the normal functioning of the organizational and economic mechanism of a project, systematic movement towards the set goal [2]. One of the main tasks solved in the field of project management is the task of construction cost managing, the solution of which is aimed at the provision of the process of the implementation of a construction project with financial resources in the required volumes and within a given time frame. Project cost management provides its implementation within the approved budget by sequentially carrying out the stages of budgeting, cost control and cost adjustments [14]. The key stage is budgeting i.e. the determination of the costs of a project. The budget for a construction project is developed on the basis of the determination of the estimated cost of the project after all the structural and technical characteristics of the future structure are identified [15]. A reliable determination of the estimated cost of construction largely determines the further implementation of the construction project, which allows project executors to focus on the solution of other issues of project management [16]. As it is noted in the work [17], this improves the efficiency of project management.

The resource method is among the methods for the determination of the estimated cost operating in the system of estimated costing, the most reliable and appropriate one to the conditions of a market economy, since it allows determining the cost of construction based on the consumption of resources and the real current cost for them in a territorial context. At the same time, in Russia in construction costing the base-index method has been used since the days of the planned economy, in the overwhelming majority of cases, which is characterized by unreasonable averaging of the cost of resources [18]. Therefore, in order to eliminate the current situation, in December 2017 the Federal State Information System for Costing in Construction (FSISCC) was created in accordance with the roadmap for the development of costing. The main task of which is to monitor the cost of construction resources in a territorial context for each region of the country.

The state monitoring system will increase the accuracy of estimate calculations due to the transition to the resource method to estimate documentation. This is relevant both for general construction and specialized construction work [19]. At the same time, the quality of implementation of investment projects is growing and target indicators are improving [20]. Unfortunately, for a number of reasons, the transition to new costing system is being delayed. The launch of FSISCC was expected in 2019, but now it is already obvious that FSISCC will fully work even later. Therefore, the use of the most reliable resource method to determine the estimated cost of construction is currently not possible. To continue the work on the introduction of this progressive method, a number of priority measures should be taken, namely it is necessary to clarify the list of building resources by types of capital construction objects "Highways" and "Artificial road structures", "High-speed railways" and organize monitoring of the current costs of building resources, used for road and bridge construction. The creation of inter-industrial center for monitoring the current costs of construction resources on the basis of the Ministry of Transport and monitoring the costs of construction resources in accordance with the list provided by industry departments will help solve this problem and timely update the estimated costs for construction resources in FSISCC necessary for this work.

2. Materials and methods

The base-index method of estimates used as the main one is based on a well-tested, but not devoid of shortcomings, estimate and regulatory framework developed long time ago. Due to scientific and technological progress, stimulating the emergence of new equipment, technologies and materials, the estimated and regulatory framework of the construction industry is rather outdated. For several years (since 2014) a costing reform has been ongoing. It includes active work on updating the existing ones and developing new regulatory and methodological documents in the field of costing. According to the analysis carried out by the authors, 42% of resource-technological models explicitly contain outdated types of construction materials. Another 26% require replacement of technological schemes and production methods. The results of the study show that the system for the determination of the cost of construction needs to eliminate a large number of identified shortcomings, which reduce the reliability to determine the cost of construction products.

3. Results

One of the most serious shortcomings negatively affecting the reliability to determine the cost of transport construction and requiring urgent elimination is the lack of industry index for recalculating the estimated cost from the basic to the current cost level. For this reason, the projected indices for the

IOP Conf. Series: Materials Science and Engineering 1103 (2021) 012036 doi:10.1088/1757-899X/1103/1/012036

cost element "Other objects" used during the calculation of the cost of work in road construction do not reflect the actual costs spent by contractors, due to the significant difference in the composition and list of construction resources used in the construction of road structures in comparison with others objects. A depressing situation is emerging when road construction contractors hardly cover the costs of construction work and excavations in accordance with the level of payment under the current estimate and regulatory framework cause significant losses. The current situation leads to the smashup of construction and installation organizations and their withdrawal from the market, which threatens the existence of a strategically important sector of the national economy and the implementation of a long-term transport strategy. In this regard, it is urgently necessary to organize the procedure for the development of indices to recalculate the estimated cost of transport construction.

One more problematic issue for the road industry is the task of the development of new estimated standards necessary to ensure the possibility to consider industry facilities of innovative technologies and materials in design documentation. Since 2016, more than 450 proposals have been formulated for the development of estimated standards necessary for road industry and only seven have been approved so far.

The development of estimated standards is carried out in the industry on the basis of current legislation. During practical application of standards, the problems of interaction between participants arise. They are associated with the uncertainty of the scope of responsibility of participants in processes and deadlines. This requires the adjustment of laws and methods in terms of the structure and composition of requirements for the participants involved in all processes of the development of estimated standards. The question of the formation of a transparent mechanism for the approval of the procedure for the development of new estimate standards, which allows the implementation of the design of facilities within the time frame established by the contract, is still open.

It is necessary to determine the financial source for the development of estimated standards, to optimize the "Procedure to approve of estimated standards" in terms of the reduction of time for their consideration and approval. Today, approval is 1.5 - 2 years, which exceeds the duration of the design of a large facility. At present, a new variant of the normative document has been prepared, taking into account the recommendations of industry experts. The approval of the document will eliminate the above mentioned problems, which will allow intensifying the development of new estimated standards.

For the purpose of technical support of work on the preparation of draft estimate standards and their subsequent examination, it is necessary to create an industry-specific software package that would systematize work on the development of estimate standards using technological processes of reuse, which reduce development time. During the solution of the tasks of the increase in the efficiency of project cost management in the industry, it is recommended to work out the issue of the harmonization of estimated standards as part of the estimate and regulatory bases of various levels (federal, industrial and territorial), on the basis of which the estimate documentation for transport infrastructure facilities is made.

The analysis of cost management at transport construction projects showed that cost management plays an increasingly important role in project management system. The classical model, in which cost management had a separate role, is less and less consistent with the reality of Russian construction. The reason for this is the incorrect explanation of the cost of work as a part of estimate and regulatory framework. Outdated costs for construction work lead to the fact that even with highly qualified engineers, it becomes impossible to ensure the compliance of the technical and economic parts of the project with the specified requirements. As a result, a company managing a project is forced to adjust the cost of a project throughout the entire engineering cycle, from investment concept to commissioning. This gives grounds to state that the management of a construction project in Russia is, first of all, the management of cost, its accuracy and reliability.

This result is quite new and even unexpected since it was previously believed that it is necessary to manage production and processes. Nowadays the tasks of financial management, implemented with the help of estimated costing tools, have become the most important. These conclusions are important both for the determination of the directions of future scientific research and for training future specialists of project management.

4. Conclusion

The relevant problems of costing considered in the article, occurred in transport construction of Russia, have led to the transformation of the project management scheme in construction. Instead of the applied role of costing as an additional control element, the subject of cost management becomes dominant in the structure of project management. Accordingly, the problems of costing, which were previously private, are becoming global.

The proposed recommendations for the solution of problems are reduced to the following main provisions: it is necessary to continue work on the implementation of FSISCC, defining the range of used resources and monitoring of current costs for them. In addition, it is necessary to organize a procedure for the determination of the indices to recalculate the estimated cost from the basic cost level to the current one. Since the final transition to the resource method for the determination of the estimated cost is delayed, the development of new estimated standards does not lose its relevance taking into account the use of modern technologies and means of production.

The implementation of the above mentioned measures objectively should not only increase the reliability of to determine the cost of road construction, but also ensure the introduction of materials with better operational properties. It will ultimately improve the efficiency of project management for their construction as a whole and thereby ensure the implementation of the transport strategy throughout the country.

References

[1] Poltava A V and Korchagin A P 2019 Problems of determining the reliable cost of construction in the framework of project management *Development of the methodology of modern economic science, management and education in the context of information and digital trends: Materials of the III Interdisciplinary all-Russian scientific conference* pp 208-213

[2] Jandová Monika, Tomeš Zdeněk and Nash Chris 2016 High-Speed Rail for Central and Eastern European Countries: A Conference Report *Review of Economic Perspectives* **16(3)** 269-275. DOI: 10.1515/revecp-2016-0016

[3] Carmen Vázquez Varela and José M Martínez Navarro 2016 High-speed railway and tourism: is there an impact on intermediate cities evidence from two case studies in castilla-la mancha (Spain) *Journal of Urban and Regional Analysis* **VIII(2)** 133-158

[4] Van Yu, Czzin` L and Van S 2014 Railway development and analysis of the situation in China *Proc. of the int. sci.-pract. conf. dedicated to the 70th anniversary of the Department, Technology of transport processes and logistics* (A S Balalaev, ed) pp 13-18

[5] Zajcev A, Kraskovskij V and Terleczkij S 2012 A platform or mound: we need a reasonable approach *Roads. Innovations in construction* (St Petersburg) **19** 18-21

[6] Silka D N and Babaeva M S 2018 Ways to improve the quality and efficiency of relationships between participants in investment and construction activities *Economy and entrepreneurship* **1(90)** 682-685

[7] Vaganova M D and Shcherbakova N A 2019 The process of transition to a new budget and regulatory framework in construction *Young scientists - development of the National technology initiative (Search)* **1-2** 158-161

[8] Kraskovskij V 2014 General issues of design of artificial structures on the roads using magnetic levitation technology *Magnetic levitation transport systems and technologies. Proc. of the 2nd Int. sci. conf.* (Yu F Antonov, ed) pp 75-80

[9] H W Yang 2015 Performance of pile foundation for the civil infrastructure of high speed rail in severe ground subsidence area *Proc. of the Int. Association of Hydrological Sciences* **372** 525-528. DOI: 10.5194/piahs-372-525-2015

[10] Solov`ev V V 2016 Industry characteristics define the integrated cost indexes of construction *Economy of railways* **6** 46-55
[11] Francesca Pagliara, Luigi Biggiero, Alessia Patrone and Francesco Peruggini 2016 An analysis of spatial equity concerning investments in high-speed rail systems: the case study of Italy *Transport Problems* **11(3)** 55-68. DOI: 10.20858/tp.2016.11.3.6
[12] Xueqiao Yu, Maoxiang Lang, Yang Gao, Kai Wang, Ching-Hsia Su, Sang-Bing Tsai, Mingkun Huo, Xiao Yu and Shiqi Li 2018 An Empirical Study on the Design of China High-Speed Rail Express Train Operation Plan—From a Sustainable Transport Perspective *Sustainability* **10(7)** 2478. DOI: 10.3390/su10072478
[13] Sazrul Leena Binti Sa'adin, Sakdirat Kaewunruen and David Jaroszweski 2016 Risks of Climate Change with Respect to the Singapore-Malaysia High Speed Rail System *Climate* **4(4)** 65. DOI: 10.3390/cli4040065
[14] Yuxiang Wang, Xueli Liu and Feng 2018 Wang Economic Impact of the High-Speed Railway on Housing Prices in China *Sustainability* **10(12)** 4799. DOI: 10.3390/su10124799
[15] Fridkin V 2006 *Principles of forming in the theory of linear-extended structures* (Moscow: Lad`ya) 512 p
[16] Solov`ev V V and Kuzneczova A E 2016 Modeling of cost standards for railway construction projects *Economy of railways* **12** 44–51
[17] Tarasenko A 2006 Retrospective of formation of indicators of an estimation of efficiency of investments *Economy and management* **3** 131-134 (In Russian)
[18] Volkov B A and Solov`yov V V 2013 Real estate of railway transport of Russia *Real estate, economy, management* **2** 128-130
[19] Solovyov V V, Korchagin A P and Abu-Haidar S B 2018 Directions of updating estimated standards in transport construction *World of transport* **16(2(75))** 116-127
[20] Majboroda V, Titov A and Morgunov M 2016 On the problem of development of systems quality assessment and management of complex objects and processes *Quality. Innovations. Education* **8-10** 84-93

IOP Conf. Series: Materials Science and Engineering 1103 (2021) 012037 doi:10.1088/1757-899X/1103/1/012037

Recursive method in assigning materials in construction projects

D Zaycev[1], A Korchagin[2], V Solovev[2] and S Guzhov[1]

[1] Department of Road Economy, Moscow Automobile State Road Transport Technical University, 64, Leningradsky Ave., Moscow, 125319, Russia

[2] Department of Economics and management in construction, Moscow State University of Civil Engineering (NRU MGSU), 26, Yaroslavskoe Highway, Moscow, 129337, Russia

E-mail: zaycevdv77@mail.ru

Abstract. The paper considers the issues of improvement of the evaluation methods of new materials in construction. The features of regulatory control over the use of resources in Russia are characterized by the presence of a dual approach to choosing a regulatory framework, which in some cases leads to incorrect choice of materials. Engineering solutions during design are closely dependent on the cost parameters laid down in the regulatory and methodological framework. As a result, the problem of improving norms and methodologies is relevant for highly significant regulation. The property of recursivity of design solutions, normative support of construction and quantitative parameters of construction materials used in the design of structures is noted. The mutual influence of the process of developing standards and their further application in projects made it possible to use the recurrent formula tool for pre-design analysis of the possibility of using certain materials. The method allows performing a separate analysis of the possibility of application in projects for different materials, which can be useful when introducing new materials. On the basis of the proposed calculation method, conclusions on the need to improve the process of developing standards were drawn. The main problems of regulating the consumption of construction resources at the present stage are formulated. The need to design new standards based on the practice of applying new materials in construction projects was noted. The paper suggests the mechanisms to solve the problem of improving the reliability of determining a set of construction materials when designing construction works in transport construction, which would reduce the time for developing new and updating the existing standards. The assessment of the applicability of construction materials should not remain within the framework of subjective decisions of the engineering practice; on the contrary, it should improve the calculation methods to justify the applicability of materials.

1. Introduction

Industry-specific features of construction products determine the need to develop and apply new methods of construction materials and technical resources, taking into account design characteristics, individual conditions of production and economic activity of contracting organizations, market conditions and other factors [1, 2]. Besides, specialists pay increased attention to the issues related to improving the mechanism for developing new and updating costing standards. In this regard, the relevance of proposals aimed at solving the problem of a reasonable choice of construction materials is beyond any doubt.

It should be noted that in the period from the beginning of the 2000s to the present, not only the normative, but also the methodological segment in the pricing of construction products is divided into federal and industrial [3]. The share of industry standards is about 40 percent in the total quantity. Methodological requirements for the accounting of transport schemes for the delivery of construction materials and for the calculation of other costs as part of the consolidated estimate also have significant differences in federal and industry bases [4, 5]. All this leads to a well-known difference in the cost of the same object, calculated according to different bases. In fact, in the practice of investment activities, the choice of the estimated regulatory framework is the determining and major factor in defining and justifying investment costs [6]. In the practice of investment projects abroad there is no unambiguous affinity to regulatory support, priority remains with the general economic selection criteria, as noted by many researchers [7-11]. For projects financed from the budget system of the Russian Federation, the procedure for applying the federal budget and regulatory framework is legally fixed, however, it should be borne in mind that there is an extensive class of specialized objects of the industry segment implemented by quasi-state structures – Russian Railways [12, 13], Rosatom Group of Companies and others [14]. In this case, the choice of the estimate and regulatory framework ceases to be unambiguous, and it opens up the possibility of a two-fold determination of the estimated cost, as noted in [15].

2. Materials and methods

The current situation, when for the specialized industry segment of construction the choice of construction materials and technologies for their processing implies a dual definition of costs, also determines the dual approach to the selection of the supplier [16]. In conditions of market interaction, the possibility of using similar types of materials with different properties under the influence of a price factor in a construction project should be considered possible. In any case, the experience of the interaction of suppliers of materials with consumers in countries with market regulation allows making such conclusions [17]. The practice of implementing investment projects in Russia shows that the choice of material of span structures of pedestrian bridges depends on the peculiarities of application of certain regulatory bases, as confirmed in [18]. If there is an accurate standard for new materials – epoxy resin-based composites with filler – it was used in the design documentation. With the relative use of standards, i.e. with their actual non-conformity, traditional materials (steel, reinforced concrete) with non-competitive technical characteristics were used.

The task of preliminary determination of material selection at the pre-design stage was considered by the authors in a discrete task. This was facilitated by studies on the selection of engineering solutions in construction [19-22], as well as conclusions drawn from the analysis of a number of investment projects in transport construction [23, 24]. The recursive function tool is used, which allows obtaining a decision on the selection of the construction material type under consideration based on the analysis of previous design solutions and corresponding normative support of the design activity.

For a classic recursive-specific factorial, the function of which can be taken as the volume or cost of material No. i, a subsequent recursive reference (for example, to the data of the design documentation) is made with a decrease in the argument by one. Depending on the initial value of the argument through n recursive references, the calculation of the function will stop due to the termination of recursion. Thus, the computed function for any argument from the domain is finite in essence.

In practice, the choice of materials in a project and their accounting in the composition of production or value standards is better described by a recurrent formula, in which each next component of the sequence is calculated as a result of a function from n previous components. This is best in line with the practical principle of the interdependence of project and normative activities.

Let us denote the generating function of the sequence f_0, f_1, ... using F(z). Build the polynomial $K(z)=C+a_1z+a_2z^2+a_3z^3+\ldots+a_mz^m$. This polynomial can be considered as the generating function of the sequence 11,a_1, a_2,…,a_m, 0,0,…. This sequence represents a set of quantitative ideas on the applied material – its target consumption, cost in pricing.

Using standard methods of working with characteristic polynomials and rational functions generated by the generating function, it is possible to present the generating function as follows:

$$F(z) = \sum_{n=0}^{\infty} \left(\sum_{i=1}^{S} p_i(n) a_i^n \right) z^n,$$

where $p_i(n)$ – polynomial consisting of actual data on the use of material in construction projects in quantity S. The variety z^n is determined by the rules and conditions for the use of construction material in design. The more varied the scope of the material, the more z is needed to assess its applicability.

The peculiarity of this approach is that it is possible to compare the values of the generating function for different materials, on the basis of which it is possible to draw conclusions on the priority of using materials in different segments of the construction industry, which indirectly contributes to the solution of previously set applied tasks [25].

The proposed method is design-theoretical, and solves the problem of improving the design standards for buildings and structures.

3. Results

On the basis of solving the set tasks, it first of all becomes possible to reasonably recommend the construction industry to carry out the following:

- to create a single information space on industry pricing issues, as well as to complete the formation of a single data pool on the list of used construction resources and their value in the context of the constituent entities of the Russian Federation;
- to create a new industry center (project office), which in the "one window" mode will consider all issues on the addition, update or elimination of outdated standards of the current estimate and regulatory framework used to determine the set of material resources during construction, reconstruction, overhaul, and maintenance of capital construction facilities;
- to simplify the process of creating standards, expand the range of specialists involved in standards' development and update;
- to create a single publicly available regulation for the development and update of resource standards.

Within the framework of the proposed events, it is necessary to constantly carry out work aimed at optimizing the processes of collecting and analyzing design and as built documentation for previously implemented projects, as well as proposed innovative materials and associated technologies for the performance of works, summarizing the received information and forming the basis for re-use projects, guided by the principles of optimizing budget expenditure. The collection and analysis of information on changes in market prices and the list of construction resources used in the industry in the context of the constituent entities of the Russian Federation will make it possible to quickly form draft changes in standards, as well as analyze data for annual update of consolidated cost indicators of construction and installation works.

In order to ensure the quality of the work performed, all organizations interested in developing new and updating existing costing standards should continue to implement the events necessary to supplement the regulatory framework with costing standards required in the industry, including:

- analysis of construction resources included in the costing standards for their replacement with more modern ones, the parameters of which will meet the requirements of existing regulatory documents;
- interaction with industry producers in terms of supporting requests for inclusion in the list of legal persons providing information necessary for the formation of estimated prices of construction resources, and for the inclusion of materials produced by them in the Classification of Construction Resources;

- collection of data on the cost of purchasing price-generating construction resources in the industry with a view to their further comparison with the prices of similar resources available in the estimated regulatory framework and further settlement of issues arising in this field with organizations responsible for the formation of estimated prices.

Besides, it is advisable to create an understandable mechanism for approving the procedure for timely development of costing standards that take into account the use of new technologies and structures, which allows the implementation of the design of the facility within the established time, since the period of their approval often exceeds the duration of the design.

It is necessary to accelerate the process of creating and implementing an information analytical system in the structures of industry customers, which will provide timely information on changes and additions to costing standards, indexes, prices for construction resources, catalogs of model projects and optimal design solutions, as well as catalogs of efficient technologies, new materials and modern equipment. It is necessary to implement comprehensive automation of the construction cost management through the implementation of the industry software complex.

At the same time, we recommend simplifying the process of developing estimates and expanding the range of participants in this process. Specialized commercial organizations should be widely involved in the development and update of costing standards, and state structures should ensure the examination of results and, if there is a positive conclusion, the order of inclusion into the Federal Register of costing standards.

The authors propose to consider the possibility of using the Cost Estimate Designer and expand its functionality. The standard designer is primarily a regulation, and secondly – a software product that allows creating costing standards according to single publicly available rules. Currently, the access to this type of activity is artificially limited and therefore is implemented only by a narrow circle of specialists, which requires changes. The implementation of this task is not as difficult as many people think, but it will require agreement on the positions of all interested parties.

The main idea of the tool is to create and accumulate a database of the simplest operations, and subsequently compile costing standards that describe a particular technological process, acting on the principle of combination. By accumulating the required number of simple (elementary) operations from year to year, the rate of development of standards for new technological processes will increase, since rationing will require not all processes related to the new production technology, but only those that have not previously found a display in the Designer.

The structure of the Standard Designer shall consist of a working module, basic directories and a number of auxiliary directories.

The working module will combine data from directories, and process chains will be formed from “elementary” operations. As a result of combinations, an estimated standard is created that describes a particular technological process, which is stored in the data storage and can subsequently be used, including as part of a more complicated complex process.

The main and auxiliary directories can include normative literature on the proposals of the developers of costing standards.

Implementing the concept of the designer it becomes possible to significantly reduce the time for the development of new and update the existing estimates, create a tool that will work on a single principle of universality, attract a wider range of specialists and reduce the budget for financing these works.

4. Conclusions

The use of recursive expressions in the analysis of problems with the use of construction materials allows solving a number of related problems of the design process and its regulatory support. It is also important to solve another problem – not all existing regulatory and methodological documents take into account the elements of the transition period to the resource method for determining the cost of construction, as well as additional types of costs that customers, design and contracting organizations actually bear when implementing construction projects. Therefore, it is necessary to continue work on updating outdated regulatory and methodological documents with their mandatory subsequent inclusion into the Federal Register of Costing Standards.

The implementation of planned activities will provide industry specialists with the necessary regulatory documentation, optimize the process of developing new and update the existing costing standards, which will ultimately increase the reliability of determining the cost of construction products and reduce the likelihood of controversial situations when calculating the performed work.

References

[1] Tarasov M M 2020 Analysis of the use of the PPP tool in the investment and construction sphere *Economics and management: problems, solutions* **1(3)** 26-31 (in Russian)

[2] Silka D N 2011 Key factors for identifying the phases of the business cycle in construction *Bulletin of the University* **21** 210-212 (In Russian)

[3] Ardzinov V D 2018 Problems and risks of reforming pricing and budget rationing in construction, in: *Proceedings of the scientific and practical conference with international participation RYASKE-2018,* S G Oparin (ed) pp 353-363 (In Russian)

[4] Yuanle Cheng and Hai Hu 2020 Analysis of the Problems, Causes and Countermeasures in the Brand Construction of the Postgraduate Entrance Examination Institutions *Open Journal of Business and Management* (Nanjing, China) **08(02)** 32-44. DOI: 10.4236/ojbm.2020.82025

[5] Hasan Mathar, Sadi Assaf, Mohammad, Hassanain, Abdullatif Abdallah, Ahmed M Z Sayed 2020 Critical success factors for large building construction projects: Perception of consultants and contractors *Built Environment Project and Asset Management* **9(5)** 29-33. DOI: 10.1108/BEPAM-07-2019-0057

[6] Zykov R A 2020 Problems of development of innovative infrastructure of the petrochemical complex of Russia, in: *Problems of development of modern society. Collection of scientific articles of the 5th all-Russian scientific and practical conference,* V M Kuzmina (ed) pp 244-246 (In Russian)

[7] Yalu Ping 2020 Analysis of the Influence of Multiple Linear Regression on Construction Price *Statistics and Application* **09(01)** 56-60

[8] Adamko P and Chutka J 2020 Company bankruptcy and its prediction in conditions of globalization *SHS Web of Conferences* **74** 05002. DOI:10.1051/shsconf/20207405002

[9] Rashid Bhutta and Angappan Regupathi 2020 Predicting Corporate Bankruptcy: Lessons from the Past *Asian J. of Multidisciplinary Studies* **8(1)** 13-21

[10] Ir. Shamsuddin Sabri, Ir. Mohd Erie Husairrie Ismail, Mohd Nazri Ali and Rohayu Abd Aziz Roads B 2020 *Towards better road contractor performance ranch* (Public Works Department (PWD), Malaysia) Preprint No. 2394 2020

[11] Yu Zhang, Dangqiang Yin and Yating Zhu 2019 Method for Construction Progress Measurement for EPC Oil and Gas Pipeline Projects *Journal of Oil and Gas Technology* **41(06)** 39-47. DOI: 10.1007/978-981-13-0107-0_109

[12] Bashirov A E, Rubchenko D S and Kurbackaya E P 2012 On pricing in the construction of railway transport facilities *Economy of Railways* **6** 34 (In Russian)

[13] Silka D N and Babaeva M S 2018 Ways to improve the quality and efficiency of relationships between participants in investment and construction activities *Economy and entrepreneurship* **1(90)** 682-685 (In Russian)

[14] Goryachkin P V 2010 On the state of issues of pricing and estimated rationing in construction *Urban planning* **6** 50-58 (In Russian)

[15] Solov'ev V V 2016 The significance of the concept "400 days" for the industry's estimated rationing *Economy of Railways* **5** 62-69 (In Russian)

[16] Solov'ev V V and Korchagin A P 2014 Ways to improve the quality of pre-project documentation in Russian Railways *Economy of Railways* **3** 21-26 (In Russian)

[17] Wang S, Zhang H, Wang C and Wu Y 2020 Cost Analysis Between Prefabricated Buildings and Traditional Buildings *IOP Conf. Ser.: Mater. Sci. Eng.* **768** 052090. DOI: 10.1088/1757-899X/768/5/052090

[18] Alekseev S A, Viktorov M Yu, Grabovyj P G, Zajceva L I, Kazanskij N V et al 2019 *Spatial restructuring. New meanings and rules of investment and construction activities* (Moscow: Delo) (In Russian)

[19] Garcia-Macia D 2020 Labor Costs and Corporate Investment in Italy *IMF Working Papers* **20(38)**. DOI: 10.5089/9781513529721.001

[20] Park D 2019 The Effect of Leverage on Labor Cost Management *Review of Accounting and Policy Studies* **24(1)** 25-63. DOI: 10.21737/RAPS.2019.02.24.1.25

[21] Do Y, Kang N and Li L 2019 Temporary Employment and Behavior of Labor Cost *Korea Association of Business Education* **34(3)** 325-344. DOI: 10.23839/kabe.2019.34.3.325

[22] Kereri J 2019 Construction Sector versus Labor Costs in Kenya: A Two Way Love Affair *International Journal of Architecture, Engineering and Construction* **8(2)**. DOI: 10.7492/IJAEC.2019.011

[23] Solov'ev V V and Korchagin A P 2020 Topical issues of improving the pricing system in construction in the transition period *Vestnik MGSU* **15(4)** 605-616

[24] Poltava A V and Korchagin A P 2019 Problems of determining the reliable cost of construction in the framework of project management, in: *Development of the methodology of modern economics, management and education in the context of information and digital trends: Materials of the III Interdisciplinary all-Russian scientific conference* pp 208-213

[25] Korchagin A P 2019 Problems and prospects of development of costing standards in construction *Financial aspects of structural transformations of the economy* **5** 83-90

AUTHOR INDEX

Institute of Physics
Dirac House, Temple Back
Bristol BS1 6BE UK

ISSN: 1757-8981
ISBN 978-1-7138-3397-0

9 781713 833970

CURRAN ASSOCIATES INC.
proceedings.com